Springer Texts in Statistics

Advisors:
George Casella   Stephen Fienberg   Ingram Olkin

Springer
New York
Berlin
Heidelberg
Hong Kong
London
Milan
Paris
Tokyo

Springer Texts in Statistics

(continued after index)

Jeffrey S. Simonoff

Analyzing Categorical Data

With 64 Figures

 Springer

Jeffrey S. Simonoff
Leonard N. Stern School of Business
New York University
New York, NY 10012-0258
USA
jsimonof@stern.nyu.edu

Editorial Board

George Casella
Department of Statistics
University of Florida
Gainesville, FL 32611-8545
USA

Stephen Fienberg
Department of Statistics
Carnegie Mellon University
Pittsburgh, PA 15213-3890
USA

Ingram Olkin
Department of Statistics
Stanford University
Stanford, CA 94305
USA

Cover illustration: The Poisson regression model (Figure 5.1).

Library of Congress Cataloging-in-Publication Data
Simonoff, Jeffrey S.
    Analyzing categorical data / Jeffrey S. Simonoff.
      p. cm. — (Springer texts in statistics)
    Includes bibliographical references and index.

    1. Multivariate analysis.   I. Title.   II. Series.
QA278.S524 2003
519.5'35—dc21   2003044946

ISBN 978-1-4419-1837-6   e-ISBN 978-0-387-21727-7

© 2010 Springer-Verlag New York, Inc.
All rights reserved. This work may not be translated or copied in whole or in part without the
written permission of the publisher (Springer-Verlag New York, Inc., 175 Fifth Avenue, New York,
NY 10010, USA), except for brief excerpts in connection with reviews or scholarly analysis. Use
in connection with any form of information storage and retrieval, electronic adaptation, computer
software, or by similar or dissimilar methodology now known or hereafter developed is forbidden.
The use in this publication of trade names, trademarks, service marks, and similar terms, even if
they are not identified as such, is not to be taken as an expression of opinion as to whether or not
they are subject to proprietary rights.

Printed in the United States of America.

9 8 7 6 5 4 3 2 1

www.springer-ny.com

Springer-Verlag New York Berlin Heidelberg
A member of BertelsmannSpringer Science+Business Media GmbH

To my parents, Pearl and Morris Simonoff

Preface

This book grew out of notes that I prepared for a class in categorical data analysis that I gave at the Stern School of Business of New York University during the Fall 1998 semester. The class drew from a very diverse pool of students, including undergraduate statistics majors, M.B.A. students, M.S. and Ph.D. statistics students, and M.S. and Ph.D. students in other fields, including management, economics, and public administration.

My task was to come up with a way of presenting the material in a way that such a heterogeneous group could grasp. I immediately hit on the idea of using regression ideas to drive everything, since all of the students would have seen regression before, at some level. This is not a new idea; many books have in recent years exploited the generalized linear model when discussing categorical data analysis. I had in mind something a little different, however—a heavily data-analytic approach that covered a broad range of categorical data problems, from the count data models common in econometric modeling, to the loglinear models familiar to statisticians and social scientists, to binomial and multinomial regression models originally used in biological applications, with linear regression at the core.

This origin has several implications for the reader of this book. First, Chapters 2 and 3 contain a more detailed overview of least squares regression modeling than is typical in books of this type, since this material is continually drawn on when describing analogous techniques for categorical data. There is also a good deal of detailed material on univariate discrete random variables (binomial, Poisson, negative binomial, multinomial) in Chapter 4. My hope is that these three chapters will make it possible for the book to stand alone more effectively, and make it useful for readers with a wide range of backgrounds. On the other hand, they make the book longer than it might have been; there is a lot to get through if a reader just sits down and attempts to read straight through.

The Poisson regression model, and its variants and extensions, is the engine for much of the material here. This is not unusual for books on econometric models for count data (for example, Long, 1997, or Cameron and Trevidi, 1998), but is not typical for categorical data analysis books written by statisticians, which tend to highlight the Poisson regression model as the basis of loglinear modeling, but do not focus very much on count data modeling problems directly (for example, Agresti, 1996, or Lloyd, 1999). On the other hand, this book also includes extensive discussion of loglinear models for contingency tables, including tables with special structure, which is common in statistical categorical data analysis books, but not count data modeling books. The close connection between these models and useful models for binomial and multinomial data makes it easy to then include material on logistic regression (and its variants and competitors) as well. The approach is classical; for a Bayesian approach to many of these problems, see Johnson and Albert (1999).

The target audience for this book is similar to the (student) audience for my original class, but extended to include working data analysts, whether they are statisticians, social scientists, engineers, or workers in any other area. My hope is that anyone who might be faced with categorical data to analyze would benefit from reading and using the book. Some exposure to linear regression modeling would be helpful, although the material in Chapters 2–4 is designed to provide enough background for the later material. The book has a strong focus on applying methods to real problems (which accounts for the active, rather than passive, nature of its title). For this reason, there is more detailed discussion of examples than is typical in books of this type, including more background material on the problem, more model checking and selection, and more discussion of implications from a contextual point of view. These discussions are set aside in the text with grey rules and titles, in order to emphasize their importance. Nothing can take the place of reading the original papers (or doing the analysis yourself), but my intention is to give the reader more of a flavor of the full data-analytic experience than is typical in a textbook. I hope that the readers will find the examples interesting on their own merits, not just as examples of categorical data analysis. Many are from recent papers in subject-area scientific journals. A more detailed description of the organization of the book is given in Section 1.2.

Many of the basic techniques for categorical data analysis are available in almost all statistical packages. A great deal (but not all) of categorical data modeling can be done using any statistical package that has a generalized linear model function. All of the statistical modeling and figures in the text are based on S-PLUS (Insightful Corporation, 2001), including functions and libraries written by myself and other people, with the following exceptions:

- ANOAS (Eliason, 1986) was used to fit the Goodman RC association model.

- Egret (Cytel Software Corporation, 2001a) was used for beta-binomial regression.

- LIMDEP (Greene, 2000a) was used for zero inflated count regression and truncated Poisson regression.

- LogXact (Cytel Software Corporation, 1999) was used to conduct conditional analyses of logistic regression models.

- SAS (SAS Institute, 2000) was used to fit some ordinal regression models.

- SPSS (SPSS, Inc., 2001) was used to construct the Hosmer–Lemeshow statistic when fitting a binary logistic regression model.

- StatXact (Cytel Software Corporation, 2001b) was used for various conditional analyses on contingency tables.

I have set up a Web site related to the material in this book at the address http://www.stern.nyu.edu/~jsimonof/AnalCatData (a link also can be found at the Springer-Verlag Web site, http://www.springer-ny.com, under "Author Websites"). The site includes computer code, functions, and macros in S-PLUS (and the free package R, which is virtually identical to S-PLUS; see Ihaka and Gentleman, 1996), and SAS for the material in the book and the data sets used in the text and exercises (these data sets are identified by name in typewriter font in the book). Answers to selected exercises are available to instructors who adopt the book as a course text. For more information, see the book's Web site or the Springer-Verlag Web site.

I would like to thank some of the people who helped me in the preparation of this book. The students in my class on categorical data analysis helped me to focus my ideas on the subject in a systematic way. Many people provided me with interesting data sets; their names are given in the book where the data are introduced. David Firth, Mark Handcock, and Gary Simon read and commented on draft versions of the text. Yufeng Ding and Zheng Sun helped with software issues, and checked many of the computational results given in the text. John Kimmel was his usual patient and supportive self in guiding the book through the publication process. I would also like to thank my family for their support and encouragement during this long process.

East Meadow, New York   Jeffrey S. Simonoff
                                                  May 2003

Contents

1
Introduction

1.1 The Nature of Categorical Data

Students whose complete exposure to statistical methods comes from an introductory statistics class can easily get the impression that, with few exceptions, data analysis is all about examining continuous data, and more specifically, Gaussian (normally) distributed data. Anyone who actually *analyzes* data, however, quickly learns that there's more to the world than just the normal distribution. In particular, many of the data problems faced in practice involve categorical data — that is, qualitative data where the possible values that a variable can take on form a set of categories. For example, a market researcher might ask respondents whether they use a particular product ("yes"/"no"), how often they use it ("never"/"less than once per week"/"between 4 and 10 times per month"/"more than 10 times per month"), how satisfied they are with the product ("completely dissatisfied"/"somewhat dissatisfied"/"neutral"/"somewhat satisfied"/"completely satisfied"), and what competitors' products they have also used ("Brand A"/"Brand B"/etc.). The first and last of these examples are *nominal* (unordered) categorical variables, while the middle two are *ordinal* (ordered). Such data often are summarized in the form of tables of counts (or *contingency tables*). Standard Gaussian-based methods are not appropriate for such data, and methods specifically designed for categorical data are necessary.

A somewhat ambiguous question is whether (quantitative) *discrete* data also should be considered categorical. For example, if a person is asked how many traffic tickets they've received in a given year, the possible responses take the form of a set of integers $(0, 1, 2, \ldots)$. Should these integers also be considered categories? If the set of possible responses is small, the same methods that are appropriate for purely categorical data are certainly appropriate. If the number of possible responses in large, there is typically little practical difference between treating the variable as discrete or continuous. Despite this, there are still advantages to recognizing the discrete nature of the data, since such analyses can be more effective in addressing the specific characteristics of such data (such as the variance of a random variable being directly related to its mean).

This book addresses data of this sort. The guiding principle will be to highlight and exploit connections with Gaussian-based methodology, especially regression modeling. The approach is heavily applied and data-driven, with the focus being on the practical aspects of data analysis rather than formal mathematical theory. For this reason, many of the issues that arise relating to, for example, identifying unusual observations, checking assumptions, model selection, and so on, are relevant to all data analysis, not just data analysis involving categorical data.

Some examples of the categorical data analyses that we will look at include the following:

1. The winning three-digit number for the Pick 3 game of the Texas Lottery is chosen using three machines, wherein one of 10 balls (numbered 0 through 9) is selected. A particular machine was pulled from use because of suspicions that it wasn't working correctly. A sample of 150 drawings from the machine was taken and examined. The balls are loaded into the machine in sequential order (low numbers on the bottom) before mixing, so there is a natural ordering to the ten categories $\{0, 1, \ldots, 9\}$. Is there evidence that the machine is not working correctly? This is a question relating to goodness-of-fit for a univariate (multinomial) categorical variable.

2. The Storm Prediction Center, in Norman, Oklahoma, is the agency charged with recording and reporting storm events in the United States. Among the information recorded and presented to the public is the number of deaths from tornadoes in the United States by month. This discrete variable can take on nonnegative integer values. Is it possible to model this discrete random variable as a function of other variables using an appropriate discrete regression model?

3. A survey was given to 364 undergraduate business students designed to examine the key factors that are involved in selecting a major. The study investigated the timing of the major decision as well as the related perceptions and preferences of marketing majors, nonmajors, and undecided students. The timing variable takes on values in the set {Before college, Freshman, Sophomore, Junior, Senior}; the

major variable takes on values in the set {Marketing, Nonmarketing, Undecided}; questions on how important different factors were in choosing a major take on values according to a five-point scale from "Not at all important" to "Extremely important"; stated opinions about the marketing major resulted in responses on a five-point scale from "Strongly disagree" to "Strongly agree." How do these different categorical variables (some where the categories have a natural ordering) relate to each other?

4. On January 28, 1986, the space shuttle *Challenger* took off on the 25th flight on the National Aeronautics and Space Administration's (NASA) space shuttle program. Less than two minutes into the flight, the spacecraft exploded, killing all on board. A Presidential Commission was appointed to determine the cause of the accident. They ultimately did so, and wrote the *Report of the Presidential Commission on the Space Shuttle Challenger Accident (1986)*. Each space shuttle uses two booster rockets to help lift it into orbit. Each booster rocket consists of several pieces whose joints are sealed with rubber O-rings, which are designed to prevent the release of hot gases produced during combustion. Each booster contains three primary O-rings, for a total of six for the craft. If an O-ring fails to recover its shape after compression is removed, a joint will not be sealed, which can result in a gas leak and explosion. In the 23 previous flights for which there were data (the hardware for one flight was lost at sea), the O-rings were examined for damage. For each flight the number of damaged O-rings can take on values in $\{0, 1, \ldots, 6\}$. What factors might be related to the probability of O-ring damage?

5. The identification of the authorship of disputed works is important in historical research, and also in criminal investigation from written evidence. Quantitative methods can be used for this based on numerical summaries of authorship styles. A study based on the counts of different function words (words with very little contextual meaning) in samples of the works of four authors is designed to build models that can efficiently distinguish the authors from one another. Which words are useful for this purpose?

1.2 Organization of This Book

This book is divided into three broad parts. The first, consisting of Chapters 2 and 3, reviews Gaussian-based data-analytic methods. After a brief discussion of univariate (one-variable) normally distributed data, the bulk of Chapter 2 examines the linear model, including least squares estimation, interval estimation and hypothesis testing, and checking of assumptions.

Virtually every introductory statistics course, at virtually any mathematical level, focuses on the normal distribution and least squares regression, and it is not at all unreasonable for the reader to consider skipping this chapter altogether. I would discourage that, however, for two reasons. First, a little review is never a bad thing. Second, and more importantly, the familiarity of Gaussian-based analyses is a crucial link to understanding the Poisson- and binomial-based analyses that are the foundation of categorical data analysis. This link is repeatedly exploited in later chapters.

Chapter 3 extends the discussion of least squares to more advanced topics related to the general area of model fitting and model selection. This includes the construction of hypothesis tests for linear contrasts, methods for choosing which predictors are needed in a model, and models that include categorical predictors (including interactions of those predictors), and regression diagnostics and assumption checking. Much of the material in this chapter is not typically covered in an introductory course, making it likely that the reader who takes a shortcut to Chapter 4 will find herself referring back to the chapter anyway.

The second part of the book, Chapters 4 through 8, examines the modeling of count data, including as a function of other variables. Chapter 4 introduces the building blocks of categorical data analysis. The chapter focuses on the important distributions for categorical data, the Poisson, binomial, multinomial, negative binomial, and beta-binomial. The formulation of tests of goodness-of-fit based on a large-sample approximation to the χ^2 distribution is described, and approaches when such an approximation is unwarranted are addressed through the use of exact inference.

Regression modeling for count data is the subject of Chapter 5. The concept of the generalized linear model allows for the fitting of loglinear regression models based on the Poisson distribution, with many direct parallels to least squares regression. We discuss the handling of count data with variation inconsistent with the Poisson (under- and overdispersion) using both model-based and non–model-based methods. Chapter 6 extends this modeling to two-dimensional contingency tables, which can be viewed as loglinear models based on categorical predictors. Model fitting when there is additional structure on the table, such as when one or both of the categorizing variables in the table have a natural ordering, or square tables (when the row and column categorizations are the same) is the focus of Chapter 7, and extensions of the models of Chapters 6 and 7 to higher dimensional tables are discussed in Chapter 8.

Chapters 9 and 10 constitute the third part of the book. Chapter 9 discusses regression analysis of binary data using logistic regression and its competitors. Chapter 10 then builds on such models to allow for response variables with multiple categories, which might or might not have a natural ordering.

All of the succeeding chapters end with the section "Background Material." Each section provides references to books or articles related to the material in the chapters, including material that is relevant but not directly covered in the main text, for those readers who might need to go further in analyzing their own data. Citations to relatively recent results in categorical data modeling that are mentioned in the chapter are also given here. All of the remaining chapters include exercises, which are worth studying. Besides providing new test beds for the methods discussed in the chapter, some also include discussion of methods and types of analyses that are not in the main text.

.

2
Gaussian-Based Data Analysis

In the next two chapters we examine univariate and regression analysis based on the central distribution of statistical inference and data analysis, the normal, or Gaussian, distribution. It is important to note that the brief overview of least squares regression given here is not a substitute for the thorough discussion that would appear in a good regression textbook. See the "Background material" section of this chapter for several examples of such books.

2.1 The Normal (Gaussian) Random Variable

2.1.1 The Gaussian Density Function

The most important continuous random variable (indeed, the most important random variable of any type) is the *normal*, or *Gaussian* random variable. A random variable is defined to be normally distributed if its density function is given by

$$f(x; \mu, \sigma^2) = \frac{1}{\sqrt{2\pi}\sigma} \; e^{-\frac{(x-\mu)^2}{2\sigma^2}}, \tag{2.1}$$

which is written $X \sim N(\mu, \sigma^2)$. Figure 2.1 gives the familiar bell-shaped density for the standard normal random variable, with $\mu = 0$ and $\sigma = 1$ (solid line). The Gaussian random variable has a density that is symmetric around its mean μ, and has variance σ^2.

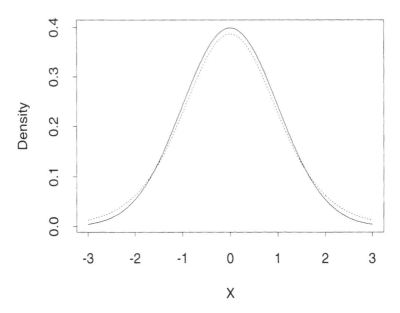

FIGURE 2.1. Density function for standard normal random variable (solid line) and t random variable on 8 degrees of freedom (dashed line).

2.1.2   Large-Sample Inference for the Gaussian Random Variable

The Gaussian density (2.1) is based on the parameters μ and σ. Typically these parameters are unknown, and would be estimated from data. Let $\{x_1, \ldots, x_n\}$ be a random sample of size n from a normally distributed population. The probability of the observed data as a function of the parameters is called the *likelihood* function. This function is the basis for the most common statistical inference strategy, *maximum likelihood estimation*. The maximum likelihood estimator (MLE) is defined as the maximizer of the likelihood function (or, equivalently, the log-likelihood function), and thus has the appealing property of being the parameter value(s) that maximize the probability of the observed data occurring. For normally distributed data, the log-likelihood has the form

$$L = \sum_{i=1}^{n} \log f(x_i; \mu, \sigma^2)$$

$$= -n \log \sigma - \sum_{i=1}^{n} \frac{(x_i - \mu)^2}{2\sigma^2},$$

where constants that are not functions of (μ, σ) are dropped out. Differentiating this function with respect to μ and σ and setting the partial derivatives to zero yields the score equations,

$$\frac{\partial L}{\partial \mu} = \frac{\sum_{i=1}^{n}(x_i - \mu)}{\sigma^2}$$

$= 0$ and

$$\frac{\partial L}{\partial \sigma^2} = -\frac{n}{2\sigma^2} + \left[\frac{\sum_{i=1}^{n}(x_i - \mu)^2}{2\sigma^4}\right] = 0.$$

Solving these equations gives the maximum likelihood estimates

$$\hat{\mu} = \overline{X} = \frac{1}{n}\sum_{i=1}^{n} x_i$$

(the sample mean), and

$$\hat{\sigma}^2 = \frac{1}{n}\sum_{i=1}^{n}(x_i - \hat{\mu})^2$$

(the sample variance multiplied by $(n-1)/n$).

Under standard conditions, the MLE is asymptotically (that is, for large samples) normally distributed. This provides the basis for data-based inference. In general, the MLE for a parameter θ is approximately Gaussian for large samples, with mean equal to θ, and variance equal to the inverse of the so-called Fisher information I, a matrix whose (i, j)th element is

$$-E\left[\frac{\partial^2 L}{\partial \theta_i \theta_j}\right] = E\left[\frac{\partial L}{\partial \theta_i} \times \frac{\partial L}{\partial \theta_j}\right] \qquad (2.2)$$

(for a brief overview of matrix algebra, see Appendix A). A large-sample $100 \times (1-\alpha)\%$ confidence interval for any scalar parameter θ based on the MLE $\hat{\theta}$ then has the form

$$\hat{\theta} \pm z_{\alpha/2}\sqrt{\hat{V}(\hat{\theta})}, \qquad (2.3)$$

where $z_{\alpha/2}$ is appropriate Gaussian-based critical value and $\hat{V}(\hat{\theta})$ is the estimated variance of $\hat{\theta}$.

For a Gaussian random sample,

$$I_{11} = -E\left(\frac{\partial^2 L}{\partial \mu^2}\right)$$

$$= n/\sigma^2$$

$$I_{12} = -E\left(\frac{\partial^2 L}{\partial \mu \partial \sigma^2}\right)$$

$$= E\left(\frac{\sum(x_i - \mu)}{\sigma^4}\right)$$

$$= 0$$

$$I_{21} = I_{12}$$

$$I_{22} = -E\left(\frac{\partial^2 L}{\partial(\sigma^2)^2}\right)$$

$$= n/(2\sigma^4)$$

The estimated asymptotic variance matrix of $(\hat{\mu}, \hat{\sigma}^2)$ is thus

$$\begin{pmatrix} n/\hat{\sigma}^2 & 0 \\ 0 & n/(2\hat{\sigma}^4) \end{pmatrix}^{-1} = \begin{pmatrix} \hat{\sigma}^2/n & 0 \\ 0 & 2\hat{\sigma}^4/n \end{pmatrix}.$$

An approximate 95% confidence interval for μ, for example, is

$$\overline{X} \pm z_{\alpha/2}\frac{\hat{\sigma}}{\sqrt{n}}. \tag{2.4}$$

Note that the asymptotic covariance (and hence the correlation) of $\hat{\mu}$ and $\hat{\sigma}^2$ is zero; in fact, these two estimates are statistically independent for finite samples as well. The so-called observed information matrix, which corresponds to (2.2) using the observed rather than the expected second partial derivatives, also can be used to construct asymptotic confidence intervals. For the Gaussian model the observed and Fisher information matrices are identical, leaving no question as to the correct form of the asymptotic intervals. Still another consistent estimator of the asymptotic covariance matrix of maximum likelihood estimators, the outer-product estimator, is based on the observed values of the first partial derivatives given in (2.2).

Say that a researcher was interested in testing a specific claim about an unknown parameter, such as that θ equals some specified value θ_0. A large-sample test of the null hypothesis

$$H_0 : \theta = \theta_0$$

versus the two-sided alternative

$$H_a : \theta \neq \theta_0$$

is also based on asymptotic normality, with the statistic

$$z = \frac{\hat{\theta} - \theta_0}{\sqrt{\widehat{V_0}(\hat{\theta})}} \tag{2.5}$$

being compared to a two-tailed standard normal reference through the use of a p-value (the probability that a standard normal random variable would be larger in absolute value than $|z|$). Here $\widehat{V_0}(\hat{\theta})$ is the estimated variance of $\hat{\theta}$ under the null hypothesis (when $\theta = \theta_0$), which is often, but not always, the same as the unrestricted estimated variance used in (2.3). In the case of a Gaussian mean, the statistic has the form

$$z = \frac{\overline{X} - \mu_0}{\hat{\sigma}/\sqrt{n}}. \tag{2.6}$$

The similarity of (2.3) and (2.5) is, of course, no accident, as confidence intervals and hypothesis tests are two sides of the same inferential coin. In particular, an observed $\hat{\theta}$ can be defined to be statistically significantly different from θ_0 if θ_0 is not in the $100 \times (1 - \alpha)\%$ confidence interval for θ, and a $100 \times (1 - \alpha)\%$ confidence interval can be defined as being those values of θ such that the observed $\hat{\theta}$ is not statistically significantly different from them at level α. If the null-restricted and unrestricted estimates of the variance of $\hat{\theta}$ are the same (as is the case for Gaussian data), these alternative constructions are equivalent, but otherwise they lead to different tests and intervals.

2.1.3 Exact Inference for the Gaussian Random Variable

The confidence interval (2.4) and hypothesis test (2.6) are valid if the sample is large enough, but what about small samples? By "valid" we mean that a 95% confidence interval, say, really does cover the true mean 95% of the time (the *coverage* of the interval), and a test conducted at a .05 level, say, really does reject the null hypothesis 5% of the time when the null is true (the *size* of the test). Large-sample intervals and tests can, in general, have true coverage and size higher or lower than the expected (nominal) values. It is easy to show that the coverage of the interval (2.4) is too low and the size of the test (2.6) is too high for small samples.

The Gaussian random variable has (among many nice properties) the property that exact (for any sample size) inference is possible, and in fact, easy. With just a slight adjustment to (2.4) and (2.6), respectively, intervals and tests can be constructed that have the exact coverage and size, respectively, for any size sample. These are based on the t-distribution, a random variable with a density function that is similarly shaped to the normal, but with fatter tails. The t-distribution is indexed by its degrees of freedom, with infinite degrees of freedom corresponding to the Gaussian distribution (refer again to Figure 2.1 on page 8, where the dashed line is the density for a t random variable on 8 degrees of freedom). The exact $100 \times (1 - \alpha)\%$ confidence interval for a Gaussian mean has the form

$$\overline{X} \pm t_{\alpha/2}^{(n-1)} \frac{s}{\sqrt{n}},$$

where $t_{\alpha/2}^{(n-1)}$ is the appropriate critical value based on a t-distribution on $n - 1$ degrees of freedom, and $s^2 = \sum(x_i - \overline{X})^2/(n - 1)$ is the sample variance. Analogously, the exact hypothesis test corresponding to (2.6) is

$$t = \frac{\overline{X} - \mu_0}{s/\sqrt{n}},$$

which is referenced to a t-distribution on $n - 1$ degrees of freedom.

2.2 Linear Regression and Least Squares

The analysis of data using regression modeling is characterized by two things: one particular variable that we are interested in understanding or modeling, the *target* variable (usually represented by y), and a set of p other variables that we think might be useful in predicting or modeling the target variable, the *predicting* variables (usually represented by x_1, x_2, etc.). Typically, a regression analysis is used for one (or more) of three purposes: prediction of the target variable (forecasting), modeling the relationship between $\mathbf{x}$ and y, and testing hypotheses.

2.2.1 The Linear Regression Model

Standard regression analysis is based on the linear model. The model can be characterized as follows. The data consist of n sets of observations $\{x_{1i}, x_{2i}, \ldots, x_{pi}, y_i\}$, and it is assumed that these observations satisfy a linear relationship,

$$y_i = \beta_0 + \beta_1 x_{1i} + \cdots + \beta_p x_{pi} + \varepsilon_i, \qquad (2.7)$$

where the $\boldsymbol{\beta}$ coefficients are unknown parameters, and the ε_i are random error terms. If the data constitute a random sample from an underlying population of this form, it is possible to make inferences about the parameters. By a *linear* model, it is meant that the model is linear in the parameters; a quadratic model

$$y_i = \beta_0 + \beta_1 x_i + \beta_2 x_i^2 + \varepsilon_i,$$

paradoxically enough, is a linear model, since x and x^2 are just versions of x_1 and x_2. Linear models are simpler to understand, and they're simpler

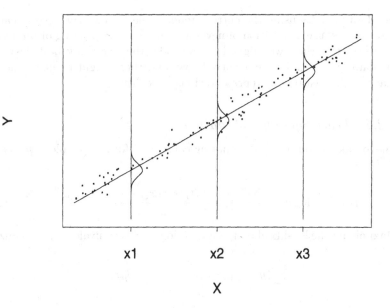

FIGURE 2.2. The simple regression model.

mathematically, but, most importantly, they work well for a wide range of circumstances (but definitely not *all* circumstances).

Making assumptions about the errors ε leads to an assumed probability structure for the target given the predictor values ($y_i | \mathbf{x}_i$, where "|" is read "given" or "conditional on"). If the ε_i are independent, normally distributed, with zero mean and constant variance σ^2, (2.7) implies that

$$y_i | \mathbf{x}_i \sim N(\beta_0 + \beta_1 x_{1i} + \cdots + \beta_p x_{pi}, \sigma^2). \tag{2.8}$$

Figure 2.2 gives a graphical representation of the linear regression model when there is only one predictor (simple regression). The true regression function is a straight line (when there are more predictors it becomes a plane or hyperplane). An observation is scattered off the line based on a Gaussian distribution with mean $\beta_0 + \beta_1 x$ and standard deviation σ. Three versions of these conditional Gaussian distributions are given for three distinct values of the predictor, but in fact the sample reflects n such distributions, one for each observation.

It's a good idea when considering this model, and any model that is discussed in this book, to remember that we do **not** believe that the linear model is an *actual* representation of reality; rather, we think that perhaps it provides a *useful* representation of reality. A model can be used to explore the relationships between variables and make accurate forecasts based on those relationships while still not being the "true" model. Further, statistical models are only provisional, representing a tentative version of beliefs about the random process being studied. They can, and should, change, based on analysis using the current model, selection among several candidate models, the acquisition of new data, and so on. Speaking of *the* model is incorrect; rather, we might have a model that we think best describes the data given what we currently know, and there might be other models that are also reasonable representations of reality.

2.2.2 Least Squares Estimation

The log-likelihood function (ignoring constants) for the sample of y's based on (2.8) is

$$L = -\frac{n}{2}\log\sigma^2 - \frac{\sum_{i=1}^{n}(y_i - \beta_0 - \beta_1 x_{1i} - \cdots - \beta_p x_{pi})^2}{2\sigma^2}. \tag{2.9}$$

Maximizing the log-likelihood as a function of $\boldsymbol{\beta}$ is equivalent to minimizing

$$\sum_{i=1}^{n}(y_i - \beta_0 - \beta_1 x_{1i} - \cdots - \beta_p x_{pi})^2. \tag{2.10}$$

Let $\hat{\boldsymbol{\beta}} = \{\hat{\beta}_0, \ldots, \hat{\beta}_p\}$ be the maximum likelihood estimates from a given set of data. Substituting these estimates into (2.7) gives the *fitted values*,

$$\hat{y}_i = \hat{\beta}_0 + \hat{\beta}_1 x_{1i} + \cdots + \hat{\beta}_p x_{pi}.$$

The set of differences between the observed and fitted target values,

$$r_i = y_i - \hat{y}_i,$$

are called the *residuals*. From (2.10) it is obvious that the MLE minimizes the sum of squares of the residuals; hence, the name least squares regression.

The model (2.7) can be written more succinctly using matrix notation. Define the following matrix and vectors as follows:

$$X = \begin{pmatrix} 1 & x_{11} & \cdots & x_{p1} \\ \vdots & \vdots & & \vdots \\ 1 & x_{1n} & \cdots & x_{pn} \end{pmatrix} \quad \mathbf{y} = \begin{pmatrix} y_1 \\ \vdots \\ y_n \end{pmatrix} \quad \boldsymbol{\beta} = \begin{pmatrix} \beta_0 \\ \beta_1 \\ \vdots \\ \beta_p \end{pmatrix} \quad \boldsymbol{\varepsilon} = \begin{pmatrix} \varepsilon_1 \\ \vdots \\ \varepsilon_n \end{pmatrix}.$$

The regression model can then be written succinctly as

$$\mathbf{y} = X\boldsymbol{\beta} + \boldsymbol{\varepsilon}.$$

The normal equations (which determine the least squares estimates of $\boldsymbol{\beta}$) can be shown (using multivariable calculus) to be

$$(X'X)\boldsymbol{\beta} = X'\mathbf{y},$$

which implies that the least squares estimates satisfy

$$\hat{\beta} = (X'X)^{-1}X'\mathbf{y}.$$

The fitted values are then

$$\hat{\mathbf{y}} = X\hat{\beta} = X(X'X)^{-1}X'\mathbf{y} \equiv H\mathbf{y},$$

where $H = X(X'X)^{-1}X'$ is the so-called "hat" matrix. The residuals $\mathbf{r} = \mathbf{y} - \hat{\mathbf{y}}$ thus satisfy

$$\mathbf{r} = \mathbf{y} - \hat{\mathbf{y}} = \mathbf{y} - X(X'X)^{-1}X'\mathbf{y} = (I - X(X'X)^{-1}X')\mathbf{y},$$

or

$$\mathbf{r} = (I - H)\mathbf{y}.$$

Two useful properties of the hat matrix H are that it is symmetric ($H = H'$) and idempotent ($H = H^2$).

The MLE of the variance of the errors σ^2 is also determined using multivariable calculus. The MLE equals

$$\tilde{\sigma}^2 = \frac{\sum_{i=1}^{n}(y_i - \hat{y}_i)^2}{n}.$$

This is not the estimate typically used, however. Instead, the residual mean square, an unbiased estimate of σ^2, is used, which equals

$$\hat{\sigma}^2 = \frac{\sum_{i=1}^{n}(y_i - \hat{y}_i)^2}{n - p - 1}.$$

2.2.3 Interpreting Regression Coefficients

The least squares regression coefficients have very specific meanings. They are often misinterpreted, so it is important to be clear on what they are (and are not). Consider first the intercept, $\hat{\beta}_0$.

$\hat{\beta}_0$: The estimated expected value of the target variable when the predictors all equal zero.

Note that this might not have any physical interpretation, since a zero value for the predictors might be meaningless. In this situation, it is pointless to try to interpret this value. This suggests the possibility of transforming the predictors so that a zero value is meaningful. If all of the predictors are centered to have mean zero, then $\hat{\beta}_0$ necessarily equals $\overline{Y}$, the sample mean of the target values.

The estimated coefficient for the jth predictor ($j = 1, \ldots, p$) is interpreted in the following way.

$\hat{\beta}_j$: The estimated expected change in the target variable associated with a one unit change in the jth predicting variable, holding all else in the model fixed.

There are several noteworthy aspects to this interpretation. Note the word *associated* — we cannot say that a change in the target variable is **caused** by a change in the predictor, only that they are associated with each other. That is, correlation does not imply causation.

Another key point is the phrase "holding all else in the model fixed," the implications of which are often ignored. Consider the following hypothetical example. A random sample of college students at a particular university is taken in order to understand the relationship between college grade point average (GPA) and other variables. A model is built with college GPA as a function of high school GPA and the standardized Scholastic Aptitude Test (SAT), with resultant least squares fit

College GPA = 1.3 + .7 × High School GPA − .0001 × SAT score.

It is tempting to say (and many people would say) that the coefficient for SAT score has the "wrong sign," because it says that higher values of SAT are associated with lower values of college GPA. This is not correct. The problem is that it is likely in this context that what an analyst would find intuitive is the *marginal* relationship between college GPA and just SAT score (ignoring all else), one that we would indeed expect to be a direct (positive) one. The regression coefficient does not say anything about that marginal relationship. Rather, it refers to the conditional relationship, which is apparently that higher values of SAT are associated with lower values of college GPA, holding high school GPA fixed. High school GPA and SAT are no doubt related to each other, and it is quite likely that this relationship would complicate any understanding of, or intuition about, the conditional relationship between college GPA and SAT score. Multiple regression coefficients should not be interpreted marginally. If you really are interested in the relationship between college GPA and just SAT score, you should simply do a regression of college GPA on only SAT score.

This does not mean that multiple regression coefficients are uninterpretable, only that care is necessary when interpreting them. In some situations it is clear that the conditional interpretation of regression coefficients is precisely what is desired. For example, consider the problem of estimating the demand for a product as a function of the price of the product, as well as other variables (such as the price of substitute products, complementary products, macroeconomic measures, etc.), using annual data. The coefficient for the price of the product in a multiple regression represents the change in demand associated with a change in price, holding all of the other economic factors fixed, which is exactly what is wanted. The marginal coefficient from a regression of demand on price alone is not very useful, since it doesn't take into account the other factors (other than price of the product) that would affect the demand.

Another common use of multiple regression that depends on this conditional interpretation of the coefficients is to explicitly include "control" variables in a model in order to try to account for their effect statistically. This is particularly important in observational data (data that are not the result of a designed experiment), since in that case the effects of other variables cannot be ignored because of random assignment in the experiment. For observational data it is not possible to physically intervene in the experiment to "hold other variables fixed," but the multiple regression framework effectively allows this to be done statistically.

2.2.4 Assessing the Strength of a Regression Relationship

The construction of the least squares estimates implies that

$$\sum_{i=1}^{n}(y_i - \overline{Y})^2 = \sum_{i=1}^{n}(y_i - \hat{y}_i)^2 + \sum_{i=1}^{n}(\hat{y}_i - \overline{Y})^2.$$

This formula says that the variability in the target variable (the left side of the equation, termed the corrected total sum of squares) can be split into two mutually exclusive parts — the variability left over after doing the regression (the first term on the right side, the residual sum of squares), and the variability accounted for by doing the regression (the second term, the regression sum of squares). This immediately suggests the desirability of a large R^2, where

$$R^2 = \frac{\sum_i(\hat{y}_i - \overline{Y})^2}{\sum_i(y_i - \overline{Y})^2} \equiv \frac{\text{Regression SS}}{\text{Corrected total SS}} = 1 - \frac{\text{Residual SS}}{\text{Corrected total SS}}.$$

The R^2 value (also called the coefficient of determination) estimates the proportion of variability in y accounted for by the predictors in the population. Values closer to 1 indicate a good deal of predictive power of the predictors for the target variable, while values closer to 0 indicate little predictive power. Another (equivalent) formula for R^2 is

$$R^2 = \text{corr}(y_i, \hat{y}_i)^2, \tag{2.11}$$

where

$$\text{corr}(y_i, \hat{y}_i) = \frac{\sum_i(y_i - \overline{Y})(\hat{y}_i - \overline{\hat{Y}})}{\sqrt{\sum_i(y_i - \overline{Y})^2 \sum_i(\hat{y}_i - \overline{\hat{Y}})^2}}$$

is the sample correlation coefficient between $\mathbf{y}$ and $\hat{\mathbf{y}}$. That is, it is a direct measure of how similar the observed and fitted target values are.

2.3   Inference for the Least Squares Regression Model

2.3.1   Hypothesis Tests and Confidence Intervals for $\boldsymbol{\beta}$

It is always possible to fit a linear regression model to a data set, but does doing so bring any understanding of the underlying process? Are patterns in scatter plots reflecting a genuine relationship, or simply due to random chance? Do some variables provide predictive power for y, while others add little of importance? These can be viewed as hypothesis testing questions.

There are two types of hypothesis tests of immediate interest.

1. Do *any* of the predictors provide predictive power for the target variable? This is a test of the overall significance of the regression,

$$H_0 : \beta_1 = \cdots = \beta_p = 0$$

versus

$$H_a : \text{some } \beta_j \neq 0, \qquad j = 1, \ldots, p.$$

The test of these hypotheses is the *F-test*,

$$F = \frac{\text{Regression MS}}{\text{Residual MS}} \equiv \frac{\text{Regression SS}/p}{\text{Residual SS}/(n - p - 1)}. \qquad (2.12)$$

This is referenced against a null F-distribution on $(p, n-p-1)$ degrees of freedom.

2. Given the other variables in the model, does a particular predictor provide additional predictive power? This corresponds to a test of the significance of an individual coefficient,

$$H_0 : \beta_j = 0, \qquad j = 0, \ldots, p$$

versus

$$H_a : \beta_j \neq 0.$$

This is tested using a *t-test*,

$$t_j = \frac{\hat{\beta}_j}{\text{s.e.}(\hat{\beta}_j)}, \qquad (2.13)$$

which is compared to a t-distribution on $n - p - 1$ degrees of freedom. Other values of β_j can be specified in the null hypothesis (say β_{j0}), with the t-statistic becoming

$$t_j = \frac{\hat{\beta}_j - \beta_{j0}}{\text{s.e.}(\hat{\beta}_j)}.$$

The values of $\widehat{s.e.}(\hat{\beta}_j)$ are obtained as the square roots of the diagonal elements of $\hat{V}(\hat{\beta}) = (X'X)^{-1}\hat{\sigma}^2$, where $\hat{\sigma}^2$ is the residual mean square,

$$\hat{\sigma}^2 = \frac{\sum_{i=1}^{n}(y_i - \hat{y}_i)^2}{n - p - 1}, \qquad (2.14)$$

the estimate of the variance of the errors, σ^2. As was noted earlier, this is almost, but not quite, the MLE of σ^2, which has n in the denominator instead of $n - p - 1$.

As always, a confidence interval provides an alternative way of summarizing the degree of precision in the estimate of a regression parameter. That is, a $100 \times (1 - \alpha)\%$ confidence interval for β_j has the form

$$\hat{\beta}_j \pm t_{\alpha/2}^{n-p-1}\widehat{s.e.}(\hat{\beta}_j).$$

F- and t-tests provide information about statistical significance, but they can't say anything about the practical importance of the model. Does knowing $x_1, \ldots, x_p$ really tell you anything of value about y? This isn't a question that can be answered completely statistically; it requires knowledge and understanding of the data. Statistics can help, though. Recall that we assume that the errors have standard deviation σ. That means that, roughly speaking, we would expect to know the value of y to within $\pm 2\sigma$ after doing the regression most (95%) of the time (since the errors are assumed to be normally distributed). The square root of the residual mean square (2.14), termed the standard error of the estimate, provides an estimate of σ that can be used in this formula.

2.3.2 Interval Estimation for Predicted and Fitted Values

An even more accurate assessment of this is provided by a *prediction interval* given a particular value of $\mathbf{x}$. This interval provides guidance as to how accurate $\hat{y}_0$ is as a prediction of y for some particular value $\mathbf{x}_0$; its width depends on both $\hat{\sigma}$ and the position of $\mathbf{x}_0$ relative to the centroid of the predictors (the point located at the means of all predictors), since values further from the centroid are harder to predict. Specifically, for a simple regression, the standard error of a predicted value based on a value x_0 of the predicting variable is

$$\widehat{s.e.}(\hat{y}_0^P) = \hat{\sigma}\sqrt{1 + \frac{1}{n} + \frac{(x_0 - \overline{X})^2}{\sum(x_i - \overline{X})^2}}$$

(we might term this more precisely the *estimated* standard error of $\hat{y}_0^P$, since an estimate of σ is used). More generally,

$$\hat{V}(\hat{y}_0^P) = [1 + \mathbf{x}_0'(X'X)^{-1}\mathbf{x}_0]\hat{\sigma}^2$$

(again, more correctly, this is the estimated variance of $\hat{y}_0^P$). Here $\mathbf{x}_0$ is taken to include a 1 in the first entry (corresponding to the constant term). The prediction interval is then

$$\hat{y}_0 \pm t_{\alpha/2}^{n-p-1} \widehat{\text{s.e.}}(\hat{y}_0^P),$$

where $\widehat{\text{s.e.}}(\hat{y}_0^P) = \sqrt{\hat{V}(\hat{y}_0^P)}$.

The prediction interval should not be confused with a *confidence interval* for a fitted value, which will be narrower. The prediction interval is used to provide an interval estimate for a prediction of y for one member of the population with a particular value of $\mathbf{x}_0$; the confidence interval is used to provide an interval estimate for the true average value of y for all members of the population with a particular value of $\mathbf{x}_0$. The corresponding standard error, termed the standard error for a fitted value, is the square root of

$$\hat{V}(\hat{y}_0^F) = \mathbf{x}_0'(X'X)^{-1}\mathbf{x}_0\hat{\sigma}^2,$$

with corresponding confidence interval

$$\hat{y}_0 \pm t_{\alpha/2}^{n-p-1} \widehat{\text{s.e.}}(\hat{y}_0^F).$$

2.4   Checking Assumptions

As was noted earlier, all of these tests, intervals, predictions, and so on, are based on believing that the assumptions of the regression hold. Thus, it is crucially important that these assumptions be checked. Remarkably enough, a few very simple plots can provide most of the evidence needed to check the assumptions.

1. A plot of the residuals versus the fitted values. This plot should have no pattern to it; that is, no structure should be apparent. Certain kinds of structure indicate potential problems:

    (a) A point (or a few points) isolated at the top or bottom, or left or right. In addition, often the rest of the points have a noticeable "tilt" to them. These isolated points are unusual points, and can have a strong effect on the regression. They need to be examined carefully, and possibly removed from the data set.

    (b) An impression of different heights of the point cloud as the plot is examined from left to right. This indicates heteroscedasticity (nonconstant variance).

2. Plots of the residuals versus each of the predictors. Again, a plot with no apparent structure is desired.

3. If the data set has a time structure to it, residuals should be plotted versus time. Again, there should be no apparent pattern. If there is a cyclical structure, this indicates that the errors are not uncorrelated, as they are supposed to be (that is, there is *autocorrelation*).

4. A normal plot of the residuals. This plot assesses the apparent normality of the residuals, by plotting the observed ordered residuals on one axis and the expected positions (under normality) of those ordered residuals on the other. The plot should look like a straight line (roughly). Isolated points once again represent unusual observations, while a curved line indicates that the errors are probably not normally distributed, and tests and intervals might not be trustworthy.

2.5   An Example

In this section we illustrate the methods of the previous sections in analyzing a real data set. In this analysis, and in all further data analyses, discussion of the data set is separated from the textual material by grey rules.

Monthly Tornado Mortality Statistics

The Storm Prediction Center, an agency of the National Oceanic and Atmospheric Administration of the U.S. Department of Commerce, tracks (among many other things) the number and characteristics of tornadoes in the United States. Medical and civil defense authorities are interested in understanding and predicting the patterns and consequences of these dangerous storms. We would expect that the number of monthly deaths from tornadoes is related to the number of tornadoes or (perhaps even more so) the number of killer tornadoes. Is it possible to build a model relating the number of deaths to the number of tornadoes? The data we will examine are the final numbers of tornadoes, killer tornadoes, and tornado deaths, for the months January 1996 through December 1999. The data are from the World Wide Web page of the Storm Prediction Center (http://www.spc.noaa.gov), and are given in the file **tornado**.

The first step in any analysis is to **look** at the data. In the regression context, that means looking at histograms of the target and predictors, and scatter plots of the target variable versus each predictor. We will not give all of the plots that would be constructed, but in an actual analysis all should be examined.

Figure 2.3 gives scatter plots of the number of tornado deaths versus each of the predictors. Neither plot is very satisfactory from the point of view of the hypothesized linear least squares model. While both plots show a direct

Number of deaths versus number of tornadoes

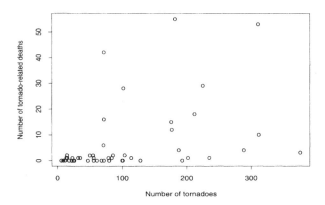

Number of deaths versus number of killer tornadoes

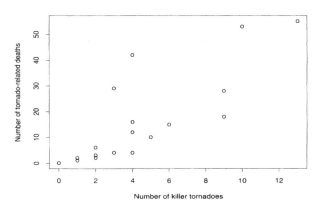

FIGURE 2.3. Scatter plots of number of tornado-related deaths versus number of tornadoes (top plot) and number of killer tornadoes (bottom plot).

relationship (as would be expected), there is noticeably higher variability of the number of deaths to the right of the plots than to the left. Further, the relationship in each plot seems more nonlinear than linear, starting out relatively flat on the left and then curving upwards to the right. Normally this would immediately suggest changing the hypothesized model, but for expository purposes we will ignore this for now.

Fitting a linear model via least squares yields the fitted equation

$$\text{Number of deaths} = -.846 - .007 \times \text{Number of tornadoes}$$
$$+ 4.005 \times \text{Number of killer tornadoes},$$

with the following output.

Analysis of Variance

Source	df	SS	MS	F	p
Regression	2	6227.2	3113.6	67.61	0.000
Residual	45	2072.4	46.1		
Total	47	8299.7			

Predictor	Coef	s.e.	t	p
Constant	-0.846	1.510	-0.56	0.578
Tornadoes	-0.007	0.013	-0.56	0.577
Killer tornadoes	4.005	0.390	10.28	0.000

The two predictors do provide significant predictive power, with a highly significant F-statistic, and an R^2 of roughly 75%. The standard error of the estimate ($\hat{\sigma}$) is 6.8, which means that 95% of the time the number of deaths due to tornadoes in a month can be predicted to within roughly ±13.6. The standard deviation of the number of deaths (before doing the regression) is roughly 13.3, so the regression has narrowed the prediction error by almost half. While the t-statistic for number of killer tornadoes shows that this predictor is highly significant, apparently the number of tornadoes adds nothing of importance after the number of killer tornadoes is given (a not unintuitive result).

At this point we could simplify the model by omitting the number of tornadoes as a predictor, but before doing that, consider Figure 2.4. This is a plot of the residuals versus the fitted values for this model, and it clearly indicates problems with this model. The point cloud in the plot widens from left to right, indicating nonconstant variance.

There is also curvature in the plot, which demonstrates that the wrong model is being fit. The dashed line is a simple scatter plot smoother that shows the dominant pattern in the plot. This *median-based regressygon* splits the data into 12 blocks of four observations each, where the blocks are determined by ordering on the horizontal axis. The regressygon then plots the medians of the residuals versus the medians of the fitted values within each block, and connects them with straight lines. If the assumed regression model held, the regressygon would be relatively straight and horizontal, since there would be no relationship between the level of the residuals and the fitted values. In this case there is noticeable curvature, with the residuals first decreasing and then increasing with increasing fitted value.

Ignoring these problems for the moment, the simple model based only on the number of killer tornadoes is appropriate, with the following summary output.

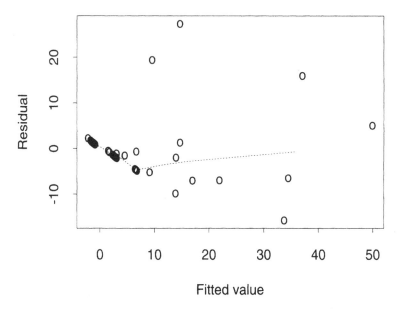

FIGURE 2.4. Scatter plot of residuals versus fitted values for linear least squares model. The dashed line is a median-based regressygon scatter plot smoother.

```
Analysis of Variance

Source        df         SS         MS         F          p
Regression     1     6212.7     6212.7     136.94     0.000
Residual      46     2087.0       45.4
Total         47     8299.7


Predictor               Coef       s.e.          t          p
Constant              -1.365      1.186      -1.15      0.256
Killer tornadoes       3.893      0.333      11.70      0.000
```

The R^2 for this model is virtually identical to that of the two-predictor model, reinforcing that the second predictor adds nothing of importance to the model. The coefficient for the number of killer tornadoes implies that an additional killer tornado is associated with almost four more tornado-related deaths in a month. Residual plots from this model are virtually identical to those from the two-predictor model, so they will not be reproduced here. A further unsatisfactory nature of this model fit is that there were 18 months with no killer tornadoes, and for each of those months the predicted number of deaths is less than zero, a clearly nonsensical result. This could be avoided by fitting a model with constant set equal to zero (consistent with the requirement that in months where there are no killer tornadoes there are no deaths due to tornadoes), but this would not address the nonlinearity and nonconstant variance already noted in the data. An alternative approach would be to transform the target variable, using $y^* = \log(y + 1)$, for example ($\log y$ would not be defined for the months where $y = 0$). While this might address the nonlinearity and nonconstant variance somewhat, this transformation suffers from the difficulty that $E[\log(y + 1)]$ (which is what is being modeled in the analysis on the transformed target variable) is not a simple function of $E(y)$ (which is what is actually of interest) when the latter value is small, as is the case for many of the observations in this data set.

The real problem with this example is that we've ignored a fundamental aspect of these data. The target variable does not satisfy the model (2.8), and, in fact, cannot possibly satisfy it, since it only can take on nonnegative integer values. We need a regression method that is designed for such count data. That is precisely what we will develop in Chapter 5.

2.6 Background Material

The outer-product estimator of the variance of maximum likelihood estimators is often called the BHHH estimator in the econometrics literature. It was proposed by Berndt et al. (1974) in the context of nonlinear models.

There are many excellent introductory textbooks covering regression analysis, including Chatterjee et al. (1999), Draper and Smith (1998), Ryan (1997), and Weisberg (1985). Chatterjee et al. (1995) contains worked-out analyses of real data sets, including ones requiring analysis using regression methods.

The term "regressygon" was coined by Simonoff and Hurvich (1993) by analogy with the "regressogram" of Tukey (1947, 1961). The latter estimate bins the data along the predictor axis, and then estimates the conditional mean of the target using a constant within each bin, yielding an estimate that looks like a histogram; the former connects mid-bin values with straight lines, yielding an estimate that looks like a frequency polygon. These are examples of nonparametric regression estimators. Although they are simple to implement, they are not very good estimators, being nonsmooth and relatively inaccurate. Many much better smoothers have been developed in recent years. Discussion of many of these estimators can be found in Chapter 5 of Simonoff (1996).

2.7 Exercises

1. Franklin (2002) examined the efficacy of various insect repellents against mosquito bites. Researchers applied repellents to the arms of 15 volunteers, which were then inserted into a cage with a fixed number of unfed mosquitoes. The elapsed time until the first bite was then recorded. They found that the mean time until the first bite when the product *OFF! Deep Woods* was applied was $\overline{X} = 301.5$ minutes, with standard deviation $s = 38.9$ minutes.

   (a) Construct a 90% confidence interval for the expected elapsed time until the first bite for this product.

   (b) Is the observed average time until the first bite significantly more than four hours? Use $\alpha = .05$ here.

   (c) What assumptions are you making in parts (a) and (b)?

2. The following table gives the number of earthquakes with either macroseismic intensity at least VI (a slightly damaging event) or a magnitude on the Richter scale of at least 4.4 in Irpinia in Italy between 1830 and 1989 by decade (Pievatolo and Rotondi, 2000). Construct a 95% confidence interval for μ, the per-decade rate of earthquakes of this severity in the region.

Decade	Earthquakes	Decade	Earthquakes
1830–1839	2	1910–1919	5
1840–1849	1	1920–1929	2
1850–1859	4	1930–1939	2
1860–1869	1	1940–1949	0
1870–1879	0	1950–1959	2
1880–1889	0	1960–1969	4
1890–1899	2	1970–1979	2
1900–1909	3	1980–1989	5

3. Lewis Fry Richardson was a British meteorologist known for investigations of numerical weather forecasting and the mathematics of war (Hayes, 2002). He compiled a list of armed conflicts around the world, and found that for a 110-year period from the mid-1800s through the mid-1900s, there were 63 years with no "magnitude 4" conflicts (between 3163 and 31,622 deaths) beginning, 35 years with one such conflict beginning, 11 years with two conflicts, and one year with three. What is the estimated rate of magnitude 4 conflict starts per year? Is this rate significantly more than one every two years? What assumptions are you making in constructing this test? Do they seem reasonable here?

4. The movie *Jaws* has left an indelible image in the minds of the public of the potential dangers of a shark attack, but just how likely is it for someone to suffer an unprovoked attack from a shark? The American Elasmobranch Society of the Florida Museum of Natural History maintains the International Shark Attack File, a record of unprovoked shark attacks (and resultant fatalities) worldwide. The file floridashark comes from this source, and gives data on shark attacks from 1946 to 1999 for the state of Florida. I thank George H. Burgess, director of the Society, for access to these data. Patterns in Florida are particularly interesting, since the peninsular nature of the state means that almost all Florida residents and visitors are within 90 minutes of the Atlantic Ocean or Gulf of Mexico. Further, since population figures for Florida are available, it is possible to account for changes in coastal population by examining patterns in attack rates, rather than simply attack counts.

   (a) Construct a scatter plot of unprovoked shark attacks per million residents versus year. Does there appear to be a time trend in attack rate? How would you explain this pattern?

   (b) Fit a linear regression relationship of shark attack rate on year. What is the interpretation of the estimated slope coefficient? What about the estimated intercept? Baum et al. (2003) reported an estimated 10 to 20% annual drop in many shark species populations in the waters around Florida for the period 1988 to 2003; is that consistent with the observed shark attack rate?

   (c) Is there a statistically significant relationship between unprovoked shark attacks and year?

   (d) What proportion of the variability in shark attack rates is accounted for by year?

   (e) According to the U.S. Census, the population of Florida in 2000 was 15,982,378. What is the estimated number of unprovoked shark attacks in Florida in 2000? There were, in fact, 34 such attacks. Is this a surprising number, given your estimate? Construct an appropriate interval estimate for the number of unprovoked shark attacks in 2000.

   (f) Construct a plot of the residuals versus the fitted values for this model. Do the assumptions of least squares regression appear to hold? Plot the residuals in time order. Is there evidence of autocorrelation in the errors?

5. The file **anscombe** contains four data sets given in Anscombe (1973).

   (a) For each of the data sets, fit a linear regression model relating Y to X. What are the fitted equations in each case? Is X a significant predictor for Y in each case? What is so striking about your results?

(b) Construct scatter plots for each of the pairs of variables. Are the plots surprising to you? Do you think linear least squares regression is appropriate for each of these data sets? Why or why not?

6. The January 11, 2001, issue of the Burlington (VT) *Free Press* contained a story on the number of homicides in Vermont in 2000, and included a table giving the number of homicides in the state for the years 1990 through 2000: 14, 24, 21, 15, 5, 13, 11, 9, 12, 17, and 11 (Stone, 2001). Is there a time trend in homicides apparent in these data? Fit a least squares regression model on year to assess this question. Are the least squares assumptions satisfied here?

7. Part of the application package for virtually all American business schools is the score from the Graduate Management Admission Test (GMAT), which is administered by the Educational Testing Service under contract from the Graduate Management Admission Council. The file **gmat** gives data based on a sample of first and second-year fulltime M.B.A. students at the Leonard N. Stern School of Business of New York University, taken in Fall 2001. I am indebted to Ada Choi and Yan Yu for gathering these data and sharing them with me. The available variables for predicting GMAT are the number of years of work experience of the respondent, the number of days spent preparing for the test, college grade point average (GPA), and gender (**Female** = 1 for women and 0 for men).

(a) Is the overall regression relationship statistically significant?

(b) Do any of the individual predictors provide significant predictive power for GMAT given the others?

(c) What proportion of the variability in GMAT is accounted for by the regression?

(d) Construct residual plots for this model. Comment on what you see. Do the assumptions of least squares regression hold here?

(e) A second year student walks in the door. Give a rough 95% interval for the accuracy within which you could expect this model to predict her GMAT. How well do you think this model would work if she had 12 years of prior work experience?

3
Gaussian-Based Model Building

In this chapter we build on the material of the previous chapter, extending the model-building and model-checking capabilities of the least squares linear regression model. Most of this material typically is not covered in an introductory statistics course. We will focus on the aspects of advanced regression modeling that are of direct relevance to the categorical data modeling methods discussed in succeeding chapters.

3.1   Linear Contrasts and Hypothesis Tests

In the example examining deaths from tornadoes in the previous chapter, the insignificance of the t-statistic for the number of tornadoes led to a simplification of the model (from two predictors to one). This type of simplification is an example of a more general question: is a simpler version of the full model, one that is a special case of it (a *subset* of it), adequate to fit the data? For example, in a regression of college grade point averages for a sample of college students on SAT verbal score and SAT quantitative score, a natural alternative to the full regression model on both predictors is a regression on the sum of the two scores, the total SAT score. The full regression model to fit to these data is

$$\text{Grade point average}_i = \beta_0 + \beta_1 \text{SAT Verbal}_i + \beta_2 \text{SAT Quantitative}_i + \varepsilon_i,$$

while the simpler subset model is

$$\text{Grade point average}_i = \beta_0 + \gamma_1 \text{SAT Total}_i + \varepsilon_i.$$

Since the total SAT score is the sum of the verbal and quantitative scores, the subset model is a special case of the full model, with $\beta_1 = \beta_2 \equiv \gamma_1$. This equality condition is called a *linear contrast*, because it defines a linear condition on the parameters of the regression model (that is, it only involves additions, subtractions and equalities of constant multiples of the parameters).

We can now state our original question about whether the total SAT score is all that is needed in the model as a hypothesis test about this linear contrast. In any regression modeling we prefer an adequate simpler model to a more complicated one; since in a hypothesis test the null hypothesis is what we believe unless convinced otherwise, in this case, that is the simpler (subset) model. The alternative hypothesis is the full model (with no conditions on β). That is, we test

$$H_0 : \beta_1 = \beta_2$$

versus

$$H_a : \beta_1 \neq \beta_2.$$

These hypotheses are tested using a *partial F-test*. The F-statistic has the form

$$F = \frac{(\text{Residual SS}_s - \text{Residual SS}_f)/d}{\text{Residual SS}_f/(n - p - 1)}, \tag{3.1}$$

where Residual SS_s is the residual sum of squares for the subset model, Residual SS_f is the residual sum of squares for the full model, p is the number of predictors in the full model, and d is the difference between the number of parameters in the full model and the number of parameters in the subset model. This statistic is compared to an F distribution on $(d, n - p - 1)$ degrees of freedom. So, for example, for the example above, $p = 2$ and $d = 3 - 2 = 1$, so the observed F-statistic would be compared to an F distribution on $(1, n - 3)$ degrees of freedom. The F-statistic to test the overall significance of the regression (2.12) is a special case of this construction (with contrasts $\beta_1 = \cdots = \beta_p = 0$), as are the individual t-statistics (2.13) that test the significance of any variable (with contrast $\beta_j = 0$, and then $F_j = t_j^2$).

An alternative form for the F-statistic (3.1) is

$$F = \frac{(R_f^2 - R_s^2)/d}{(1 - R_f^2)/(n - p - 1)},$$

where R_s^2 is the R^2 for the subset model and R_f^2 is the R^2 for the full model. That is, if the fit of the full model (as measured by R^2) isn't much better than the fit of the subset model, the F-statistic is small, and we do not reject using the subset model. If, on the other hand, the difference in R^2 values is large (indicating much better fit for the full model), we do reject the subset model in favor of the full model.

3.2   Categorical Predictors

3.2.1   One Categorical Predictor with Two Levels

The most common way to include a categorical predictor that takes on two levels (for example, male and female for gender) is to use an indicator $(0/1)$ variable. For example, if a multiple regression model has the form

$$y_i = \beta_0 + \beta_M \text{Male}_i + \beta_1 x_{1i} + \cdots + \beta_p x_{pi} + \varepsilon_i, \tag{3.2}$$

where the variable Male equals one for males and zero for females, the coefficient β_M has the appealing interpretation as the expected difference in y between men and women given the other predictors. This is called a "constant shift" model, since it models the relationship between y and $\{x_1, \ldots, x_p\}$ as two parallel hyperplanes shifted by a constant (one hyperplane for men and one for women). The intercept is the expected value of y for females if all of the other predictors equal zero. The indicator variable is treated as any other predictor is when modeling the data, with the t-statistic associated with it assessing whether there is significant additional predictive power from knowing gender given the other predictors.

This t-statistic has a particularly familiar form when the indicator variable is the only predictor in the model. It is then testing whether the average y for one group is significantly different from the average y for the other group, and is just the ordinary two-sample t-statistic. If a regression is fit on only an indicator variable that equals zero for group 1 and one for group 2, the resultant t-statistic for the slope is equivalent to

$$t = \frac{\overline{Y}_2 - \overline{Y}_1}{s_{\text{pooled}} \sqrt{\frac{1}{n_1} + \frac{1}{n_2}}}, \tag{3.3}$$

where $\overline{Y}_1$ and $\overline{Y}_2$ are the means of y for group 1 and 2, respectively, and

$$s_{\text{pooled}}^2 = \frac{(n_1 - 1)s_1^2 + (n_2 - 1)s_2^2}{n_1 + n_2 - 2}$$

is the pooled estimate of variance, a weighted average of the group variances s_1^2 and s_2^2, weighted by (one less than) the group sample sizes n_1 and n_2.

In fact, there is nothing inherently special about using a $0/1$ variable to code two groups, as using any two distinct numerical values would lead to the same t-statistics and overall F-statistic. The only advantage of using the indicator variable is that its slope coefficient then has the natural interpretation as the expected difference in the target between the two groups given the other predictors. An alternative way to represent the two groups is using *effect codings*, where one group is coded 1 and the other is coded -1. Say model (3.2) is fit using an effect coding for gender (1 for males, -1 for females),

$$y_i = \gamma_0 + \gamma_{GM} \text{Gender}_i + \gamma_1 x_{1i} + \cdots + \gamma_p x_{pi} + \varepsilon_i. \tag{3.4}$$

All of the measures of fit and association (R^2, overall F-statistic, t-statistics for all of the predictors) will be the same in (3.2) and (3.4). Further, the coefficients for $\{x_1, \ldots, x_p\}$ will be identical in the two models (that is, $\beta_j = \gamma_j$ for $j = 1, \ldots, p$). The name "effect coding" comes from the interpretation of γ_0 and γ_{GM}. In (3.2) the expected values of y (given the other predictors equal zero) are β_0 for females and $\beta_0 + \beta_M$ for males, but the effect coding form (3.4) instead centers the equation at $\gamma_0 = \beta_0 + \beta_M/2$, which corresponds to the expected y (given the other predictors equal zero) if gender is accounted for in the model, but unknown. This is different from the intercept in a regression model on only $\{x_1, \ldots, x_p\}$, where gender is not accounted for. The expected y then moves up or down $\gamma_{GM} = \beta_M/2$ corresponding to being male or female. That is, γ_{GM} is the effect of being male, relative to gender being unknown, while $-\gamma_{GM}$ is the effect of being female, relative to gender being unknown, given that the values of all of the other predictors are fixed. Note that implicitly a slope coefficient for females, γ_{GF}, is also defined, but it is constrained to be $-\gamma_{GM}$. While using an effect coding adds little compared with using an indicator variable in the case of predictors with two levels, such variables can be very useful when dealing with categorical predictors with more than two levels, as is done in the next section.

3.2.2   One Categorical Predictor with More Than Two Levels

Consider now the generalization to a categorical predictor with $K > 2$ levels (or groups). The generalization of the constant shift model (3.2) or (3.4) is a model with K different parallel hyperplanes, one for each level. That is,

$$y_{ij} = \beta_0 + \alpha_i + \beta_1 x_{1ij} + \cdots + \beta_p x_{pij} + \varepsilon_{ij}, \quad i = 1, \ldots, K, \ j = 1, \ldots, n_i, \tag{3.5}$$

where y_{ij} is the value of y for the jth member of the ith group, α_i is the effect of being in the ith group, $\{x_1, \ldots, x_p\}$ are numerical predictors (with associated slope parameters $\{\beta_1, \ldots, \beta_p\}$), ε_{ij} is the error term, and there are n_i observations in the ith group. Note that the α values must sum to zero to make the model well-defined; this corresponds to the requirement that the implicit coefficient γ_{GF} from (3.4) be constrained to equal $-\gamma_{GM}$.

Since the α_i term is defined as the effect of being in the ith group, (3.5) is of the same form as (3.4), and would be fit using $K - 1$ effect codings. Pick one level as a reference level; say it's level K. For $\ell = 1, \ldots, K - 1$, define a predictor corresponding to the $\{x_{1ij}, \ldots, x_{pij}, y_{ij}\}$ observation as

$$z_{\ell,ij} = \begin{cases} 1 & \text{if } i = \ell \\ -1 & \text{if } i = K \\ 0 & \text{otherwise.} \end{cases}$$

That is, the effect coding for level i equals one for all observations from level i, -1 for all observations from level K, and zero otherwise. Now, the regression

$$y_{ij} = \beta_0 + \alpha_1 z_{1,ij} + \cdots + \alpha_{K-1} z_{K-1,ij} + \beta_1 x_{1ij} + \cdots + \beta_p x_{pij} + \varepsilon_{ij},$$

with implicit $\alpha_K = -\sum_{i=1}^{K-1} \alpha_i$, is equivalent to (3.5). Model (3.5) is called an *analysis of covariance* model, and if there are no numerical predictors $\{x_1, \ldots, x_p\}$ it is a *one-way analysis of variance* model. The usefulness of the categorical predictor in the model is assessed statistically using the partial F-test that compares the model (3.5) to a version that omits the α terms. One advantage of using effect codings to fit (3.5) is that the t-test for each coefficient α_i tests whether the expected target variable, given the other predictors, is higher or lower in group i relative to not knowing the group at all. That is, the individual t-statistics are a useful way to explore for which groups the expected target values are distinct from the overall level determined by the numerical predictors.

3.2.3 More Than One Categorical Predictor

Analysis of variance and covariance models can be generalized easily to more than one categorical variable. Say there are two such variables: one representing rows having I categories, and one representing columns having J categories. The two-way analysis of variance model has the form

$$y_{ijk} = \mu + \alpha_i + \beta_j + (\alpha\beta)_{ij} + \varepsilon_{ijk}, \tag{3.6}$$
$$i = 1, \ldots, I, \ j = 1, \ldots, J, \ k = 1, \ldots, n_{ij}.$$

The α and β parameters represent the *main effects* of rows and columns, respectively, and have the same general interpretation that the effect in a one-way analysis of variance does. The $(\alpha\beta)$ represents an *interaction effect*. As in model (3.5) the α, β, and $(\alpha\beta)$ terms sum to zero in order to make the model well-defined.

What does the interaction effect mean? Consider the following example. Say it was desired to investigate viewership of television shows. One way of measuring viewership is using Nielsen ratings, which are based on diaries kept by a sample of "Nielsen families." The rating is (an estimate of) the percentage of televisions tuned to a particular show at a particular time out of all televisions. Table 3.1 represents average prime time audience ratings separated by network (ABC, CBS, Fox, or NBC) and type of show (comedy, drama, or news). The table is based on actual data from the 1998–1999 television season. Overall averages do not correspond to the average of cell means because the cell means are not based on the same number of observations.

Without information about the variability of individual ratings around their means, and the number of shows represented in each cell, it's not pos-

Network		Type of show			
	Comedy		*Drama*		*News*	*Average*
ABC	7.0	$\xleftrightarrow{-1.1}$	5.9	$\xleftrightarrow{2.2}$	8.1	6.8
	$\updownarrow$ .2		$\updownarrow$ 1.8		$\updownarrow$ .1	
CBS	7.2	$\xleftrightarrow{.5}$	7.7	$\xleftrightarrow{.5}$	8.2	7.6
	$\updownarrow$ −2.3		$\updownarrow$ −2.1		$\updownarrow$ −4.1	
Fox	4.9	$\xleftrightarrow{.7}$	5.6	$\xleftrightarrow{-1.5}$	4.1	5.2
	$\updownarrow$ 2.9		$\updownarrow$ 1.6		$\updownarrow$ 4	
NBC	7.8	$\xleftrightarrow{-.6}$	7.2	$\xleftrightarrow{.9}$	8.1	7.7
Average	6.7		6.7		7.9	6.9

TABLE 3.1. Average Nielsen ratings separated by Network and Type of show.

sible to assess the significance of any effects in the table, but the numbers do provide the ability to informally determine what these effects would mean. The mean ratings for the networks suggest a network effect of CBS and NBC being the highest-rated networks, followed by ABC, and then Fox. Type of show also apparently affects the ratings, as news shows get considerably higher average ratings than comedies and dramas. The interaction effect, however, says that the network effect is different for different types of shows; equivalently, it says that the type of show effect is different for different networks. That is, knowing which network a show is on changes the pattern of ratings for different types of shows, or knowing the type of show changes the pattern of ratings for different networks.

The numbers above the horizontal arrows and next to the vertical arrows reflect the changing effects of one variable given the level of the other. The interaction seems to be driven by two major variations on the main effects: the relatively poor performance of dramas on ABC (and, to a lesser extent, NBC), and the relatively poor performance of news shows on Fox. Figure 3.1 gives an interaction plot, a graphical representation of an interaction effect. In this plot the vertical axis represents the group averages of the target variable, with the categories of one effect along the horizontal axis, and different lines representing the categories of the other effect. If there is no interaction effect in the data, the lines in an interaction plot will be roughly parallel. This is clearly not true here, reflecting the problems apparently faced by ABC and NBC in dramas and by Fox in news shows.

Model (3.6) can be fit using effect codings. The main effects are fit using $I - 1$ codings for rows, and $J - 1$ codings for columns, respectively. The interaction effect is fit by constructing and including all $(I - 1)(J - 1)$ pairwise products of the codings for rows and columns. The statistical significance of each effect (given the presence of the other two) then can be tested using the partial F-test based on dropping the codings from the model (for example, the test for the significance of the interaction compares

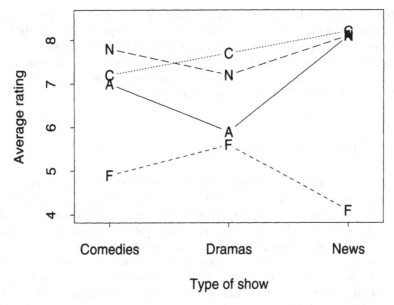

FIGURE 3.1. Interaction plot for Nielsen rating data. Average ratings for ABC (solid lines connecting "A"s), CBS (dotted lines connecting "C"s), Fox (short dashed lines connecting "F"s), and NBC (long dashed lines connecting "N"s) for three types of shows are given.

the model with all of the variables to the one that only includes the codings for the main effects). Such tests are sometimes referred to being based on "Type III sums of squares" in this context.

It is important to note that even though each of the three F-statistics (two main effects and the interaction effect) is well-defined, that doesn't mean that they are useful. Since an interaction effect implies that each main effect is not unambiguously defined (each main effect changes depending on the level of the other variable), tests for the significance of the main effects are rarely meaningful in the presence of an interaction. Thus, models with interactions can be thought of as *hierarchical*, in the sense a model that includes an interaction should include the main effects that "comprise" it, even if the partial F-test for such a main effect is not statistically significant.

Interactions between categorical and numerical predictors also can be defined using effect codings. In this context an interaction is present if the slope of the numerical predictor changes depending on the level of the categorical predictor. So, for example, model (3.5) generalizes to

$$y_{ij} = \beta_0 + \alpha_i + \beta_{1i}x_{1ij} + \cdots + \beta_{pi}x_{pij} + \varepsilon_{ij}, \quad i = 1, \ldots, K, \ j = 1, \ldots, n_i,$$
(3.7)

where each slope coefficient has (potentially) K different values. Model (3.7) is fit by regressing on the $K - 1$ effect codings, the p numerical predictors, and the $p(K - 1)$ pairwise products of the effect codings and the numerical predictors. Various partial F-tests can then be constructed to test interesting hypotheses. Comparing the fit of (3.7) to the model after dropping each set of products of the effect codings with the ℓth numerical predictor (for $\ell = 1, \ldots, p$) tests whether different slopes for predictor x_ℓ add significant predictive power to the model.

3.3 Regression Diagnostics

An implicit assumption in fitting regression models is that all of the data follow the hypothesized linear model; that is, there aren't any cases far off the regression function (*outliers*). In addition, cases that have unusual values of the predictors (*leverage points*) are also problematic, as they can have a strong effect on estimated regression parameters, measures of fit, and so on. A point might be a leverage point because it has an unusual value for one or more predictors, or because it has an unusual combination of values for several predictors (even if none of the values are that unusual by themselves). Once you've identified such a case, it's very important to identify what it is that makes the case unusual (for example, y is surprisingly large for the given predictor values, the observed values for two predictors don't typically occur together, etc.). Further, you need to try to determine what might have happened in the random process under study that would result in such a case.

Residual plots are very useful to detect outliers and leverage points. In a plot of residuals versus fitted values, or residuals versus each predictor, points by themselves on the top or bottom are potential outliers, while points by themselves on the left or right are potential leverage points. In a normal plot of the residuals, outliers show up as distinct at the bottom left (negative outliers) or top right (positive outliers).

Still, it is sometimes the case (particularly for multiple regression data sets) that these plots don't identify unusual observations very well. For this reason, diagnostics have been developed to help identify unusual cases. These diagnostics fall into three broad categories: (i) variations on residuals, (ii) designed to identify outliers, measures of leverage, designed to identify leverage points, and (iii) broader measures of influence on the regression if an observation is omitted from the data set.

3.3.1 Identifying Outliers

By definition, an outlier is a point off the regression hyperplane. Thus, its residual should be large (in absolute value). There are a few different versions of residuals that can be used in model checking (we cannot use the raw residuals $\mathbf{r}$, since they depend on the scale of y). The simplest version says that since $V(\varepsilon_i) = \sigma^2$, a simple estimate of the variance of the residual (the estimate for ε_i) is $\hat{\sigma}^2$. Thus, a simple standardized residual (one where all residuals have roughly standard deviation one, and so can be compared with one another easily) is $r_i^* = r_i/s$, where s is the standard error of the estimate.

Recall that $\mathbf{r} = (I - H)\mathbf{y}$, where H is the hat matrix. Basic matrix probability calculations then yield

$$
\begin{aligned}
V(\mathbf{r}) &= V[(I - H)\mathbf{y}] \\
&= (I - H)V(\mathbf{y})(I - H)' \\
&= (I - H)\sigma^2 I(I - H)' \\
&= (I - H)(I - H)'\sigma^2 \\
&= (I - H)\sigma^2,
\end{aligned}
$$

since H is symmetric and idempotent. Thus, $V(r_i) = (1 - h_{ii})\sigma^2$, where h_{ii} is the ith diagonal element of H. A "better" standardized residual (one that has standard deviation exactly equal to one) is therefore

$$
\tilde{r}_i = \frac{r_i}{s\sqrt{1 - h_{ii}}}. \tag{3.8}
$$

Recalling that the (unknown) errors ε are assumed to be normally distributed, we can see that the standardized residuals can be expected to be (roughly) standard normal. For example, we would expect about 95% of them to be within ± 2. A good guideline for standardized residuals is that a case with a standardized residual larger than about ± 2.5 should be investigated as a potential outlier, since that would only be expected to occur randomly about 1% of the time. Note, however, that for very large samples, that means that many observations could have standardized residuals outside ± 2.5 while still not being outliers.

3.3.2   Identifying Leverage Points

Looking at residuals doesn't help in the detection of leverage points, since they don't necessarily fall off the regression surface (and can, in fact, draw it towards them, thereby *reducing* their residuals). What is needed is a measure of how far a case is from a "typical" value. This is provided by the so-called *leverage value* (also sometimes called the "hat" value, for reasons that will be obvious in a moment). The leverage value is designed to measure the potential for an observation to have a large effect on the fitted regression line, and is simply a measure of how far a particular case is (based on only predictor values) from the average of all cases, with distance being measured in such a way that the correlations between the predictors is taken into account.

The formula for the fitted values gives a justification for the definition of the leverage value. Recall that $\hat{\mathbf{y}} = H\mathbf{y}$. Writing this out observation by observation reveals that

$$\hat{y}_i = h_{i1}y_1 + h_{i2}y_2 + \cdots + h_{ii}y_i + \cdots h_{in}y_n.$$

That is, the ith diagonal element of the hat matrix measures how much of an effect an observation has on its own fitted value. The symmetry and idempotence of H implies that $0 < h_{ii} < 1$; the closer h_{ii} gets to 1, the more a fitted value is completely determined by its own observed value. This means that the regression line (or plane, or hyperplane) has been drawn towards that point, and it is a leverage point.

It can be shown that the sum of the n leverage values equals $p + 1$, where p is the number of predicting variables in the regression. That is, the average leverage value is $\frac{p+1}{n}$. A good guideline for what constitutes a large leverage value is $2.5 \left(\frac{p+1}{n}\right)$; cases with values greater than that should be investigated as potential leverage points.

3.3.3 Identifying Influential Points

A different way to look at the unusual observation problem is to focus on the effect an observation has on the regression. An observation that, if it were removed, would result in a large change in the regression is an *influential point*, and obviously dangerous to ignore. A common measure of influence (although not the only one) is *Cook's distance*, which measures the change in the fitted regression coefficients if an observation were dropped from the regression, relative to the inherent variability of the coefficient estimates themselves. This value equals

$$D_i = \frac{\tilde{r}_i^2 h_{ii}}{(p+1)(1-h_{ii})}, \tag{3.9}$$

so it obviously combines the outlier and leverage point diagnostics into one measure. A value of Cook's D over one or so is flagging a point that should probably be studied further.

Any good statistical package provides these diagnostics as a standard option from the regression. These values should always be determined and looked at. That can mean simply printing them out to look at, using univariate pictures of them, such histograms, stem-and-leaf plots or boxplots, or plotting an index plot of them versus the case number (for time series data, this is a time series plot of the diagnostics, but it provides information in any case). Graphical methods are particularly useful, as they might make values that are unusually high (or low, in the case of standardized residuals) relative to the others stand out more (even if they are not flagged by the somewhat arbitrary guidelines given earlier).

It is worth noting what these diagnostics don't do very well. Specifically, they are all sensitive to the so-called *masking effect*. This occurs when several unusual observations are similar to each other. When this happens, the diagnostics, which all focus on changes in the regression when a single observation is deleted, fail, since the presence of the other outlier(s) means that the fitted regression changes very little. The problem of identifying multiple outliers in regression data is one of the hardest problems in statistics, and is a topic of ongoing research.

Note that if you have omitted an observation or observations from your data set, you have effectively created a new data set. What that means is that results from the analysis with the observation(s) included in the data are no longer valid, and you need to start from the beginning in deciding the variables you want in the model, fitting the model, checking assumptions, and so on.

Monthly Tornado Mortality Statistics (continued)

We will continue here the analysis of the monthly tornado statistics from Section 2.5, ignoring for now the violations of assumptions apparent in Figure 2.4. Two additional (categorical) predictors are added to the model, the year (1996, 1997, 1998, or 1999) and the month (January through December). Figure 3.2 gives side-by-side boxplots of the number of tornado-related deaths separated by year (top) and month (bottom). There does not appear to be very much difference in the number of deaths by year, although there is more variability in the latter two years. On the other hand, there is a clear effect related to month, with far more deaths in the first five months of the year than in the latter seven.

There are, in fact, generally more tornadoes in the United States in June and July than there are in February or March, but there is a crucial difference in where the tornadoes occur. Early spring tornadoes tend to occur in the southeastern part of the United States, while summer tornadoes tend to occur in the southern Plains states (the so-called "Tornado Alley"). The former location has much higher population density, so when a tornado occurs there, it is much more likely to lead to (multiple) deaths than in the latter location.

The following output gives partial-F tests for each of the effects in a model based on the two categorical main effects and the two numerical predictors used earlier.

Source	df	SS	MS	F	P
Tornadoes	1	180.77	180.77	4.09	0.052
Killer tornadoes	1	2846.92	2846.92	64.35	0.000
Year	3	76.99	25.66	0.58	0.633

Number of deaths by year

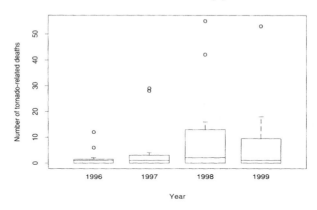

Number of deaths by month

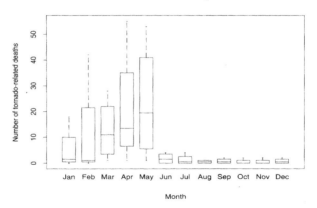

FIGURE 3.2. Side-by-side boxplots of number of tornado-related deaths separated by Year (top plot) and Month (bottom plot).

Month	11	643.52	58.50	1.32	0.259
Error	31	1371.56	44.24		
Total	47	8299.67			

Note that the four-level categorical predictor Year is based on three effect codings (hence the three degrees of freedom associated with it), and the 12-level predictor Month is based on 11 effect codings. The F-tests for the two numerical predictors are equivalent to t-tests for whether the slope coefficient equals zero. As in Section 2.5, the number of killer tornadoes is highly significant, but now the number of tornadoes is also marginally significant. It is perhaps not too surprising that the Year effect is not statistically significant, but the apparent lack of information from the month effect is a bit surprising. Of course, the F-statistic tests the importance of the month effect given the other variables in the model; perhaps knowing the month adds little given the number of killer tornadoes. Based on the F-tests, the next step would seem to be to omit the year and month effects, and use the model with the two tornado predictors, but as we already saw in Section 2.5, if this is done the number of tornadoes becomes quite insignificant. This seeming contradiction is due to several factors: the nonlinearity and heteroscedasticity noted earlier, the possible overspecification of the month effect (Figure 3.2 suggests that many of the months might have similar tornado death experiences, implying the need for fewer than 12 distinct month effects), and an inherent problem with the use of t- and F-tests to select the variables in a regression model when the predictors are related to one another (we will discuss ways of addressing this last problem in Section 3.4). The R^2 for this model is 83.5%, an improvement over the 75% of the model based on only the number of killer tornadoes. The standard error of the estimate is $\sqrt{44.24} = 6.65$, implying predictions to within ± 13.3 deaths roughly 95% of the time.

The following output gives the least squares coefficients of the numerical predictors and effect codings.

Predictor	Coef	s.e.	t	p
Constant	3.671	2.676	1.37	0.180
Killer tornadoes	4.392	0.548	8.02	0.000
Tornadoes	-0.057	0.028	-2.02	0.052
Year				
1996	-0.763	1.717	-0.44	0.660
1997	-0.675	1.680	-0.40	0.691
1998	2.229	1.693	1.32	0.198
1999	-0.791	1.680	-0.47	0.641
Month				
January	-6.684	3.330	-2.01	0.053
February	2.612	3.712	0.70	0.487
March	-4.190	3.626	-1.16	0.257
April	-0.004	3.642	-0.00	0.999
May	14.181	5.079	2.79	0.009
June	6.667	5.304	1.26	0.218
July	0.228	3.453	0.07	0.948
August	-1.140	3.281	-0.35	0.730
September	-2.031	3.274	-0.62	0.540
October	-1.483	3.306	-0.45	0.657
November	-3.725	3.664	-1.02	0.317
December	-4.431	3.904	-1.14	0.265

Each of the coefficients corresponding to the year effect is close to zero, reinforcing the lack of additional predictive power of year in this model. While only one coefficient for the month effect codings is strongly significantly different from zero (which could just be because of random chance, given the many t-tests being constructed), the coefficients do generally follow the pattern Figure 3.2 suggested of higher-than-expected deaths in May, and lower-than-expected deaths in the summer, autumn, and winter. Investigation into models with potentially different slope coefficients for different months is postponed until Section 3.4.

A plot of residuals versus fitted values for this model displays the same nonlinearity and heteroscedasticity as Figure 2.4 and is not reproduced here. Regression diagnostics aren't very meaningful given these model violations, but we produce them in Figure 3.3 for expository purposes (we will also refer to them again in Chapter 5). Each plot is an index plot of the diagnostics. The guideline of ± 2.5 is given in the top plot for standardized residuals as a pair of dotted lines; none of the leverage values or Cook's distances reach their guideline values, so no lines corresponding to them are given. Observation 26, corresponding to February 1998, shows up as an outlier in the plot of standardized residuals, and also has the largest Cook's distance. There were 42 tornado-related deaths in that month, high for a February with only four killer tornadoes and 72 total tornadoes (compare, for example, the next month, which also had four killer tornadoes and 72 total tornadoes, and only 16 deaths). The guideline for leverage values for these data is $(2.5)(17)/48 = .89$, and none of the observed values are close to that (more importantly, none stand out as very unusual relative to the others). Several months are noticeable in the plot of Cook's distances, especially the outlier month (February 1998).

At this point we should attempt to investigate what makes this month so unusual. In fact, all four killer tornadoes that month occurred in a two-hour period on the night of February 22, 1998, in central Florida, with the latter two tragically causing 38 deaths. These were two of the deadliest tornadoes during the entire 1996–1999 time period. We could now consider removing the observation from the data set to see what effect this would have on inferences. We will not pursue that here, given the previously noted nonlinearity and heteroscedasticity in the residuals.

3.4 Model Selection

As more information is gathered about a process, the need to choose among competing models becomes more severe. Using an inappropriate model can

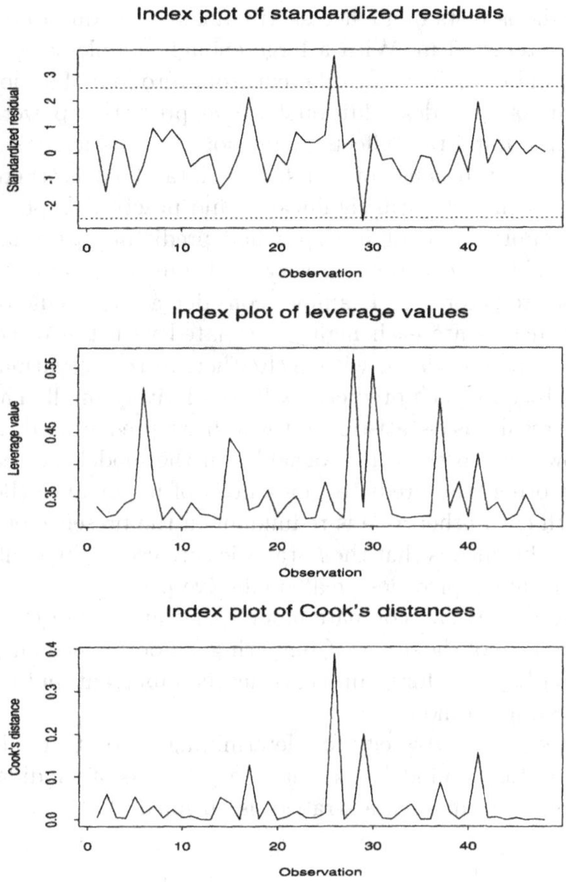

Index plot of standardized residuals

Index plot of leverage values

Index plot of Cook's distances

FIGURE 3.3. Index plots of standardized residuals (top plot), leverage values (middle plot) and Cook's distances (bottom plot) for tornado deaths data.

lead to misleading inferences. Omitting important effects (*underfitting*) reduces predictive power, biases estimates of effects for included predictors, and results in less understanding of the process being studied. Including unnecessary effects (*overfitting*) complicates descriptions of the process, and tends to lead to poorer predictions because of the additional unnecessary noise. Great advances in computing power have led to the mistaken belief that these choices can be completely automated. In fact, model selection is as much of an art as a science, and involves subjective decision-making.

We have already seen how hypothesis tests (t-tests and partial F-tests) can be used to try to decide which predictors are needed in a model. Unfortunately, there are several reasons why such tests are not adequate for the task of choosing the appropriate model to use.

First, hypothesis tests don't necessarily answer the question a data analyst is most interested in. With a large enough sample, almost any estimated slope will be significantly different from zero, but that doesn't mean that the predictor provides additional *useful* predictive power. Similarly, in small samples, important effects might not be statistically significant at typical levels simply because of insufficient data. That is, there is a clear distinction between statistical significance and practical importance.

A second important point is that when predictors are related to one another, t- or F-tests can provide very misleading indications of the importance of a predictor. For example, consider a two-predictor situation where the predictors are each highly correlated with the target variable, and are also highly correlated with each other. In this situation it is likely that the t-statistic for each predictor will be relatively small. This is not an inappropriate result, as t-statistics test whether a predictor adds significant predictive power given the other variables in the model. For each predictor, given the other, the predictor adds little of importance (being highly correlated with each other, one is redundant in the presence of the other). Unfortunately this means that the t-statistics are useless in identifying that either predictor alone provides great predictive power.

Partial F-tests only can compare models that are nested (that is, where one is a special case of the other). Comparing a model based on $\{x_1, x_3, x_5\}$ to one based on $\{x_2, x_4\}$, for example, is clearly important, but is impossible using these testing methods.

What is needed is a strategy for determining a "best" model (or even better, a set of "best" models) among a larger class of candidate models. In the next few sections such a strategy is outlined.

3.4.1   Choosing a Set of Candidate Models

In recent years it has become commonplace for databases to be constructed with hundreds (or thousands) of variables and hundreds of thousands (or millions) of observations. It is tempting to consider all possible sets of variables as potential predictors in a regression model, limited only by available computing power. This would be a mistake. If too large a set of possible predictors is allowed, it is very likely that variables will be identified as important just due to random chance. This sort of overfitting is known as "data dredging," and is probably the most serious danger when selecting regression predictors.

The set of possible models should ideally be chosen before seeing any data based on as thorough an understanding of the underlying random process as possible. Potential predictors should be justifiable on theoretical grounds if at all possible. This is by necessity at least somewhat subjective, but good basic principles exist. Potential models to consider should be based on the scientific literature and previous relevant experiments. In particular, if a model simply doesn't "make sense," it shouldn't be considered among the possible candidates.

3.4.2   Choosing the "Best" Model

The title of this section is actually a misnomer, as there is often no single "best" model to choose. Rather, the goal is to find the (hopefully small) set of models that best describe the data.

We need to be clear what we mean by the (or a) "best" model. As was stated on page 13, any model is viewed as an approximation of reality. In this sense, there is no "true" model at all (or, perhaps, what is basically the same, the true model is too complex to be useful). Our goal is not to find the "true" model, but rather to find a model, or set of models, that best balances fit and simplicity (the so-called principle of parsimony, using as few parameters as possible while still adequately representing the relationships in the data). This should result in a model that provides useful descriptions of the process being studied from estimated parameters, and which can be used for predictions of future events.

A general way of evaluating the worth of a model is through the use of statistical information. A detailed discussion of the determination of information measures is beyond the scope of this book, but see the references in Section 3.6 for several good books on the subject. The Akaike Information Criterion AIC has the form

$$AIC = -2L + 2\nu, \qquad (3.10)$$

where L is the log-likelihood function, and ν is the number of estimated parameters in the model ($p + 2$ for the regression models discussed thus far: $\beta_0, \beta_1, \ldots, \beta_p$, and σ^2). Using (2.9) and omitting constants that are not functions of the parameters gives

$$AIC = n \log(\hat{\sigma}^2) + 2\nu \qquad (3.11)$$

as the AIC form for least squares regression models, where $\hat{\sigma}^2$ is the MLE of σ^2.

The criterion used to select a model is to minimize AIC. It is clear from (3.11) that this achieves the goal of balancing goodness-of-fit (through smaller $\hat{\sigma}^2$) with parsimony (through smaller $\nu = p + 2$). When comparing models, those with smaller values of AIC are preferred over those with larger values. Another criterion that is similar to AIC that is often used in regression modeling is Mallows' C_p, but as this measure does not generalize to models for count data, we will not discuss it here.

It is well known that AIC has a tendency to lead to overfitting, particularly in small samples. That is, the penalty term in AIC designed to guard against too complicated a model (2ν) is not strong enough. A corrected version of AIC that helps address this problem is the corrected AIC,

$$AIC_C = -2L + 2\nu \left(\frac{n}{n - \nu - 1} \right) \qquad (3.12)$$

$$= AIC + \frac{2\nu(\nu + 1)}{n - \nu - 1} \qquad (3.13)$$

(Hurvich and Tsai, 1989). Equations (3.12) and (3.13) show that (especially for small samples) models with fewer parameters (smaller ν) will be more strongly preferred when minimizing AIC_C than when minimizing AIC, providing stronger protection against overfitting. In large samples the two criteria are virtually identical, but in small samples, or models with a large number of parameters, AIC_C is the better choice. From a practical point of view, situations where a few strong effects account for most of the variability in the target variable are easier for a model selection method to identify than those with many effects of only moderate strength.

AIC and AIC_C have the desirable property that they are efficient model selection criteria. What this means is that as the sample gets larger, the error obtained in making predictions using the model chosen using these criteria becomes indistinguishable from the error obtained using the best possible model among all candidate models. That is, in this large-sample predictive sense, it is as if the best approximation was known to the data analyst. Other criteria, such as the Bayesian Information Criterion BIC (which substitutes $\log(n)\nu$ for 2ν in (3.10)) do not have this property.

Although an ordering of candidate models by AIC or AIC_C value provides a single "best" model, it is really comparisons of AIC or AIC_C values that are more interesting. If two models have values that differ by less than 2 or 3, they are effectively equivalent as possible model choices. On the other hand, if the values differ by more than 10, the model with larger AIC or AIC_C has considerably poorer fit, and would normally not be considered further.

It seems obvious to note that model comparisons are only sensible when based on the same data set, but there is a subtlety here that can trap the unwary. Most statistical packages drop any observations that have missing data in any of the variables in the model. If a data set has missing values scattered over different variables, the set of observations with complete data will change depending on which variables are in the model being examined, and the set of AIC (or AIC_C) values will not be comparable. One way around this is to only use observations with complete data for all variables under consideration, but this can result in discarding a good deal of available information for any particular model.

3.4.3   Model Selection Uncertainty

Once a "best" model is chosen, all of the usual inference tools then can be brought to bear to explain the process being studied. Unfortunately, doing this while ignoring the model selection process can lead to problems. Since the model was chosen to be best, in some sense, it will tend to exhibit less variability than would be expected by random chance. That is, conducting inference based on the chosen model ignores an additional source of variability, that of actually choosing the model (model selection based on a different sample from the same population could very well lead to a different "best" model). As a result, confidence intervals can have lower coverage than the nominal value, hypothesis tests can reject the null too often, prediction intervals can be too narrow, and so on.

Identifying and correcting for this uncertainty is a difficult problem, and an active area of research, and is beyond the scope of this book. From a practical point of view, the guidelines given in the previous section on differences in AIC or AIC_C values should be considered. Any model that has a value very close (within 2 or 3) to that of the best model should be treated as being one that might have easily been chosen based on a different sample from the same population, and any implications of such a model should be viewed as being as valid as those from the best model.

There is a simple way to get a sense of certain aspects of this effect if enough data are available. The predictive accuracy of a chosen model can be evaluated by holding out some data, applying the selected model to the new data (based on the previously estimated parameters, not estimates based on the new data), and then examining the errors made using the model. If the standard deviation of the errors from this prediction is not very different from the standard error of the estimate in the original regression, chances are that making inferences based on the chosen model will not be too dangerous or misleading. If validating the model this way is not possible, one simple adjustment that is helpful is to estimate the variance of the errors as

$$\tilde{\sigma}^2 = \frac{\sum_{i=1}^n (y_i - \hat{y}_i)^2}{n - \max(p) - 1}, \tag{3.14}$$

where $\hat{y}$ is based on the chosen "best" model, and $\max(p)$ is the number of predictors in the most complex model examined (in the sense of most predictors). Clearly if very complex models are included among the set of candidate models, $\tilde{\sigma}$ will be much larger than the standard error of the estimate from the chosen model, with correspondingly wider prediction intervals. This reinforces the benefit of limiting the set of candidate models (and the complexity of the models in that set) from the start.

Monthly Tornado Mortality Statistics (continued)

We will examine least squares estimation for the tornado mortality data once more in order to illustrate the use of AIC and AIC_C, although the nonlinearity and heteroscedasticity issues noted earlier are of course still

Predictors	AIC difference	AIC$_C$ difference
(a) Models using only main effects		
None	64.26	63.99
Killer tornadoes	0.00	0.00
Tornadoes	57.66	57.66
Year	67.34	68.22
Month	64.05	74.21
Killer tornadoes, Tornadoes	1.66	2.05
Killer tornadoes, Year	4.68	6.18
Killer tornadoes, Month	9.62	21.81
Tornadoes, Year	61.13	62.63
Tornadoes, Month	59.13	71.31
Year, Month	65.32	82.32
Killer tornadoes, Tornadoes, Year	6.32	8.57
Killer tornadoes, Tornadoes, Month	6.47	20.93
Killer tornadoes, Year, Month	13.79	33.65
Tornadoes, Year, Month	61.78	81.63
Killer tornadoes, Tornadoes, Year, Month	9.85	32.89
(b) Different slopes by month		
Killer tornadoes	16.51	0.00
Killer tornadoes × Month	0.00	42.03
(c) Models based on tornado season		
Killer tornadoes	40.38	30.22
Killer tornadoes, Tornado season	38.60	31.59
Killer tornadoes × Tornado season	0.00	0.00

TABLE 3.2. AIC and AIC_C differences of models for tornado mortality data.

present. Table 3.2 summarizes model selection using AIC and AIC_C. The first part of the table gives results for all 16 possible models using the four predictors discussed earlier (including the model with no predictors). In fact, looking at all possible regression models this way is not generally advisable, as it can greatly increase the chances of overfitting, particularly if there are many possible predictors (say, ten or more).

Rather than give the AIC and AIC_C values, the table gives the difference in values compared to the model that minimizes the value over the set of candidate models. Reporting the values this way removes the confusion from additive constants that do not affect model selection, makes it easier to identify the best model (it has difference value exactly equal to zero), and makes it easier to see which models are effectively equivalent to the best model. Both AIC and AIC_C choose the simple regression model on just the number of killer tornadoes as the best model. The two-predictor model that also includes the number of tornadoes is a viable candidate as well, but given our preference for simple models, the small t-statistic for the number of tornadoes in this model (see page 23), and the difficulty in understanding this model, the one-predictor model is clearly better. In this case using either AIC or AIC_C leads to similar choices, but the larger penalty in AIC_C for complicated models is apparent in the models that include the 11-parameter Month effect.

The second portion of the table investigates the possibility of different slopes in the model. There is no reason to think that the effect of killer tornadoes on deaths would be different for different years, but an effect related to month is plausible. This model is labeled Killer tornadoes × Month in the table, although it should be remembered that the main effects are included in the model as well. AIC strongly supports the idea of different slopes based on month, as shown by an AIC value that is 16.51 lower than the previously best model using only the number of killer tornadoes. This is a situation, however, when the stronger penalty of AIC_C is needed. With only 48 observations, a model based on 23 predictors is inadvisable, and AIC_C shows that clearly, giving a value 42.03 higher than that for the model on number of killer tornadoes alone.

The lack of any effects related to month could very well come from the requirement of distinct shifts and slopes for each of 12 months. From climatological considerations, this is probably too general a model, as there are few killer tornadoes between June and December. The last part of Table 3.2 attempts to address this point by changing the time of year variable from 12 distinct months to six distinct time periods: January through May, and all of the other months. This is referred to as "Tornado season" in the table. This representation of time of year is certainly more effective than using all 12 months, as the model using the number of killer tornadoes and tornado season is only slightly worse than the model on killer tornadoes alone. More importantly, the model with only six different slopes for the number of killer tornadoes (rather than 12) is clearly preferred over the other models examined.

We will come back to these data in Chapter 5 where their character as count data will be recognized and exploited.

3.5 Heteroscedasticity and Weighted Least Squares

A violation of the assumptions of ordinary least squares regression that is particularly relevant in models for categorical data (as we've seen when examining the tornado deaths data) is the occurrence of heteroscedasticity, or nonconstant variance. Say $V(\varepsilon_i) = \sigma_i^2$, not all equal. A general approach to deriving accurate estimates of $\boldsymbol{\beta}$ is *weighted least squares*.

The idea behind weighted least squares (WLS) is that least squares is still a good thing to do if the target and predicting variables are transformed to give a model with errors with constant variance. To keep things simple, consider simple regression. The regression model is

$$y_i = \beta_0 + \beta_1 x_i + \varepsilon_i.$$

Dividing both sides of this equation by σ_i gives

$$\frac{y_i}{\sigma_i} = \beta_0 \left(\frac{1}{\sigma_i}\right) + \beta_1 \left(\frac{x_i}{\sigma_i}\right) + \frac{\varepsilon_i}{\sigma_i}.$$

This can be rewritten

$$u_i = \beta_0 z_{1i} + \beta_1 z_{2i} + \delta_i,$$

where u_i, z_{1i}, z_{2i}, and δ_i are the obvious substitutions from the previous equation, and $V(\delta_i) = 1$ for all i. Thus, ordinary least squares estimation (without an intercept term) of u on z_1 and z_2 gives fully efficient estimates of β_0 and β_1. Note that using a constant multiple of σ_i works just as well, since the only requirement is that $V(\delta_i)$ be constant for all i. By convention, the values $1/\sigma_i^2$ are considered the weights when fitting a WLS model. Ordinary least squares (OLS) is a special case of WLS with unit weights for all i (and, in fact, most regression packages only include code for WLS, with OLS the default special case). The problem is that in general σ_i^2 is unknown, and must be estimated.

It is important to recognize the reasons behind the use of weighted least squares. The goal is not to improve measures of fit like R^2 or F; rather, the goal is to analyze the data in an appropriate fashion. There are several advantages to doing this:

1. The estimates of $\boldsymbol{\beta}$ are more efficient. That is, on average, the WLS estimates will be closer to the true regression coefficients than are the OLS estimates.

2. Hypothesis tests, such as t- and F-tests, follow the assumed distributions, and are thus proper tools for inference. The same is true for t-based confidence intervals.

3. Predictions are more sensible. If the underlying variability of a certain type of observation is larger than that for another type of observation, the prediction interval should reflect that. This is not done under OLS, but it is under WLS.

Weighted least squares can be formulated more generally using matrix notation. The model is

$$\mathbf{y} = X\boldsymbol{\beta} + \boldsymbol{\varepsilon},$$

with $E(\boldsymbol{\varepsilon}) = \mathbf{0}$ and $V(\boldsymbol{\varepsilon}) = V\sigma^2$, where V is the $n \times n$ matrix with σ_i^2/σ^2 on the diagonal and zeroes elsewhere. We will treat σ^2 as unknown, but σ_i^2/σ^2 known; in practice, this ratio must be estimated for each observation from the data. Matrix manipulations are then as follows:

$$V^{-1/2}\mathbf{y} = V^{-1/2}X\boldsymbol{\beta} + V^{-1/2}\boldsymbol{\varepsilon}$$
$$\Leftrightarrow \mathbf{u} = Z\boldsymbol{\beta} + \boldsymbol{\delta}$$

with $V(\boldsymbol{\delta}) = I\sigma^2$. Thus,

$$\hat{\boldsymbol{\beta}} = (Z'Z)^{-1}Z'\mathbf{u} \tag{3.15}$$
$$= (X'V^{-1}X)^{-1}X'V^{-1}\mathbf{y}. \tag{3.16}$$

Equation (3.15) makes clear how inference for weighted least squares models proceeds. In all of the earlier formulas, the columns of Z take the place of the predictors (that is, the ith value of the jth predictor is x_{ij}/σ_i, with the first predictor corresponding to the constant term) and $\mathbf{u}$ takes the place of $\mathbf{y}$. So, for example, t-statistics for individual coefficients are based on $V(\hat{\boldsymbol{\beta}}) = (Z'Z)^{-1}\hat{\sigma}^2$, where $\hat{\sigma}^2$ satisfies

$$\hat{\sigma}^2 = \frac{\sum_{i=1}^n (u_i - \hat{u}_i)^2}{n - p - 1}$$
$$= \frac{\sum_{i=1}^n (y_i/\sigma_i - \hat{\beta}_0/\sigma_i - \hat{\beta}_1 x_{1i}/\sigma_i - \cdots - \hat{\beta}_p x_{pi}/\sigma_i)^2}{n - p - 1}.$$

Similarly, the hat matrix is defined as

$$Z(Z'Z)^{-1}Z' = \hat{V}^{-1/2}X(X'\hat{V}^{-1}X)^{-1}X'\hat{V}^{-1/2},$$

where $\hat{V}^{-1/2}$ is the diagonal matrix with ith diagonal element equaling $\hat{\sigma}/\sigma_i$. Regression diagnostics are also formed by treating $\mathbf{u}$ as the target variable and Z as the set of predictors.

3.6   Background Material

The books mentioned in Section 2.6 also discuss much of the material in this chapter. Chatterjee and Hadi (1988) gives a thorough discussion of many regression diagnostics, and the connections between them. Barnett and Lewis (1994) discusses outlier identification and robust estimation for many data problems, while Hadi and Simonoff (1993) specifically discusses a method for identifying multiple outliers in regression data. Robust estimation for regression models is an extremely challenging problem, with the goal of developing a method that is efficient for clean data, yet resistant to multiply outlying observations in either the predictors or target variable (or both), being a highly elusive one. Given that little work has been done adapting these approaches to general count regression models, we will not address these methods here.

Burnham and Anderson (2002), Linhart and Zucchini (1986), and McQuarrie and Tsai (1998) discuss in great detail model selection for regression models (and other models), including methodology and philosophical issues. An important point to realize is that while the derivation of AIC and AIC_C assume the existence of a "true" model, their efficiency property is based on the assumption that there is no such thing, a point that has been a cause of confusion in the literature. Ye (1998) described a general framework for measuring and correcting for the effects of data dredging and model selection; some of the results there roughly correspond to correcting for model selection as in (3.14). Little and Rubin (2002) discusses many different approaches to data analysis when there are missing values in the data.

When fitting analysis of variance and covariance models, the coefficient corresponding to the reference group can be obtained as the negative of the sum of the other coefficients. This cannot provide a t-test for the significance of the coefficient, however. The easiest way to obtain this test when using software that constructs the effect codings automatically is to refit the model, specifying another group as the reference group.

3.7 Exercises

1. Consider again the data on shark attacks in Florida (page 27). Construct regression diagnostics for the model of shark attack rates versus year. Are any of the years unusual?

2. The owners of a six-acre vineyard on South Bass Island (a small island off the Ohio coast of Lake Erie) recorded production information for the vineyard over several years. Production is measured by the number of baskets (termed lugs) of grapes gathered for each of the 52 rows of the vineyard. The file **vineyard** refers to data for 1983 through 1991 (Chatterjee et al., 1995, page 304).

   (a) Build a regression model for the number of lugs on year and row number, treating each of these variables as numerical. Does the model fit the data well?

   (b) Now, build a model treating the predictors as categorical. Is this model an improvement over the model in part (a)? Answer

this question using both the appropriate hypothesis test and
information measures. Is there a consistent pattern in annual
production? Are there differences in production in different regions of the vineyard? What do you think might account for
these differences? Would it surprise you to learn that there is a
farmhouse next to rows 28–42?

(c) The file also includes a variable Group that splits the 52 rows
of the vineyard into 13 groups of four contiguous rows each.
Use this variable as a categorical predictor. Is this model an
improvement over the model in part (b)?

3. Consider again the four data sets in the file anscombe (page 27). Construct regression diagnostics for each of the four data sets. What do
these diagnostics say about the appropriateness of the least squares
model for each data set? Do the diagnostics identify all of the problems with the respective models? If not, why not?

4. The file braziltourism contains data from a survey of Brazilian
tourists in the Atlantic Coastal Forest of Brazil (Englin et al., 2003).
I am indebted to Tom Holmes and Erin Sills for making these data
available here. The goal of the survey was to study how the number
of trips to adventure tourism areas in the region could be modeled
as a function of characteristics of the tourists. Information gathered
included the age, sex (male = 0, female = 1), monthly household
income, and estimated travel cost (in U.S. dollars) of the respondent,
the number of adventure activities (for example, mountaineering and
hiking) and passive activities (for example, sightseeing and picnick-
ing) the respondent participated in, and the importance to the respondent of paving the access road into the region. The data given
here are for people surveyed outside of the adventure region.

(a) Use information measures to choose a model for number of trips
as a function of the predictors.

(b) Construct residual plots for your chosen model. Does the model
fit the data? What model violation(s) are suggested by the plots?
What would you do to address the violation(s)?

(c) Construct regression diagnostics for your chosen model. Do they
flag any observations as unusual? Do you think they can be
trusted in this situation?

5. The ith *standardized deleted residual* r_i^* is a standardized residual
that measures the difference between the observed y_i and the fitted
$\hat{y}_i$ if the ith observation had been omitted from the data. It equals

$$r_i^* = \tilde{r}_i \sqrt{\frac{n - p - 2}{n - p - 1 - \tilde{r}_i^2}}.$$

These residuals have the desirable property that if the assumed model holds, they are independent and identically distributed, with $r_i^* \sim t^{(n-p-2)}$. Calculate standardized deleted residuals for the tornado mortality data. Do they lead to different conclusions from the standardized residuals $\tilde{r}$?

6. The file `michiganacc` contains information on monthly traffic accidents in Michigan for the years 1979–1987. The data are a 0.1% random sample of all accidents, and include the number of accidents, the noninstitutional unemployment rate (as a proxy for economic activity), as well as monthly and seasonal identifiers (Lenk, 1999).

   (a) Fit a regression model for the number of accidents as a function of time, unemployment rate, and season. Do the variables provide significant predictive power? What would you consider the "best" model for these data?

   (b) Construct regression diagnostics and residual plots for your chosen model. Do they indicate any problems? If so, what would you do to correct them?

7. Consider again the data on GMAT performance of M.B.A. students (page 28).

   (a) Use the information measures AIC and AIC_C to choose a model for these data. Do you get the same model using each measure? Is your model choice consistent with what t-tests would suggest?

   (b) Examine plots of residuals versus each of the potential predictors. Is there evidence of possible heteroscedasticity? In fact, there appears to be more variability in the residuals for women than for men. Estimate the variance of the errors for each of these two groups by calculating the sample variances of the residuals in each group separately. Using these variance estimates, construct a weighted least squares fit to these data. Are the results similar to those of an OLS fit? How would prediction intervals differ for a man versus for a woman if you used the WLS model for the prediction?

4
Categorical Data and Goodness-of-Fit

This chapter covers the building blocks of the analysis of categorical data. First we discuss the important random variables that are the basis of analysis — the binomial, Poisson, and multinomial distributions.

4.1 The Binomial Random Variable

Many categorical data problems are consistent with the following scenario. A set of situations occurs, where in each situation one of two possibilities can arise. For example (the classic example), you flip a coin ten times, where each flip can come up heads or tails. Other (more interesting) examples would be a hospital treating 100 people in its emergency room in a day, and some patients live while others die; you pick twenty stocks to invest in, and the prices of some go up while others go down; you ask 200 people in a survey if they are in favor of capital punishment or not. More formally, a *binomial process* is characterized by four things:

1. there is a fixed, finite number of trials, n

2. each trial has two outcomes (arbitrarily called success and failure)

3. the probability of a success, p, is the same on all trials

4. the trials are independent.

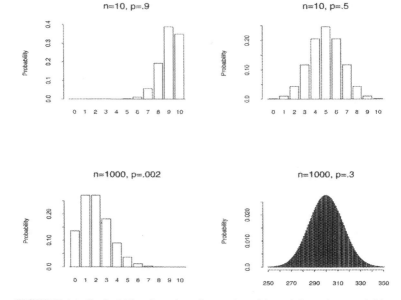

FIGURE 4.1. Probability functions for various binomial random variables.

The random variable X = number of successes in the n trials is called a *binomial random variable*, which we will write $X \sim \text{Bin}(n, p)$. The probability function for this random variable is

$$f(k; n, p) \equiv P(X = k) = \binom{n}{k} p^k (1 - p)^{n-k}, \qquad k = 0, 1, 2, \ldots, n,$$

where

$$\binom{n}{k} = \frac{n!}{k!(n-k)!} = \frac{(n)(n-1)\cdots(1)}{(k)(k-1)\cdots(1)(n-k)(n-k-1)\cdots(1)}$$

is the so-called binomial coefficient. The binomial random variable has mean $E(X) = np$ and variance $V(X) = np(1 - p)$.

Figure 4.1 gives graphs of the binomial probability function for different values of n and p (values of k with probabilities virtually equal to zero are not given for the $n = 1000$ plots). It is apparent that the probability function can be quite asymmetric, even for large n, but becomes more symmetric (and Gaussian looking) as the mean number of successes np and mean number of failures $n(1 - p)$ both get larger.

4.1.1 Large-Sample Inference

Say a binomial random variable X is observed, with x successes in n trials the result. The MLE for the true probability of success p maximizes

$f(x; n, p)$ as a function of p, and is $\hat{p} = x/n$, the observed proportion of successes in the n observed trials (that is, the frequency estimate). Since $V(X) = np(1 - p)$, $V(\hat{p}) = p(1 - p)/n$, and an approximate $100 \times (1 - \alpha)\%$ confidence interval for p based on (2.3) is

$$\hat{p} \pm (z_{\alpha/2})\sqrt{\hat{p}(1 - \hat{p})/n} \tag{4.1}$$

(this standard interval is sometimes called the Wald interval).

A better way to use the Central Limit Theorem to construct a confidence interval for p is as follows. The Central Limit Theorem implies that for large n

$$P\left(\left|\frac{\hat{p} - p}{\sqrt{p(1 - p)/n}}\right| \leq z_{\alpha/2}\right) \approx 1 - \alpha.$$

Squaring both sides of the inequality in the parentheses gives

$$P\left(\left[\frac{\hat{p} - p}{\sqrt{p(1 - p)/n}}\right]^2 \leq z_{\alpha/2}^2\right) = P\left([\hat{p} - p]^2 \leq z_{\alpha/2}^2 p(1 - p)/n\right)$$

$$\approx 1 - \alpha.$$

The zeroes of the quadratic equation $[\hat{p} - p]^2 = z_{\alpha/2}^2 p(1 - p)/n$ then define the endpoints of the confidence interval for p. Note that the only difference between this interval and (4.1) is that the standard error of $\hat{p}$ is estimated in the Wald interval by $\sqrt{\hat{p}(1 - \hat{p})/n}$, while the exact value $\sqrt{p(1 - p)/n}$ is used for this interval. Some algebra ultimately gives the form of the interval (sometimes called the score interval, the q-interval, or the Wilson interval),

$$\hat{p}\left(\frac{n}{n + z_{\alpha/2}^2}\right) + \frac{1}{2}\left(\frac{z_{\alpha/2}^2}{n + z_{\alpha/2}^2}\right)$$

$$\pm \sqrt{\left[\frac{\hat{p}(1 - \hat{p})}{n}\right]\left[\frac{n^2 z_{\alpha/2}^2}{(n + z_{\alpha/2}^2)^2}\right] + \frac{1}{4}\left[\frac{z_{\alpha/2}^4}{(n + z_{\alpha/2}^2)^2}\right]}. \tag{4.2}$$

This looks quite daunting, but what's actually going on is some "fudging" with the estimates of both p and the standard error of the estimate of p. The first part of the interval shows that the estimate of p is "shrunk" from $\hat{p}$ towards $1/2$, the center of the possible values for p. The second part of the interval shows that the estimated standard error is slightly adjusted, with the largest effect coming when $\hat{p}$ is close to 0 or 1. The score interval typically achieves its nominal coverage level for considerably smaller n than is true for the Wald interval unless p is very close to 0 or 1 (when the Wald interval is even worse), and should probably be used routinely as a replacement for (4.1).

Daily Water Consumption

A survey of 2818 people done for Rockefeller University and the International Bottled Water Association, discussed in the August 8, 2000, issue of *Newsday*, found that 34% reported that they drink the recommended eight or more 8-ounce servings of water daily (Ochs, 2000). An approximate 95% confidence interval for the true proportion of people who report that they drink this amount of water is thus

$$.34 \pm \sqrt{(.34)(.66)/2818} = .34 \pm .017 = (.323, .357).$$

Note that although the newspaper referred to 34% as an estimate of the percentage "of Americans that drink the recommended" amount, this is not quite true; rather, as indicated above, it is an estimate of the percentage that *report* that they drink the recommended amount.

The score interval for these data is $(.323, .358)$, which, given the large sample size, is unsurprisingly very close to the Wald interval.

The large-sample test (2.5) of the hypotheses

$$H_0 : p = p_0$$

versus

$$H_a : p \neq p_0$$

illustrates a case where the standard test is not equivalent to the standard confidence interval. Since the variance of $\hat{p}$ is a function of p (being $p(1 - p)/n$), the estimated variance of $\hat{p}$ under the null uses the null value p_0 rather than the estimate $\hat{p}$ (this is called a score test, or a Lagrange multiplier test; the confidence interval (4.2) is based on the same construction, which is why it is called the score interval). Thus, an approximate test statistic is

$$z = \frac{\hat{p} - p_0}{\sqrt{p_0(1 - p_0)/n}}, \tag{4.3}$$

which is compared to a Gaussian reference distribution.

The Effectiveness of Needle Exchange Programs

The idea of needle exchange programs, where intravenous drug users turn in dirty hypodermic needles for clean ones, is a very controversial one. Proponents argue that it encourages drug addicts to not share dirty needles, thereby lowering the chances of transmissions of diseases such as hepatitis and AIDS. Opponents argue that it provides government support for an illegal and unhealthy lifestyle.

Do needle exchange programs actually reduce disease risk? A study, results of which were reported at the annual meeting of the American Public Health Association in November 1994, investigated this question. Previous studies had shown that in situations where no needle exchange programs were used, between 4% and 8% of addicts who were not infected with the human immunodeficiency virus (that is, were HIV-negative) would become newly infected over a two-year period. In this study, 2500 HIV-negative intravenous drug users in New York City were enrolled in a needle exchange program. Two years later, 2% of the people in the study had become newly infected. Do these results indicate that the proportion of new infections using the needle exchange program is different from that without a needle exchange program?

The hypotheses being tested are

$$H_0 : p = .04$$

versus

$$H_a : p \neq .04.$$

We use $p = .04$ as the null value because it is the value in the null $(.04 \leq p \leq .08)$ that is closest to the observed proportion; this makes it as difficult as possible to reject the null, resulting in a conservative test. We define the alternative hypothesis to be two-sided, since it was possible that providing free needles might have worsened, rather than helped, the situation. The test statistic is

$$z = \frac{.02 - .04}{\sqrt{(.04)(.96)/2500}} = -5.10,$$

which has a tail probability less than .00001. That is, there is very strong evidence that the probability of new infection is lower in the needle exchange program than the previous standard of 4%.

4.1.2 Sample Size and Power Calculations

The hypothesis test (4.3), and the Gaussian approximation upon which it is based, can be used to guide the data analyst in roughly determining the amount of data necessary to identify a specified effect. Suppose that it is desired that a true (alternative) probability p_a that is δ away from the null value p_0 will be detected with probability β based on a test of size α. To keep things simpler, consider a one-sided test designed to identify $\delta > 0$. The desired condition is

$$P\left(\frac{\bar{p} - p_0}{\sqrt{p_0(1 - p_0)/n}} > z_\alpha\right) = \beta \qquad (4.4)$$

when the true $p_a = p_0 + \delta$. The value β is the *power* of the test to detect an effect of size δ. By the Central Limit Theorem, $\bar{p} \sim N(p_a, p_a(1-p_a)/n)$, implying from (4.4)

$$\begin{aligned}
\beta &= P\left(\frac{\bar{p} - p_0}{\sqrt{p_0(1 - p_0)/n}} > z_\alpha\right) \\
&= P(\bar{p} > p_0 + z_\alpha\sqrt{p_0(1 - p_0)/n}) \\
&\approx P\left(Z > \frac{p_0 + z_\alpha\sqrt{p_0(1 - p_0)/n} - p_a}{\sqrt{p_a(1 - p_a)/n}}\right) \\
&= P\left(Z > z_\alpha\sqrt{\frac{p_0(1 - p_0)}{p_a(1 - p_a)}} - \frac{\delta}{\sqrt{p_a(1 - p_a)/n}}\right).
\end{aligned}$$

Since the probability on the right side of the equation equals β, by definition

$$z_\alpha\sqrt{\frac{p_0(1 - p_0)}{p_a(1 - p_a)}} - \frac{\delta}{\sqrt{p_a(1 - p_a)/n}} = z_\beta.$$

Solving for n gives the appropriate sample size,

$$n = \left(\frac{z_\alpha\sqrt{p_0(1 - p_0)} - z_\beta\sqrt{p_a(1 - p_a)}}{\delta}\right)^2. \qquad (4.5)$$

The same sample size results when attempting to detect a negative effect.

The Effectiveness of Needle Exchange Programs (continued)

It was noted previously that a sample size of 2500 was used in the needle exchange program study. How might a number like this be determined? Say the researchers felt that a one percentage point drop in new infection rate was clinically important, and they would want to identify such an effect with probability .7 based on a .025 level test (corresponding to a two-sided .05 level test). Then $p_0 = .04$, $p_a = .03$, $\delta = -.01$, $z_\alpha = 1.96$, and $\beta = .7$ (implying $z_\beta = -.5244$). Substituting these values into (4.5) gives

$n = 2242.3$, or (rounding up) $n = 2243$, a rough sample size reasonably close to the value actually used.

4.1.3 Inference from a Sample of Binomial Random Variables

What if the data constitute a sample of binomial random variables from a $\text{Bin}(n,p)$ population? Say a sample of n realizations from a binomial random variable is taken. Each $x_i \sim \text{Bin}(n_i, p)$; that is, the number of trials for each realization might be different, but the true probability of success for each realization is assumed to be the same. The likelihood function is then

$$\prod_{i=1}^{n} f(x_i; n_i, p) = \prod_{i=1}^{n} \left[\binom{n_i}{x_i} p^{x_i} (1-p)^{n_i - x_i} \right]$$

$$= \left[\prod_{i=1}^{n} \binom{n_i}{x_i} \right] p^{\sum_i x_i} (1-p)^{\sum_i (n_i - x_i)}.$$

The MLE for p maximizes this function, and is easily shown to be

$$\hat{p} = \frac{\sum_i x_i}{\sum_i n_i}; \tag{4.6}$$

that is, to estimate p, pool all of the sample realizations and use the overall empirical success proportion. The asymptotic variance of $\hat{p}$ is $p(1-p)/\sum_i n_i$. This result also can be justified using the fact that the sum of two independent binomial random variables, $X \sim \text{Bin}(n,p)$ and $Y \sim \text{Bin}(m,p)$, is also binomial $X + Y \sim \text{Bin}(n+m, p)$. This latter fact also explains why the binomial probability functions plotted in Figure 4.1 look increasingly Gaussian as np increases. If $X \sim \text{Bin}(n,p)$, then $X = \sum_i B_i$, where the B_i are independent $\text{Bin}(1,p)$ (sometimes called *Bernoulli* distributed). The Central Limit Theorem implies that this sum has a distribution increasingly Gaussian as n gets larger.

4.1.4 Exact Inference

As we saw in Section 2.1.3, one of the nice properties of the Gaussian distribution is that exact (small-sample) inference is easy to implement using the t-distribution. The need for such methods is even more important for the binomial random variable than it is for the Gaussian, since (as Figure 4.1 shows) the distribution of $\hat{p}$ is discrete and potentially asymmetric, and potentially far from Gaussian (since $\hat{p} = x/n$, it has a binomial distribution scaled by the sample size).

It is perhaps easier to see how exact inference proceeds for hypothesis testing than for confidence intervals. Consider first the one-sided hypothesis test of

$$H_0 : p = p_0$$

versus

$$H_a : p > p_0$$

(the test for $p < p_0$ proceeds in the obvious way). From the form of the alternative hypothesis we see that the test is based on rejecting the null when $\hat{p}$ is large enough, or equivalently when the observed number of successes x is large enough. The size of the test α is the probability of rejecting the null when it is in fact true, so the test tries to find the value c such that

$$P(X \geq c \,|\, p = p_0) \leq \alpha. \tag{4.7}$$

One immediate difference from the approximate (Gaussian-based) inference that we're used to is that the size α is not necessarily achieved; we can only guarantee that the probability of falsely rejecting the null is no more than α. The value c is chosen so that the probability $P(X \geq c \,|\, p = p_0)$ is as close to α as possible without exceeding it.

Say we are testing hypotheses based on $p_0 = .4$ based on $n = 10$ binomial trials, with $\alpha = .05$. The following table gives the null binomial probabilities.

x	$P(X = x \,|\, p = .4)$	$P(X \geq x \,|\, p = .4)$
0	0.0060	1.0000
1	0.0403	0.9940
2	0.1209	0.9537
3	0.2150	0.8328
4	0.2508	0.6178
5	0.2007	0.3670
6	0.1115	0.1663
7	0.0425	0.0548
8	0.0106	0.0123
9	0.0016	0.0017
10	0.0001	0.0001

The hypothesis test is based on (4.7) with $\alpha = .05$, so the test rejects the null if the number of successes is at least 8. Despite a nominal size of .05, the actual size of the test is only $.0106 + .0016 + .0001 = .0123$; that is, the true size is much less than the nominal size, making the test very *conservative*. This is unavoidable, given the form of the null binomial distribution. Rejecting the null when the number of successes is at least 7 would not be appropriate at a .05 level, since the exact p-value for that value is $.0425 + .0106 + .0016 + .0001 = .0548$, which is greater than the nominal .05.

Part of the problem here is the rigid adherence to a reject/don't reject decision rule. This is overcome by reporting a p-value rather than rejection or lack of rejection of the null at a specific α level, but while this is straightforward for statistics with continuous distributions, for discrete statistics there is still a problem. If the null is true, the average p-value should be .5, but for discrete statistics it is typically higher than .5. That is, there is a tendency not to reject often enough (exact p-values are too high). This can be corrected using the *mid-p-value*, which equals half the probability of the observed event, plus the probabilities of more extreme events. So, for example, for the binomial situation above, the mid-p-value if 8 successes is observed is $.0106/2 + .0016 + .0001 = .007$ (compared to the exact p-value of .0123), while the mid-p-value if 7 successes is observed is $.0425/2 + .0106 + .0016 + .0001 = .0336$ (compared to the exact p-value of .0548), reflecting stronger evidence against the null hypothesis in each case. Using the mid-p-value does not guarantee that the true significance level will be less than α, but it usually performs well, and is useful to combat the conservativeness that comes from discreteness in exact tests for discrete data.

What if the alternative hypothesis of interest is two-sided? That is, the hypotheses being tested are

$$H_0 : p = p_0$$

versus

$$H_a : p \neq p_0.$$

The test now would naturally reject for either small $\hat{p}$ or large $\hat{p}$, or equivalently, if $X \leq c_1$ or $X \geq c_2$. There is no unique way to define c_1 and c_2. One approach is the so-called Clopper–Pearson method, which defines an α level test as two one-sided tests, each at an $\alpha/2$ level. A table of the needed binomial calculations is given at the top of the next page. The two-sided $\alpha = .05$ test is a combination of a lower .025 level test (reject the null if there are zero successes) and an upper .025 level test (reject the null if there are at least 8 successes), which gives a total true size of $.0060 + .0123 = .0183$, once again highly conservative. The two-sided p-value is taken to be twice the one-sided value. So, for example, if 8 successes were observed, the one-sided p-value testing if $p > .4$ is .012, so the two-sided p-value would be .024.

x	$P(X = x \mid p = .4)$	$P(X \geq x \mid p = .4)$	$P(X \leq x \mid p = .4)$
0	0.0060	1.0000	0.0060
1	0.0403	0.9940	0.0463
2	0.1209	0.9537	0.1672
3	0.2150	0.8328	0.3822
4	0.2508	0.6178	0.6330
5	0.2007	0.3670	0.8337
6	0.1115	0.1663	0.9452
7	0.0425	0.0548	0.9877
8	0.0106	0.0123	0.9983
9	0.0016	0.0017	0.9999
10	0.0001	0.0001	1.0000

An alternative approach is the Wilson–Sterne rule, which defines the "tail" of the distribution of the statistic based on ordering the null probabilities of X (or, equivalently, those of $\hat{p}$).

x	$P(X = x \mid p = .4)$	Cumulative probability
10	0.0001	0.0001
9	0.0016	0.0017
0	0.0060	0.0077
8	0.0106	0.0183
1	0.0403	0.0586
7	0.0425	0.1011
6	0.1115	0.2126
2	0.1209	0.3335
5	0.2007	0.5342
3	0.2150	0.7492
4	0.2508	1.0000

This table orders the observed number of successes from least to most likely based on null probabilities. An $\alpha = .05$ level test then rejects the null if the number of successes is zero, or at least 8, yielding exact p-value .0183. In this case, for $\alpha = .05$ the Clopper–Pearson and Wilson–Sterne tests are identical, but for other values of α they are not. For example, if $\alpha = .06$, the Clopper–Pearson test is unchanged, but the Wilson–Sterne test adds $X = 1$ to the rejection region, yielding an exact p-value of .0586. The reason for the difference in the two tests is the asymmetry of the $\text{Bin}(10, .4)$ probability function, since for the Wilson–Sterne rule .0463 of the null size comes from having a small number of successes, while only .0123 comes from having a large number of successes. Note that since the Wilson–Sterne rule provides a well-defined ordering of events, the mid-p-value is easily defined. For example, if one success if observed, the mid-p-value is $.0403/2 + .0183 = .0385$, which can be compared to the exact p-value of .0586.

As was pointed out in Section 2.1.2, a $100 \times (1 - \alpha)\%$ confidence interval for p can be defined to include those values of p such that the observed $\hat{p}$ is not statistically significantly different from them at level α. Typically the Clopper–Pearson test (which aims at having size $\alpha/2$ in each tail) is used to form a symmetric confidence interval (it is symmetric in terms of having lack of coverage $\alpha/2$ in each tail, not in terms of symmetry around $\hat{p}$). If x successes are observed in n trials, the confidence interval is then (p_ℓ, p_u), where p_ℓ satisfies

$$P(X \geq x \mid p = p_\ell) = \alpha/2$$

and p_u satisfies

$$P(X \leq x \mid p = p_u) = \alpha/2,$$

where $X \sim \text{Bin}(n, p)$. These definitions don't apply if $x = 0$ (no successes) or $x = n$ (no failures); in the former case, $p_\ell = 0$, while in the latter case $p_u = 1$, as would be expected. It turns out that p_ℓ and p_u can be calculated in closed form,

$$p_\ell = \begin{cases} 0 & \text{if } x = 0 \\ \beta_{1-\alpha/2, x, n-x+1} & \text{if } x \neq 0, \end{cases}$$

and

$$p_u = \begin{cases} 1 & \text{if } x = n \\ \beta_{\alpha/2, x+1, n} & \text{if } x \neq n, \end{cases}$$

where $\beta_{\alpha, \nu_1, \nu_2}$ is the upper α percentile of a Beta distribution on (ν_1, ν_2) degrees of freedom. Note that a confidence region for p based on the Clopper–Pearson rule is, in fact, an interval, which is not necessarily the case if a region based on the Wilson–Sterne rule is formed (the region can be an interval with "holes").

Small-sample inference problems become more difficult as the sampling structure becomes more complex, but all reflect the basic issues that arise for the binomial random variable. The discreteness of the distribution of the statistic of interest leads to conservative tests, and if that distribution takes on only a few values, the conservativeness can be serious. Further, while one-sided tests are often straightforward to define, two-sided tests don't necessarily have a unique definition. Another consequence of the discreteness of the binomial distribution is that the exact power function does not increase monotonically with increasing sample size, as is implied by the approximate power (4.4), which is based on a continuous (Gaussian) distribution.

Testing for Lyme Disease

Lyme disease is a tick-borne infection caused by the spirochete *Borrelia burgdorferi*, that, when left untreated, can cause arthritis, nervous system damage, and damage to the heart. Treatment using antibiotics in the early stages of Lyme disease is generally highly effective, but treatment is more difficult in later stages. The standard test for infection requires a time lag of up to several weeks, and can be inconclusive. In February 1999, the Centers for Disease Control and Prevention (CDC) conducted a small study of the effectiveness of a simple diagnostic immunologic assay test for infection with *B. burgdorferi* (Schriefer et al., 2000; Schutzer et al., 1999). In this study, samples were taken from 13 patients with early, culture-confirmed Lyme disease, and from 25 patients with no history of Lyme disease, and the assay test was applied to the samples. The estimated sensitivity for the immune complex assay test (the proportion of tests on patients with Lyme disease that came back positive) was $8/13 = 61.5\%$. A large-sample 95% confidence interval for the sensitivity of this test is thus

$$.615 \pm (1.96)\sqrt{\frac{(.615)(.385)}{13}} = (.351, .880).$$

Similarly, the estimated specificity for the immune complex assay test (the proportion of tests on patients without Lyme disease that came back negative) was $19/25 = 76\%$, yielding a large-sample 95% confidence interval for the specificity of

$$.76 \pm (1.96)\sqrt{\frac{(.76)(.24)}{25}} = (.593, .927).$$

The score intervals for these data are $(.355, .823)$ and $(.566, .885)$, respectively, and are noticeably different from the Wald intervals, especially at the upper end. These are relatively low sensitivity and specificity values; for example, the Wellcome Elisa test, the standard test for HIV, has sensitivity roughly .993 and specificity roughly .9999.

With such small samples, the Gaussian approximation is suspect here, so we should consider exact intervals. They turn out to be $(.316, .861)$ and $(.549, .906)$, respectively. In both cases the Wald intervals are shifted upwards, with the lower ends off by roughly 10%. The score intervals are 10–15% narrower than the Clopper–Pearson intervals, which could very well reflect the conservativeness of the exact intervals.

The study also studied the application of two-tiered serologic testing, the test recommended by the CDC, to the same samples. For this test the estimated sensitivity was $10/13 = 76.9\%$, while the estimated specificity was $22/25 = 88\%$. The large-sample 95% confidence intervals are thus

$$.769 \pm (1.96)\sqrt{\frac{(.769)(.231)}{13}} = (.540, .998)$$

(score interval $(.497, .918)$) and

$$.88 \pm (1.96)\sqrt{\frac{(.88)(.12)}{25}} = (.753, 1.007)$$

(score interval $(.700, .958)$), respectively. The Wald interval for the specificity is absurd, of course, since the specificity cannot exceed 1, so the interval would typically be truncated to $(.753, 1)$ (the score interval cannot include impossible values). The exact intervals are $(.462, .950)$ and $(.688, .975)$, respectively. Again the Wald intervals are shifted up, with the exact and score intervals for the specificity excluding values of p close to 1.

In order to assess reproducibility of testing, 10 repeat samples were taken from the blood samples (5 from patients with Lyme disease, and 5 from patients without Lyme disease). When the two-tiered serologic test was applied to the repeat samples, all 10 gave the same result as in the original samples. Thus, the estimate of reproducibility is $10/10 = 100\%$, but we don't seriously think this is the true reproducibility rate. The large-sample confidence interval (4.1) is useless here, since $\hat{p} = 1$, but the exact interval is still defined and is $(.692, 1)$, which reflects the high variability in the estimate of the reproducibility rate based on only 10 samples. The score interval is similar, being $(.722, 1)$.

Note, by the way, that if zero successes or failures are observed, a sensible alternative is to construct one-sided intervals, since one limit of the confidence interval would necessarily be 0 or 1, as appropriate. This would correspond to using the Wilson–Sterne rule for exact intervals, rather than the Clopper–Pearson rule, and is the approach used in some statistical packages. If that is done here, the exact interval is $(.741, 1)$, while the score interval is $(.787, 1)$.

4.2   The Multinomial Random Variable

A natural generalization of the binomial random variable is to allow for more than two possible outcomes in a trial (K, where $K > 2$). For example, respondents to a sample survey might be asked if they agree or disagree with a statement (yes/no), but allowance is made for the response "don't know." Patients arriving at an emergency room might be classified by degrees of response (death/hospitalization required/serious treatment but not hospitalization/minor treatment). Data of this type can be modeled using the *multinomial* distribution. Say n respondents are sampled, and n_i of them result in a response in the ith category (clearly $\sum_i n_i = n$). Then the probability function for the multinomial vector ($\mathbf{n} \sim \text{Mult}(n, \mathbf{p})$) is

$$f(\mathbf{n}; n, \mathbf{p}) = \frac{n!}{\prod_i n_i!} \prod_{i=1}^{K} p_i^{n_i}. \tag{4.8}$$

It is apparent that each cell count $n_i \sim \text{Bin}(n, p_i)$ (simply pool all categories other than the ith category together). Thus, the expected value of the number of observations falling in the ith category is $E(n_i) = np_i$, while the variance is $V(n_i) = np_i(1 - p_i)$. Since we know that the counts must sum to n, it is not surprising that they are negatively correlated with one another (if many observations fall in one cell, fewer must fall in another cell); in fact,

$$\text{corr}(n_i, n_j) = -\sqrt{\frac{p_i p_j}{(1 - p_i)(1 - p_j)}}.$$

4.2.1   Large-Sample Inference

The maximum likelihood estimator of the probability of falling in the ith category is the frequency estimator, $\hat{p}_i = n_i/n$. Since each n_i is binomially distributed, confidence intervals and hypothesis tests for an individual p_i based on (4.1), (4.2), and (4.3) are still valid. A simultaneous confidence region for *all* of the values in $\mathbf{p}$, however, requires a correction to account for the fact that there are K different intervals being formed (if each was constructed at a 95% confidence level, for example, the chance that at least one p_i would be outside its confidence interval in repeated sampling would be considerably higher than 5%). A standard correction for this is to use a *Bonferroni* correction, where each interval is constructed at a $100 \times (1 - \alpha/K)\%$ level. This will result in an overall confidence level of at least $100 \times (1 - \alpha)\%$. Since this widens each individual interval, it is typically applied to the narrower score interval (4.2).

We postpone discussion of hypothesis tests for the multinomial distribution, and exact inference, to the discussion of goodness-of-fit tests in Section 4.4.

___

Draws for the Louisiana Lottery

The winning four-digit number for the Pick 4 game of the Louisiana Lottery is chosen using four machines, wherein one of 10 balls (numbered 0 through 9) is selected. The following table gives the observed numbers of chosen digits in the 517 draws of the lottery from its start on March 1, 1999, through the drawing on August 1, 2000, the frequency estimates of the probability of each digit being chosen, and simultaneous 95% confidence intervals for **p**.

Digit	Number of draws	Frequency estimate	Confidence intervals
0	206	.0996	(.0826, .1196)
1	200	.0967	(.0800, .1165)
2	206	.0996	(.0826, .1196)
3	223	.1078	(.0902, .1285)
4	176	.0851	(.0694, .1039)
5	215	.1040	(.0866, .1243)
6	202	.0977	(.0809, .1176)
7	223	.1078	(.0902, .1285)
8	213	.1030	(.0857, .1233)
9	204	.0986	(.0817, .1186)

If the lottery is fair, each digit should have an equal chance of being chosen (that is, $p_i = .1$ for all digits). There is no evidence here to suggest that the game is not fair, as .1 is well within each of the confidence intervals. Of course, we cannot be sure of this based on only 517 draws, since the probabilities are only estimated to within roughly $\pm.02$.

___

4.3   The Poisson Random Variable

The multinomial random variable is a reasonable model for discrete data where observations are categorized over a (perhaps limited) predetermined set of categories, but what about discrete data where that is not the case? For example, how many speeding tickets will be given on the New York State Thruway this week? We know that the value is a nonnegative integer,

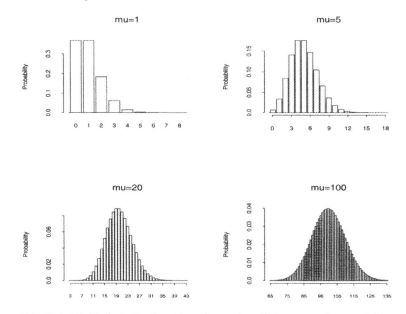

FIGURE 4.2. Probability functions for various Poisson random variables.

but we cannot say what the maximum possible value is, and if a year's worth of data (for example) are available, chances are that each specific number of tickets given will occur only once.

A distribution that is often useful when the random variable is a count of the number of occurrences of some event is the *Poisson* distribution. The probability function for a Poisson random variable $(X \sim \mathrm{Pois}(\mu))$ is

$$f(x; \mu) = \frac{e^{-\mu}\mu^x}{x!}, \qquad x = 0, 1, \ldots \qquad (4.9)$$

where μ is the mean (expected number of counts) of the random variable. It can be shown that the variance of a Poisson random variable also equals the mean μ.

The Poisson random variable is closed under summation; that is, the sum of K independent Poisson random variables, each with mean μ_i, is Poisson with mean $\sum \mu_i$. Since a Poisson random variable with mean $\mu \to \infty$ is equivalent to a sum of μ Poisson(1) random variables, the Central Limit Theorem implies that a Poisson random variable with large mean μ is approximately normally distributed with mean and variance both equal to μ. This can be seen in Figure 4.2, which gives graphs of the Poisson probability function for different values of μ. As the mean increases, the asymmetry of the probability function decreases, while the variance of the random variable increases.

The Poisson random variable arises naturally in several ways. One is as a result of a general model for the occurrence of random events in time or space. Let ν be the rate at which events occur per unit time. If the probability that exactly one event occurs in a small time interval of length h is approximately equal to νh, the probability that more than one event occurs in a small time interval is virtually zero, and if the numbers of events that occur in nonoverlapping time intervals are independent, then the total number of event occurrences in a period of time of length t has a Poisson distribution with mean $\mu = \nu t$.

Another way that the Poisson distribution arises is as an approximation to the binomial distribution. If the number of successes in n trials is binomially distributed, with the number of trials $n \to \infty$ and the probability of success $p \to 0$ such that $np \to \mu$, the distribution of the number of successes is roughly Poisson with mean μ. That is, the Poisson distribution is a reasonable candidate model for the number of occurrences of "rare" events (ones that are very unlikely to occur in any one situation, but might occur because of a large number of repeated, independent, trials).

4.3.1 Large-Sample Inference

Say a sample of n realizations from a Poisson random variable is taken. The likelihood function is then

$$\prod_{i=1}^{n} f(x_i; \mu) = \prod_{i=1}^{n} \left[\frac{e^{-\mu}\mu^{x_i}}{x_i!} \right]$$

$$= \left[\prod_{i=1}^{n} \frac{1}{x_i!} \right] e^{-n\mu} \mu^{\sum_i x_i} \tag{4.10}$$

The MLE for μ maximizes this function, and is easily shown to be the sample mean, $\hat{\mu} = \overline{X}$, with estimated standard error $\sqrt{\overline{X}/n}$. Note that this is *not* the estimated standard error for $\hat{\mu}$ from a Gaussian sample ($\sqrt{\hat{\sigma}^2/n}$), even though both estimates are the sample mean. For Poisson data the variance of X equals the mean, so the estimated standard error of $\hat{\mu}$ is based on $\hat{\mu}$ itself. Large-sample confidence intervals and hypothesis tests are then based on the usual Gaussian approximation, with a $100 \times (1-\alpha)\%$ confidence interval having the form

$$\overline{X} \pm z_{\alpha/2}\sqrt{\overline{X}/n}, \tag{4.11}$$

and a test of the null hypothesis $\mu = \mu_0$ based on the statistic

$$z = \frac{\overline{X} - \mu_0}{\sqrt{\mu_0/n}}. \tag{4.12}$$

As was true for the binomial random variable, since the estimated standard error of $\hat{\mu}$ is based on $\hat{\mu}$, the asymptotic confidence interval (4.11) and test (4.12) do not have the same form, and the true coverage of the standard interval can be much smaller than the nominal coverage. A better confidence interval (a score interval) can be constructed by inverting the test (4.12), giving

$$\overline{X} + \frac{z_{\alpha/2}^2}{2n} \pm \frac{z_{\alpha/2}}{\sqrt{n}} \sqrt{\overline{X} + z_{\alpha/2}^2/(4n)}.$$

This interval alters the standard interval by increasing slightly both the estimate of the mean and the estimated standard error, an adjustment that has a larger effect for smaller $\overline{X}$ and smaller n, as would be expected.

Goal Scoring in the National Hockey League

The Poisson distribution is a natural one for the number of goals in games such as hockey and soccer, since both games are characterized by multiple opportunities for goal scoring with low probabilities of scoring on any one opportunity. Table 4.1 summarizes the goals scored for (GF) and against (GA) the New Jersey Devils during the 1999–2000 National Hockey League season. The Devils won the Stanley Cup that year, emblematic of the league championship. The table gives the number of games wherein the Devils scored zero goals, one goal, and so on, as well as the figures for goals against. Since these data represent the entire offensive and performance of the Devils during the season, there is no need to estimate their season performance. Rather, we could view these results as being typical of the performance of a high-quality National Hockey League team during this time period, and could use these data to try to understand the offensive and defensive patterns of such teams.

The average goals scored per game by the Devils was $\overline{GF} = 3.061$, while the average goals given up per game was $\overline{GA} = 2.476$. The standard large-sample 95% confidence interval for the true average number of goals scored by high-quality teams is $3.061 \pm (1.96)(.193) = (2.682, 3.44)$, while that for goals given up is $2.476 \pm (1.96)(.174) = (2.135, 2.816)$. Given the success of the team, it seems a little surprising to see how close these two intervals are. The score intervals are $(2.705, 3.464)$, and $(2.158, 2.84)$, which are very similar to the standard intervals.

Of course, the important number isn't really goals scored or goals given up, but rather the difference between them, which depends on the joint distribution of goal scoring. We will examine joint distributions of this type in Section 7.1.4.

Goals	Goals scored	Goals against
0	3	6
1	9	21
2	24	17
3	18	16
4	14	12
5	7	9
6	3	1
7	2	
8	1	
9	1	

TABLE 4.1. Goals scored and given up by the New Jersey Devils during the 1999–2000 season. Entries are the number of games where the given number of goals was scored or given up.

4.3.2 Exact Inference

Figure 4.2 showed that the Poisson distribution looks increasingly Gaussian as μ increases because of the Central Limit Theorem. This implies that the Gaussian-based interval (4.11) and test (4.12) should work reasonably well even for small samples if μ is large enough. The interval and test should also work well for small μ if the sample size is large, since then the asymptotic normality of $\overline{X}$ applies (that is, the interval and test require large $n\mu$).

If both n and μ are small, however, the standard interval and test can be misleading. Exact tests and confidence intervals can be defined in the same way as was done for the binomial random variable in Section 4.1.4. Recall that the sum X of n observations X_i drawn from a Poisson random variable with mean μ is Poisson distributed with mean $n\mu$. Say the observed sum equals x. A Clopper–Pearson-type interval is formed by finding the values (μ_ℓ, μ_u) that satisfy

$$P(X \geq x \,|\, \mu = \mu_\ell) = \alpha/2$$

and

$$P(X \leq x \,|\, \mu = \mu_u) = \alpha/2,$$

respectively, with $\mu_\ell = 0$ if $x = 0$. It turns out that these limits can be derived analytically, and are

$$\mu_\ell = \begin{cases} 0 & \text{if } x = 0 \\ \chi^2_{1-\alpha/2,2x}/(2n) & \text{if } x \neq 0, \end{cases}$$

and

$$\mu_u = \chi^2_{\alpha/2, 2x+2}/(2n),$$

where $\chi^2_{\alpha,\nu}$ is the upper α percentile of a chi-squared distribution with ν degrees of freedom.

Severe U.S. Mainland Hurricane Strikes

The National Hurricane Center Tropical Prediction Center, a part of the National Oceanic and Atmospheric Administration, is the U.S. agency charged with issuing watches, warnings, forecasts, and analyses of hazardous tropical weather. Historical data on tropical storms are also archived at the Prediction Center's web site (http://www.nhc.noaa.gov).

There were six severe hurricanes in the United States during the years 1957–1996: in 1957 (Audrey), 1960 (Donna), 1961 (Carla), 1969 (Camille), 1989 (Hugo), and 1992 (Andrew). We classify a hurricane as severe if it is rated four or above on the Saffir–Simpson hurricane scale (sustained winds of more than 130 miles per hour) when making U.S. landfall. Thus, the MLE of the average rate of severe hurricanes per year is $\hat{\mu} = 6/40 = .15$. The standard large-sample 99% confidence interval for μ is $(-.008, .308)$, but this interval (which includes impossible negative values) is based on a suspect Gaussian approximation. The score interval reflects this, as it is $(.055, .411)$, noticeably asymmetric around .15. The exact interval is $(.038, .391)$, again reinforcing that the asymmetry of the underlying Poisson distribution implies a larger value for the upper end of the interval than the Gaussian approximation would imply.

4.3.3   The Connection Between the Poisson and the Multinomial

Say we observe a sample of K independent Poisson random variables n_i, each with mean μ_i. Naturally, the total count $n = \sum_i n_i$ is itself a random variable, but say we knew a priori that this would be n. The n_i's would no longer be Poisson, since there would be a maximum restriction on each ($n_i < n$), and they would not be independent, since a larger value of n_i would imply a smaller value for other values n_j.

The definition of conditional probability of event A given event B,

$$P(A|B) \equiv \frac{P(A \text{ and } B)}{P(B)},$$

and the Poisson likelihood function (4.10), directly imply the distribution of $\mathbf{n} = \{n_1, \ldots, n_K\}$, as follows:

$$P(i\text{th value equals } n_i, i = 1, \ldots, K \mid \textstyle\sum_i n_i = n)$$

$$= \frac{P(i\text{th value equals } n_i, i = 1, \ldots, K \text{ and } \sum_i n_i = n)}{P(\sum_i n_i = n)}$$

$$= \frac{\prod_i e^{-\mu_i} \mu_i^{n_i} / n_i!}{e^{-\sum_i \mu_i} (\sum_i \mu_i)^n / n!}$$

$$= \left(\frac{n!}{\prod_i n_i!}\right) \prod_i \left(\frac{\mu_i}{\sum_i \mu_i}\right)^{n_i}$$

$$= \left(\frac{n!}{\prod_i n_i!}\right) \prod_i p_i^{n_i},$$

where $p_i = \mu_i / (\sum_j \mu_j)$. This probability function, of course, is the multinomial probability function (4.8). That is, Poisson sampling conditional on the total sample size is equivalent to multinomial sampling, a result that will prove very useful in analyzing categorical data. Note that a special case of this result is that given two independent Poisson random variables X_1 and X_2 with means μ_1 and μ_2, respectively, the distribution of X_1 conditional on their sum $n = X_1 + X_2$ is $X_1 \sim \mathrm{Bin}(n, \mu_1/[\mu_1 + \mu_2])$.

4.4 Testing Goodness-of-Fit

Parametric modeling based on maximum likelihood is a very powerful tool, but that power comes with a price. While maximum likelihood estimates have many desirable properties when the hypothesized model is true, they can be worse than useless when the model is not true. A well-known example of this is the great utility of the sample mean and sample variance for data that are (reasonably) Gaussian, combined with their extremely poor performance for data with extreme outliers or long tails.

Modeling for categorical data suffers from the same problem, but the structure of categorical data does provide a simple tool that can sometimes help in checking whether a hypothesized model is reasonable or not — tests of *goodness-of-fit*. Goodness-of-fit tests are based on comparing the observed number of observations falling in each cell to the number that would have been expected if the hypothesized model was true. If the observed and expected numbers differ greatly, that is evidence that the hypothesized model is not correct; that is, it does not fit the data.

4.4.1 Chi-Squared Goodness-of-Fit Tests

Consider a K-cell multinomial vector $\mathbf{n} = \{n_1, \ldots, n_K\}$, with $\sum_i n_i = n$. Say that it was hypothesized that this multinomial was generated from an underlying probability vector $\mathbf{p}_0 = \{p_{10}, \ldots, p_{K0}\}$; that is,

$$H_0 : \mathbf{p} = \mathbf{p}_0.$$

The alternative hypothesis is the so-called omnibus alternative, which just states that $\mathbf{p}$, whatever it might be, is not $\mathbf{p}_0$; that is,

$$H_a : \mathbf{p} \neq \mathbf{p}_0.$$

Let $e_i = Np_{i0}$ be the expected cell counts under the null hypothesis. Karl Pearson, in a paper in 1900 that in many ways opened the door to the modern era of statistical inference, proposed the test statistic

$$X^2 = \sum_{i=1}^{K} \frac{(n_i - e_i)^2}{e_i} \tag{4.13}$$

to test the goodness-of-fit hypotheses. It is clear that large discrepancies between the observed and expected cell counts will result in larger values of X^2. If the null hypothesis is true, and n is large (so that each of the expected counts e_i is large), X^2 has (approximately) a χ^2 distribution on $K - 1$ degrees of freedom (one degree of freedom is "lost" because $\sum_i p_i = 1$), so observed values of X^2 can be compared to this distribution to determine if they indicate an unusual difference from the null. Since $V(n_i) \approx e_i$ for large N, X^2 is roughly the sum of squares of standardized distances from the expected counts under the null. The fact that this variance is a function of the mean is what makes the statistic work, since differences between the observed and expected counts can be ascribed to potential lack of fit, rather than inherent variability that is unrelated to the mean.

An alternative goodness-of-fit test is based on likelihood functions. Recall that the likelihood function gives the probability of seeing the observed sample values given specific values of unknown parameters (and hence justifies the MLE as the values of the parameters that maximize the probability of seeing what was actually observed). The *likelihood ratio* statistic (also called the *deviance*) is based on the following idea. The maximized likelihood under the null hypothesis must be less than or equal to that when the vector $\mathbf{p}$ is unrestricted (or else the MLE $\hat{\mathbf{p}}$ for $\mathbf{p}$ would be $\mathbf{p}_0$). If the null is true, $\hat{\mathbf{p}}$ won't be "too far" from $\mathbf{p}_0$, and

$$\Lambda = \frac{\text{Maximized likelihood under } H_0}{\text{Maximized likelihood for unrestricted } \mathbf{p}}$$

won't be "too far" below 1. On the other hand, if the null is not true, the maximized likelihood without restriction will be much larger than that under H_0, and Λ will be small.

The distribution for $G^2 = -2 \log \Lambda$ is easier to use than that of Λ, and this statistic equals

$$G^2 = 2 \sum_{i=1}^{K} n_i \log \left(\frac{n_i}{e_i} \right). \tag{4.14}$$

Just as is true for X^2, G^2 takes on its minimum value of 0 when $n_i = e_i$ for all i, and larger values imply greater discrepancy from the null. If the null hypothesis is true, and n is large (so that each of the expected counts e_i is large), G^2 has (approximately) a χ^2 distribution on $K - 1$ degrees of freedom; in this case X^2 and G^2 will be very similar to each other, although they can be very different if e_i values are small. By convention,

$$n_i \log \left(\frac{n_i}{e_i} \right) \equiv 0$$

when $n_i = 0$.

Both of these goodness-of-fit statistics are special cases of a general class of statistics called *power divergence* statistics, which take the form

$$2nI^\lambda = \frac{2}{\lambda(\lambda + 1)} \sum_{i=1}^K n_i \left[\left(\frac{n_i}{e_i} \right)^\lambda - 1 \right], \quad -\infty < \lambda < \infty. \tag{4.15}$$

While (4.15) is undefined for $\lambda = 0$ or -1, these forms can be defined as the limits of (4.15) as $\lambda \to 0$ and $\lambda \to -1$, respectively. X^2 is (4.15) with $\lambda = 1$, and G^2 is (4.15) with $\lambda = 0$. The statistic with $\lambda = 2/3$ has desirable properties when individual n_i's are small. All members of the power divergence family are asymptotically χ^2_{K-1} under the null when $n \to \infty$.

Sometimes the null vector $\mathbf{p}_0$ comes from a parametric family based on a q-dimensional parameter vector $\boldsymbol{\theta}$ (for example, testing if observed data are consistent with a Poisson distribution with mean μ, where $q = 1$). In this case the test statistics are modified by replacing e_i with $\hat{e}_i$, estimates based on estimating the unknown parameters of the parametric family (from the observed categorized data) using an efficient estimator such as maximum likelihood. This affects the distribution of the test statistics, in that the power divergence goodness-of-fit statistics are asymptotically χ^2_{K-q-1} under the null as $n \to \infty$.

Asymptotic p-values are based on having a large sample, but a problem can occur if the sample size is very large: goodness-of-fit tests will reject almost any null hypothesis, but the observed deviation from the null might be small, and of little practical importance. The same problem occurs in other hypothesis tests; for example, a sample mean of 0.01 is statistically significantly different from 0 at a .05 level when a sample of size 40,000 is drawn from a Gaussian distribution with unit standard deviation, but it is unlikely that this is of any practical importance. A statistic (or set of statistics) that quantifies the practical importance of observed deviations from the null should be examined to assess whether (for large samples) a significant lack of fit is in fact unimportant, or, for that matter, whether (for small samples) an apparently adequate fit in fact displays noteworthy lack of fit.

One possible set of statistics to report is the percentage difference between the observed and expected counts,

$$\frac{n_i - \hat{e}_i}{\hat{e}_i},$$

although this gives more weight to cells with small expected counts. A useful summary measure is the *index of dissimilarity*,

$$D = \sum_{i=1}^{K} \frac{|n_i - \hat{e}_i|}{2n}.$$

This value equals the percentage of underlying observations that would have to be moved from one cell to another in order for the observed counts to equal the expected counts. The minimum value of D is 0, while the maximum is less than or equal to 1 (the maximum value depends on the underlying model). Unfortunately, there is little guidance as to how large a value of D suggests practical importance, but values over $0.10 - 0.15$ are worth noting.

Benford's Law

What are the "typical" values of digits? When faced with numbers in a newspaper, for example, what numbers are "typical"? Let's be a little more specific here — what is typical for the first (leading) digit of all the numbers you run across in your travels?

You might think that the most likely choice of a distribution here is (discrete) uniform; that is,

$$P(D_1 = i) = \frac{1}{9}, \quad i = 1, \ldots, 9,$$

where D_1 is the leading digit of a number (note that the leading digit of a number cannot be zero). You might think it, but it's not true. Frank Benford, in a 1938 paper in *Proceedings of the American Philosophical Society*, proposed a general law for what he called *anomalous numbers* — "those outlaw numbers that are without known relationship rather than those that individually follow an orderly course." It first occurred to Benford that a uniform distribution was not appropriate when he noticed that tables of logarithms in libraries tended to be dirtier at the beginning than at the end (which meant that people had more occasion to look up numbers beginning with low digits than ones beginning with high digits). In fact, Simon Newcomb, in an 1881 paper in *American Journal of Mathematics*, noted the same pattern of use in tables of logarithms and inferred the same general law.

Benford's Law states that the probability distribution of the leading digit of anomalous numbers is

$$P(D_1 = i) = \log_{10}\left(\frac{i+1}{i}\right), \quad i = 1, \ldots, 9.$$

One important property of this distribution is that it is the only possible distribution that is unchanged if the underlying numbers are multiplied by a constant. This is desirable, since anomalous numbers could be reported in different scales (for example, degrees Fahrenheit or degrees Celsius, or U.S. dollars or Japanese yen), which should not change the observed pattern of leading digits.

Does Benford's Law actually work? Let's look at daily returns on the Standard and Poor's (S&P) 500 Index for January 2, 1986, through December 29, 1995 (there were 2524 trading days with nonzero returns during these 10 years). The observed distribution of leading digits for the absolute daily return, along with the expected counts based on Benford's Law, are as follows:

Digit	Observed count	Expected count
1	735	759.8
2	432	444.5
3	273	315.3
4	266	244.6
5	200	199.9
6	175	169.0
7	169	146.4
8	148	129.1
9	126	115.5
Total	2524	

All three of the goodness-of-fit tests suggest a lack of fit of Benford's Law to these data, as $X^2 = 16.2$, $G^2 = 16.1$, and $2nI^{2/3} = 16.1$, all on 8 degrees of freedom, with p-values of .04. The lack of fit apparently comes from too few returns with lower leading digits (1–3) and too many returns with higher leading digits (4–9). Note, however, that the index of dissimilarity $D = .032$, implying that the observed differences are probably of little practical importance.

Eduardo Ley (Ley, 1996) noted that the deviation from Benford's Law for stock market data comes in large part from days with very small returns (less than .1% in absolute value), as the distribution of returns for such days appears uniform. Here is the distribution of leading digits for these S&P returns, separated by the magnitude of the return:

Digit	$|Return| \geq .1\%$	$|Return| < .1\%$
1	698	37
2	387	45
3	232	41
4	228	38
5	162	38
6	136	39
7	124	45
8	108	40
9	92	34
Total	2167	357

The goodness-of-fit statistics support the idea that for stock returns Benford's Law is mixed with a uniform distribution for very small returns. The fit of Benford's Law for days with absolute returns greater than .1% is good ($X^2 = 12.1$, $p = .15$; $G^2 = 12.3$, $p = .14$; $2nI^{2/3} = 12.1$, $p = .15$), with $D = .032$, while the uniform distribution fits very well for days with absolute return less than .1% (all three statistics equal 2.6, with $p = .96$), with $D = .035$. Interestingly, the uniformity of leading digits for days with small returns effectively implies the existence of some sort of structure on those days, since it is Benford's Law, not uniformity, that is consistent with numbers that are "without known relationship."

These data also illustrate the dangers of relying solely on the statistical significance of goodness-of-fit tests to assess lack of fit. If almost 30 years worth of data are used, rather than 10, and Benford's Law is fit to digits from days with absolute return greater than .1%, the index of dissimilarity hardly changes ($D = .029$), but the goodness-of-fit statistics more than double to roughly 28 ($p \approx .0004$). If samples get large enough, small deviations from a null model can be identified, even if they aren't important.

4.4.2 Partitioning Pearson's X^2 Statistic

As omnibus tests, χ^2 goodness-of-fit tests possess two clear weaknesses. First, since they are looking for violations of the null in any direction, they lack power for specific types of alternatives. Second, they only can flag the existence of lack of fit, not the nature of that lack of fit.

Under some circumstances, these weaknesses can be addressed. Consider a K-cell multinomial vector where the cells have a natural ordering. In this situation, lack of fit can be partitioned into separate components of different types, allowing more focused information and more powerful tests.

The Pearson statistic X^2 can be partitioned into $K - 1$ components V_j^2, where

$$X^2 = V_1^2 + \cdots + V_{K-1}^2.$$

The V_j terms are defined as

$$V_j = \frac{1}{\sqrt{n}} \sum_{i=1}^{K} n_i g_j(x_i),$$

where $g_j(x_i)$ is the jth orthogonal polynomial of the random variable that takes the value x_i with probability p_i, $i = 1, \ldots, K$. In many circumstances the values x_i would simply be the cell position i. The orthogonal polynomials are constructed such that the V_j terms are asymptotically independent and identically standard normal distributed (so each V_j^2 is asymptotically χ_1^2). V_j will be large when the observed counts are (roughly) linearly related to $g_j(x_i)$. The first orthogonal polynomial represents a linear relationship with the scores x_i, as it equals

$$g_1(x_i) = \frac{x_i - \mu}{\sqrt{\mu_2}},$$

where $\mu = \sum_j x_j p_{0j}$ and $\mu_r = \sum_j (x_j - \mu)^r p_{0j}$, and $\mathbf{p}_0$ is the vector of null probabilities. Thus, V_1^2 assesses how much lack of fit can be attributed to a location effect (the counts are shifted down or up relative to the null). The second orthogonal polynomial represents a quadratic relationship with the scores, as it equals

$$g_2(x_i) = a \left[(x_i - \mu)^2 - \frac{\mu_3(x_j - \mu)}{\mu_2} - \mu_2 \right],$$

where $a = 1/\sqrt{\mu_4 - \mu_3^2/\mu_2 - \mu_2^2}$. A large value of V_2^2 summarizes a dispersion effect (the counts are too concentrated at the extreme cells (more variability) or the central cells (less variability) relative to the null). Higher-order polynomials also can be defined, although they are increasingly less interpretable. It is also possible to assess if, for example, a location (linear) effect accounts for all of the lack of fit, since the sum of the higher-order components

$$V_2^2 + \cdots + V_{K-1}^2 = X^2 - V_1^2$$

is a χ^2 test (on $K - 2$ degrees of freedom) of fit given the location effect.

Draws for the Texas Lottery

In 150 draws from a machine used in the Pick 3 game of the Texas Lottery, the observed numbers of draws for each number (0 through 9) were $\{18, 17, 16, 18, 20, 15, 15, 11, 9, 11\}$ (Eubank, 1997). If the machine is fair the numbers should be equiprobable (implying an expected 15 draws for each number), and χ^2 goodness-of-fit tests give no indication of a lack of uniformity ($X^2 = 7.7$, $G^2 = 8.1$, and $2nI^{2/3} = 7.8$ on 9 degrees of freedom, $p \approx .5$). A little more information suggests further study, however. This was not a machine chosen at random, but rather one that had been removed from use because of unusual selection patterns in the daily drawings. The balls are loaded into the machine in sequential order (low numbers on the bottom) before mixing, so there is a natural ordering to the categories. The linear component of X^2 is $V_1^2 = 5.0$, which clearly indicates a linear pattern ($p = .02$). Examination of the observed counts shows that the lower numbers were drawn too often while the higher numbers were drawn too infrequently, no doubt due to insufficient mixing of the balls in the machine. There is no indication of any lack of fit past the linear component ($X^2 - V_1^2 = 2.7$ on 8 degrees of freedom, $p = .95$). Thus, the focused linear component is able to clearly identify the existence and nature of lack of fit that the omnibus tests are unable to detect.

4.4.3   Exact Inference

As was noted earlier, the null properties of the χ^2 goodness-of-fit statistics (and hence associated p-values based on the χ^2 distribution) are based on large samples (or, more precisely, large enough fitted values in each cell). When expected counts are small, inferences from the tests based on this distribution are questionable. Of course, the question then becomes "How small is small?" There is no unambiguous answer to this question; common recommendations are that most cells should have fitted values greater than 3, and all should have fitted values greater than 1. Generally speaking, X^2 and $2nI^{2/3}$ follow the χ^2 distribution more closely under the null than does G^2, although G^2 can be more powerful when the null does not hold.

The exact distribution of these statistics can be derived in the same way that exact tests and confidence intervals were derived in Sections 4.1.4 and 4.3.2. The exact p-value is calculated by determining the set of counts $\mathbf{n}$ with $\sum_i n_i = n$ such that the resultant goodness-of-fit statistic based on those counts is at least as large as the observed statistic, and then summing the associated multinomial probabilities (4.8). This is clearly computationally intensive, but sophisticated search algorithms can be used to speed up

computation. Note, by the way, that even if the correct distribution for a
test is derived, the power of the test still can be quite low for small samples.

Major Hurricane Strikes in South Carolina

Hurricane season in the United States is defined by the National Hurricane Center as the months June through November. During the years
1900–1996, there were four major hurricane strikes in South Carolina: three
in September and one in October (a major hurricane is one that is at least
category 3 on the Saffir–Simpson hurricane scale, with sustained winds
above 110 miles per hour). Is this pattern of hurricane strikes consistent
with a uniform probability distribution over hurricane season? Equivalently,
would a Poisson distribution with constant mean number of hurricanes in
each month fit the observed data? This is a goodness-of-fit comparison of
observed counts $\mathbf{n} = (0, 0, 0, 3, 1)$ to expected counts $e_i = .8$ for five cells
(there were no major hurricane strikes anywhere in the United States during 1900–1996 in November, so we do not include November). The Pearson
goodness-of-fit statistic $X^2 = 8.5$, and comparing to a χ^2 distribution on
four degrees of freedom gives a tail probability of .075, suggesting some
lack of fit. Results are similar for the likelihood ratio statistic ($G^2 = 8.38$,
$p = .079$).

With only four observations and five cells, it is not difficult to enumerate
all of the possible values of X^2 and G^2 for testing uniformity, and thereby
get the exact p-value. Table 4.2 gives this exact distribution, and the associated exact p-values for the two statistics can be seen to be .128+.008 = .136,
considerably different from the χ^2-based values. That is, the tests suggest
considerably less evidence of a lack of uniformity than the asymptotic p-values would imply. This isn't necessarily the "right" answer, of course —
over the entire United States there is a distinct lack of uniformity in major
hurricane strikes, with almost 80% of them occurring in August and September — but it is an accurate reflection of the evidence that is available
in the data to test uniformity.

X^2	G^2	Probability
1.0	1.79	.192
3.5	4.56	.576
6.0	7.33	.096
8.5	8.38	.128
16.0	12.88	.008

TABLE 4.2. Exact distribution of X^2 and G^2 for a test of uniformity based on
four observations and five cells.

4.5   Overdispersion and Lack of Fit

The Poisson and binomial random variables are inherently restrictive as models for discrete data because they are determined by only one parameter. Consider, for example, the Poisson random variable. Since for this random variable the variance equals the mean, it cannot adequately model data where the variance is different from the mean. Typically this lack of fit is reflected in the variance being larger than the mean, and is termed *overdispersion*. Such overdispersion can arise in different ways. If there is heterogeneity in the population that is not accounted for in the model (underlying Poisson distributions with different means), the observed variance will be larger than expected. Depending on the assumed form of the heterogeneity of the populations, this extension of the Poisson model yields several useful models for overdispersion.

4.5.1   The Zero-Inflated Poisson Model

One common source of non-Poisson behavior that leads to overdispersion is when the observed counts exhibit an excess of zeroes. This can be modeled as a mixture of two subpopulations, representing that an observation occurs with probability p from a Poisson distribution, while with probability $1 - p$ the observed number of counts is zero. This *zero-inflated Poisson* (ZIP) model is consistent with a process where $100p\%$ of a population is at risk for some event, while $100(1 - p)\%$ have no risk. The probability function for this random variable is

$$P(X = x) = \begin{cases} 1 - p + pe^{-\mu} & \text{if } x = 0 \\ pe^{-\mu}\mu^x/x! & \text{if } x \neq 0. \end{cases}$$

Direct calculation gives that $E(X) = p\mu$ and $V(X) = p\mu[1+\mu(1-p)]$, with an *index of dispersion* equal to

$$\frac{V(X)}{E(X)} = 1 + \mu(1 - p),$$

which is greater than one if $p \neq 1$. Maximum likelihood estimates are not available in closed form, but can be calculated iteratively, and asymptotic confidence intervals for p and μ can be calculated as in (2.3). The ZIP model maximum likelihood fit to $P(X = 0)$ is exact, in that $n\widehat{P(X = 0)} = n_0$.

The choice between the ZIP model and the Poisson model can be put in the general model selection framework discussed in Section 3.4. For example, the adequacy of the Poisson model for a data set can be compared to that of the ZIP model using AIC, defined as in (3.10) as $AIC = -2L + 2\nu$, where L is the log-likelihood and ν is the number of estimated parameters. Define for a sample of n observations the vector $\mathbf{n}$ with $n_x, x = 0, 1, \ldots$.

equaling the number observations equal to x. For the Poisson model $\nu = 1$ and

$$L = -n\mu + \sum_{x=0}^{\infty} n_x[x \log \mu - \log x!].$$

Under the ZIP model $\nu = 2$ and

$$L = n_0 \log[1 - p + pe^{-\mu}] + \sum_{x=1}^{\infty} n_x \log[pe^{-\mu}\mu^x/x!].$$

For each model the maximum likelihood estimates of any parameters are substituted into the log-likelihood function. For small samples, the bias-corrected version of AIC (AIC_C) is preferable, where

$$AIC_C = AIC + \frac{2\nu(\nu + 1)}{n - \nu - 1},$$

as in (3.13). It should be noted that although AIC_C was derived under the least squares regression conditions of Section 3.4, this more general use of the criterion has been found empirically to be useful for these categorical data models as well.

The Decayed, Missing, and Filled Teeth Index

A standard indicator of the dental status of a person is the decayed, missing, and filled teeth (DMFT) index. The Belo Horizonte caries prevention (BELCAP) study was undertaken in an urban area of Belo Horizonte (Brazil) to examine the dental health of 797 school children (Böhning et al., 1999). The children were all 7 years of age at the beginning of the two-year study. The DMFT index for these data is based on only the eight deciduous molars, implying possible values between zero and eight, inclusive. In all of the models fit here, probabilities for more than eight decayed, missing or filled teeth are combined into the eight teeth category.

Table 4.3 summarizes the fit of Poisson and ZIP models to these data, separated into data at the beginning and at the end of the study (after one of several caries prevention interventions was administered to each child), respectively. It is apparent that a Poisson fit to the data at the beginning of the study (based on sample mean of 3.32) is not adequate, which is supported by a high index of dissimilarity ($D = .31$) and goodness-of-fit statistics that strongly reject the model ($X^2 = 957.6$ and $G^2 = 549.8$, each on 8 degrees of freedom, with tail probability effectively 0). The top plot in Figure 4.3 demonstrates that the expected count for zero damaged teeth under the Poisson model (solid line) is far too small, possibly reflecting children with no caries risk. Note that this is also reflected in overdispersion

DMFT Index	Observed Count	Poisson exp. Count	ZIP exp. Count
(a) Data at the beginning of the study			
0	172	28.7	172.0
1	73	95.4	40.8
2	96	158.6	85.1
3	80	175.7	118.4
4	95	146.0	123.6
5	83	97.0	103.1
6	85	53.8	71.7
7	65	25.5	42.8
8	48	16.4	39.4
(b) Data at the end of the study			
0	231	124.8	231.0
1	163	231.4	138.7
2	140	214.5	164.1
3	116	132.6	129.4
4	70	61.5	76.6
5	55	22.8	36.2
6	22	9.5	21.0

TABLE 4.3. DMFT index data from Belo Horizonte caries prevention study.

relative to the Poisson assumption of equal means and variances, since the sample variance here is 6.64, twice the sample mean.

Table 4.3 and Figure 4.3 (dotted line) also summarize the fit of a ZIP model to these data. Here $\hat{p} = .8$, implying that 20% of the children have no caries risk. The other 80% of the children are fit as having Poisson risk for damaged teeth with mean $\hat{\mu} = 4.17$. The ZIP model is a considerably better fit than the Poisson model, although there is still some lack of fit ($D = .11$; $X^2 = 65.7$, $G^2 = 61.4$, each on six degrees of freedom, $p < 10^{-10}$). AIC and AIC_C agree that the ZIP model is greatly preferred for the observed vector of counts, as the values for the ZIP model are roughly 486 lower for it compared to the Poisson model.

The DMFT index values at the end of the study show a similar pattern. The mean index value is 1.85, but the data are overdispersed (the variance equals 2.91), with an excess of zeroes. A Poisson model fits these data poorly ($D = .2$, $X^2 = 201.9$, $G^2 = 172.0$). The fitted ZIP model has $\hat{p} = .78$ and $\hat{\mu} = 2.37$. The closeness of the $\hat{p}$ values for the data at the beginning and end of the study is encouraging, since it supports the inference that roughly 20% of the Belo Horizonte children were at no caries risk, and would thus be unaffected by any caries prevention intervention. Goodness-of-fit tests for the model fit are still statistically significant ($X^2 = 19.5$, $G^2 = 18.2$, each with tail probability around .001), although the index of dissimilarity

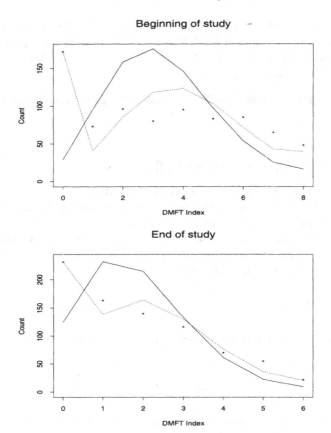

FIGURE 4.3. Plots of DMFT index values at the beginning of the BELCAP study (top plot) and at the end of the study (bottom plot). Stars represent observed counts, solid lines connect fitted counts under a Poisson model, and dotted lines connect fitted counts under a ZIP model.

is small ($D = .055$), indicating a reasonable fit from a practical point of view. Again the information measures prefer the ZIP model, being roughly 152 lower for that model.

One puzzling aspect of these data is worth comment. The values of $\hat{\mu}$ for the ZIP model at the beginning and end of the study imply that the dental health of the children improved over the two-year time period (in fact, for 60% of the children the ending DMFT index was less than the starting index). This would seem to be impossible, as a tooth that is missing would seemingly remain missing, one that is filled would remain filled or become missing, and one that is decayed would either remain decayed, become filled, or become missing. In fact, the explanation for this is that the DMFT index used here considered a tooth decayed even if it was not decayed, but only had lesions of the tooth. Since a tooth with a low grade of lesion can recover in time, smaller DMFT index values later in time are possible. This is less likely in older populations, where only tooth deterioration or maintenance might be expected.

4.5.2 The Negative Binomial Model

A more complex heterogeneity model for count data is as follows. Say each observation x_i is Poisson distributed with potentially different mean μ_i, where $\mu_i = \mu \delta_i$. If δ_i is a nonnegative random variable with $E(\delta_i) = 1$, then unconditionally (without knowledge of δ_i) each observation has mean μ. If $g(\delta_i)$ is the probability function for δ_i, then the unconditional distribution of x_i is no longer Poisson, but rather

$$f(x_i; \mu) = \int_0^\infty \frac{e^{-\mu \delta_i}(\mu \delta_i)^{x_i}}{x_i!} g(\delta_i) d\delta_i, \qquad x = 0, 1, \ldots$$

If δ_i is assumed to follow a one-parameter Gamma distribution with parameter ν,

$$g(\delta_i) = \frac{\nu^\nu}{\Gamma(\nu)} \delta_i^{\nu-1} e^{-\delta_i \nu},$$

then x_i has a *negative binomial* distribution with probability function

$$f(x_i; \mu, \nu) = \frac{\Gamma(x_i + \nu)}{x_i! \Gamma(\nu)} \left(\frac{\nu}{\nu + \mu} \right)^\nu \left(\frac{\mu}{\nu + \mu} \right)^{x_i}. \qquad (4.16)$$

Here the gamma function satisfies

$$\Gamma(a) = \int_0^\infty t^{a-1} e^{-t} dt.$$

The negative binomial probability function approaches the Poisson as $\nu \to \infty$. Define $\alpha = 1/\nu$. Then $E(x_i) = \mu$, while $V(x_i) = \mu(1 + \alpha\mu)$. Thus, the negative binomial provides an alternative to the Poisson that allows for overdispersion.

The MLE of μ, the population mean under the negative binomial model, equals the sample mean $\overline{X}$, but the MLE of ν is not available in closed form. A simple moment-based estimate of ν (if a computer is not available to calculate the MLE $\hat{\nu}$) equates $E(X) = \mu$ to $\overline{X}$ and $V(X)$ to the sample variance s^2. This yields the estimate $\tilde{\nu} = \overline{X}^2/(s^2 - \overline{X})$. Note that if $s^2 < \overline{X}$ then $\tilde{\nu} < 0$; since the negative binomial is consistent with overdispersion relative to the Poisson assumption of equal mean and variance, this would indicate that the distribution is not appropriate for those data. Putting restrictions on the parameters of the distribution leads to other mean/variance relationships. For example, taking $\nu = \mu/\alpha$ in (4.16) leads to the probability function

$$f(x_i; \mu, \alpha) = \frac{\Gamma(x_i + \mu/\alpha)}{x_i! \Gamma(\mu/\alpha)} \left(\frac{1}{1+\alpha}\right)^{\mu/\alpha} \left(\frac{\alpha}{1+\alpha}\right)^{x_i}, \tag{4.17}$$

which has $V(x_i) = \mu(1 + \alpha)$. Since the probability function (4.17) leads to a multiplicative (linear) inflation of variance, while (4.16) leads to a quadratic inflation, random variables with probability function (4.17) are sometimes referred to as following a *negative binomial I* distribution, while random variables with probability function (4.16) follow a *negative binomial II* distribution. In this book we will take "negative binomial distribution" to mean (4.16) unless otherwise specified.

The inflation of variance of the negative binomial distribution relative to the Poisson suggests that a comparison of the sample mean and variance would provide a way to test the appropriateness of the Poisson assumption. This is in fact the case, as a test based on the variance/mean ratio is, among all locally unbiased tests for testing the Poisson against a mixture of Poissons, asymptotically locally most powerful against a negative binomial distribution. Specifically, the test statistic

$$Z = \left(\frac{s^2}{\overline{X}} - 1\right) \sqrt{\frac{n-1}{2}} \tag{4.18}$$

is compared to a standard normal reference to test the Poisson assumption. This statistic still can be applied even if a negative binomial model is not the appropriate alternative, and can even be used to identify significant underdispersion compared to the Poisson.

The Number of Goals Scored in Soccer Games

In the space of less than 150 years, soccer (or Association football) has become arguably the world's most popular sport. The rules of soccer were codified in England in December 1863, and have changed relatively little since then. Scoring a goal in soccer is intimately connected to possession of the ball, since whenever a team has possession of the ball it has an opportunity to mount an attack on the opposing team's goal and score. There are many such attacks in a game, although most do not result in a goal. If it is assumed that there is a constant (small) probability p of scoring on any attack, and attacks are independent of each other, the number of goals scored will be binomially distributed with the number of trials being the number of attacks in a game. As was pointed out in Section 4.3, this binomial is well-approximated with a Poisson distribution.

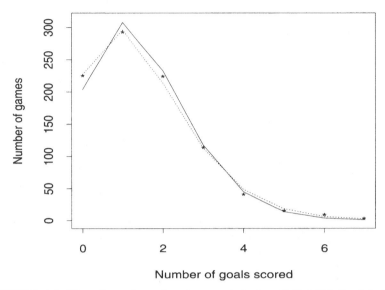

FIGURE 4.4. Plot of number of goals scored during 1967–1968 English First Division Football League season. Stars represent observed counts, the solid lines connect fitted counts under a Poisson model, and dotted lines connect fitted counts under a negative binomial model.

Figure 4.4 refers to the goals scored by each team in 462 games (yielding 924 observations) played during the 1967–1968 English First Division Football League season (Norman, 1998). The observed counts of number of teams scoring each number of goals are given by stars, while the expected counts based on a Poisson fit (with $\overline{X} = \hat{\mu} = 1.514$) are connected by solid lines (counts for seven goals are actually based on seven or more goals). It is apparent that the Poisson does not provide a good fit to these data, there being too many games with zero goals and too few with one goal. The lack of fit is supported by goodness-of-fit statistics, which strongly reject the Poisson model ($X^2 = 18.0, p = .006$; $G^2 = 13.2, p = .039$, both with $df = 6$). The data exhibit significant overdispersion, with $s^2 = 1.765$ yielding $Z = 3.56$ based on (4.18), with $p < .001$ rejecting the null of equality of mean and variance. The components of X^2 (Section 4.4.2) reinforce this, as the location component indicates no lack of fit of the Poisson ($V_1^2 = .002$), while the variance component of lack of fit is highly significant ($V_2^2 = 10.8$, $p = .001$). A negative binomial fit ($\hat{\mu} = 1.514$, $\hat{\nu} = 9.626$), with expected counts connected by dotted lines given in the figure, provides a much better fit ($X^2 = 3.8, p = .58$; $G^2 = 3.7, p = .60$, both with $df = 5$). The mixture model justification for the negative binomial model makes it seem particularly appropriate for these data, in several ways. The probability of a specific attack resulting in a goal would be expected to differ based on which player or players were controlling the ball during the attack. In addition, teams have different rates of scoring that might vary from match to match, and might vary within a match based on the game situation. See Norman (1998, pages 106–107) for more discussion of these points.

The next example shows how several of the mixture models previously described can be used together to help summarize the structure of a data set, highlighting the separation of a population into well-defined and identifiable subgroups.

Weekly Consumer Use of Shampoo

A market research company collected information from 533 men aged 16–65 years old (Romaniuk et al., 1999). These "panelists" were asked (among other things) how many times in a given week they used shampoo. The analysis here is based on data derived from Figure 1 of that paper. Figure 4.5 summarizes the results. The observed counts of number of panelists reporting each number of uses are represented by stars in the figure. The observed mean number of uses ($\overline{X} = 3.71$) is much smaller than the variance ($s^2 = 8.02$), strongly rejecting a Poisson fit to the counts of number of uses ($Z = 18.9$). We might hypothesize a mixture model based on random shampoo rates, but the negative binomial model (solid lines connecting fitted counts in the figure) with $\hat{\mu} = 3.71$ and $\hat{\nu} = 2.52$ does not fit the model well either ($X^2 = 330.4$ and $G^2 = 246.2$ on 12 degrees of freedom, and the index of dissimilarity, $D = .25$, is quite high).

Examination of the observed counts shows that there are pronounced spikes at zero uses (corresponding to nonusers) and seven uses (corresponding to people who use shampoo once each day), and a small spike at 14 uses (people who shampoo twice each day). This suggests fitting a negative binomial model (corresponding to people with a random number of uses each week, and potentially different rates of use) inflated by percentages of people who are nonusers, once-a-day users, and twice-a-day users. This is, of course, a generalization and modification of the ZIP model of Section 4.5.1.

This model can be fit using a version of the EM (Expectation/Maximization) algorithm. We start by fitting a negative binomial distribution to the data (using maximum likelihood), yielding estimates $\hat{p}_j$ of the probability of j uses of shampoo per week. This is the M-step of the EM algorithm. Under the inflated model the expected number of panelists using shampoo

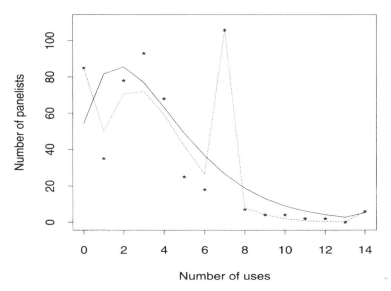

FIGURE 4.5. Plot of number of uses of shampoo in a week by panelists. Stars represent observed counts, the solid lines connect fitted counts under a negative binomial model, and dotted lines connect fitted counts under a negative binomial model mixed with point masses at 0, 7, and 14 uses.

j times is

$$E(n_j) = \begin{cases} nq_j + n(1 - q_0 - q_7 - q_{14})p_j & \text{if } j = 0, 7, 14 \\ n(1 - q_0 - q_7 - q_{14})p_j & \text{otherwise.} \end{cases}$$

Here p_j is the true probability of j shampoo uses for people from the negative binomial population, q_j $(j = 0, 7, 14)$ is the probability of a panelist being a nonuser, once-a-day, or twice-a-day user, respectively, and n is the total number of panelists. The E-step of the EM algorithm equates the expected counts to the observed estimated counts, yielding estimates of q_0, q_7, and q_{14}. Let $\tilde{p}_j = n_j/n$ be the observed frequencies of j shampoo uses. The E-step is then

$$n_j = n\hat{q}_j + n(1 - \hat{q}_0 - \hat{q}_7 - \hat{q}_{14})\hat{p}_j, \qquad j = 0, 7, 14.$$

This results in three equations in three unknowns, which yield the solutions

$$\hat{q}_j = \tilde{p}_j - (1 - \hat{Q})\hat{p}_j,$$

where

$$\hat{Q} = \frac{(\tilde{p}_0 + \tilde{p}_7 + \tilde{p}_{14}) - (\hat{p}_0 + \hat{p}_7 + \hat{p}_{14})}{1 - (\hat{p}_0 + \hat{p}_7 + \hat{p}_{14})}.$$

Based on the three estimates $\hat{\mathbf{q}}$, observations that are considered inflated from the underlying negative binomial model are removed from the sample, resulting in effective counts

$$n_j^* = n_j - n\hat{q}_j, \qquad j = 0, 7, 14$$

(in practice, these counts are rounded to the nearest integer so that the negative binomial model can be fit). Considering the modified data the new data set, the negative binomial model is fit (repeating the M-step), yielding new values of $\hat{\mathbf{p}}$. The E- and M-steps are alternated in this way until convergence (when the fitted counts change very little).

The final estimates for the shampoo use data are $\hat{\mu} = 3.4$, $\hat{\nu} = 10.7$, $\hat{q}_0 = .123$, $\hat{q}_7 = .17$, and $\hat{q}_{14} = .011$. Dotted lines connect the expected counts based on this model in Figure 4.5. This model fits the data far better than the negative binomial. While there is still some lack of fit, with the observed counts too low for $j = 1$ and too high for $j = 3$ ($X^2 = 33.3$ and $G^2 = 30.5$ on 9 degrees of freedom, $p < .001$), the index of dissimilarity $D = .079$ suggests that this lack of fit is not of great practical importance. The model estimates that 12.3% of the target population never use shampoo, 17% use it exactly once per day, 1.1% use it exactly twice per day, while the remaining 69.6% use it an average 3.4 time per week (roughly every other day), with moderate overdispersion relative to a model of independent panelists with constant shampoo usage rates (the Poisson model).

4.5.3   Overdispersed Binomial Data and the Beta-Binomial Model

A binomial assumption is just as vulnerable to apparent overdispersion as a Poisson assumption, since it also restricts the variance to be a specified function of the mean. In the same way that overdispersion relative to the Poisson assumption can arise from unmodeled heterogeneity, unmodeled heterogeneity also can lead to overdispersion relative to the binomial distribution. Say observation $x_i \sim \text{Bin}(n_i, p_i)$, and p_i is a random variable with density $g(p_i)$. The unconditional distribution of x_i is then

$$f(x_i; n_i, p_i) = \int_0^1 \binom{n_i}{x_i} p_i^{x_i} (1 - p_i)^{n_i - x_i} g(p_i) dp_i, \qquad x = 0, 1, \ldots, n_i.$$

If p_i is assumed to follow a beta distribution with parameters (α, β),

$$g(p_i) = \frac{p_i^{\alpha-1}(1 - p_i)^{\beta-1}}{B(\alpha, \beta)},$$

where

$$B(\alpha, \beta) = \int_0^1 x^{\alpha-1}(1-x)^{\beta-1}dx = \frac{\Gamma(\alpha)\Gamma(\beta)}{\Gamma(\alpha+\beta)},$$

then x_i has a *beta-binomial* distribution with probability function

$$f(x_i; n_i, \alpha, \beta) = \binom{n_i}{x_i}\frac{B(\alpha+x_i, \beta+n_i-x_i)}{B(\alpha, \beta)}.$$

Define $\pi \equiv \alpha/(\alpha+\beta)$. The moments of this distribution are then

$$E(x_i) = n_i\left(\frac{\alpha}{\alpha+\beta}\right) \equiv n_i\pi,$$

and

$$V(x_i) = \frac{n_i\alpha\beta[1+(n_i-1)(\alpha+\beta+1)^{-1}]}{(\alpha+\beta)^2} \equiv n_i\pi(1-\pi)[1+(n_i-1)\theta],$$

where $\theta = (\alpha+\beta+1)^{-1}$. Thus, the beta-binomial can be parameterized in a way to highlight its overdispersion relative to the binomial, as long as $n_i > 1$. Note that a sample of binary trials ($n_i = 1$ for all i) with random p_i cannot be distinguished from one with fixed probability p (which would be modeled using a binomial random variable). The maximum likelihood estimates of α and β based on a sample $\{x_i, n_i\}$ cannot be determined in closed form.

A different violation of the binomial assumptions leads to similar overdispersion. Let $\{y_1, \ldots, y_n\}$ be Bernoulli trials with the same underlying probability; that is, $y_j \sim \text{Bin}(1, p)$. If the trials are independent, $X = \sum_j y_j$ is, of course, binomially distributed. If, however, the trials are correlated with pairwise correlation ρ, the variance of X satisfies

$$V(X) = \sum_{j=1}^n V(y_j) + \sum_{i\neq j} \text{Cov}(y_i, y_j)$$

$$= \sum_{j=1}^n p(1-p) + \sum_{i\neq j} p(1-p)\rho$$

$$= np(1-p) + n(n-1)p(1-p)\rho = np(1-p)[1+(n-1)\rho].$$

This is identical to the variance of the beta-binomial distribution, with ρ taking the place of θ. Note that this also provides a plausible model for underdispersion, since negative correlation ($\rho < 0$) induces a variance less than the binomial variance. Even though the mean and variance of X based on this correlated Bernoulli model are the same as those of the beta-binomial distribution, the actual distributions differ (one easy way to see this is that the beta-binomial model cannot lead to underdispersion, since

α and β must be greater than zero). Note that since $V(X)$ must be positive, the amount of negative correlation allowable in this model is restricted to $\rho > -1/(n-1)$.

Overdispersion naturally leads to lack of fit, but the goodness-of-fit statistics of Section 4.4 need to be adapted from the multinomial/Poisson formulation to one for a sample of binomial observations. Let $\{x_1, \ldots, x_N\}$ be a sample of counts of successes out of trials $\{n_1, \ldots, n_N\}$ hypothesized to be binomially distributed with common success probability p_0 (that is, $x_i \sim \text{Bin}(n_i, p_0)$). Since under the null hypothesis $E(x_i) = n_i p_0$ and $V(x_i) = n_i p_0 (1 - p_0)$, the Pearson goodness-of-fit statistic is not (4.13), but rather has the form

$$X^2 = \sum_{i=1}^{N} \frac{(x_i - n_i \hat{p}_0)^2}{n_i \hat{p}_0 (1 - \hat{p}_0)}, \tag{4.19}$$

where $\hat{p}_0$ is the MLE from a sample of binomial random variables given in (4.6). Similarly, the likelihood ratio goodness-of-fit statistic is not (4.14), but is instead

$$G^2 = 2 \sum_{i=1}^{N} \left[x_i \log \left(\frac{x_i}{n_i \hat{p}_0} \right) + (n_i - x_i) \log \left(\frac{n_i - x_i}{n_i (1 - \hat{p}_0)} \right) \right]. \tag{4.20}$$

Each of these statistics would be compared to a χ^2_{N-1} reference, as long as the n_i values are large enough.

The tests in (4.19) and (4.20) are omnibus tests, aimed at identifying any violation from the binomial. If the specific alternatives of the beta-binomial or correlated binomial are of interest, more powerful tests can be developed. The asymptotically optimal test of the binomial against these alternatives is

$$Z = \frac{\sum_{i=1}^{N} \frac{(x_i - n_i \hat{p}_0)^2}{\hat{p}_0 (1 - \hat{p}_0)} - \sum_{i=1}^{N} n_i}{\left[2 \sum_{i=1}^{N} n_i (n_i - 1) \right]^{1/2}}. \tag{4.21}$$

Since the beta-binomial distribution is only consistent with overdispersion, the test Z is a one-sided test, and would be compared to the upper tail of a standard normal distribution. Correlated Bernoulli trials can lead to either over- or underdispersion, so the test is appropriately two-sided (or, equivalently, Z^2 is compared to a χ^2_1 distribution). Even if the beta-binomial and correlated binomial models are not appropriate, this statistic still provides a reasonable test for over- and underdispersion relative to the binomial.

The fit of alternative models, such as the beta-binomial, would be assessed using tests based on the appropriate properties of the model. For example, the Pearson goodness-of-fit statistic would equal

$$X^2 = \sum_{i=1}^{N} \frac{(x_i - n_i \hat{\pi})^2}{n_i \hat{\pi} (1 - \hat{\pi})[1 + (n_i - 1)\hat{\theta}]},$$

where $\hat{\pi} = \hat{\alpha}/(\hat{\alpha} + \hat{\beta})$ and $\hat{\theta} = (\hat{\alpha} + \hat{\beta} + 1)^{-1}$, and would be compared to a χ^2_{N-2} reference, since two parameters are being estimated.

The Success of Broadway Shows

The Broadway theater is one of the most important arts and entertainment industries in the world. Located in and around Times Square in New York City, the roughly 35 legitimate Broadway theaters form the backbone of one of the most highly concentrated entertainment districts in the world (along with London's West End). Besides its cultural importance, the production of dramas, comedies, and musicals on Broadway is also big business. More than 11 million tickets were sold for Broadway shows in 1999, leading to more than $550 million in gross revenues.

The Broadway stage has the unusual characteristic that demand is unpredictable, since audiences don't know if they will like a product until they actually experience it. For this reason, information transfer to the potential audience is crucial in the ultimate success of a show. The potential audience wishes to assess the quality of a show without attending it, which makes outside, presumably objective, measures of quality crucially important.

One such measure is the number of awards a show wins. The most important awards for a Broadway show are the Antoinette Perry (Tony) awards. Shows that open before the end of the Broadway season (typically at the end of April) are eligible for nomination in that season. While being nominated for such awards is nice, it is winning them that really matters (Simonoff and Ma, 2003, investigated the factors associated with the longevity of Broadway shows, and found that in general it is Tony awards, not Tony nominations, that are associated with a long run). What can we say about the chances of winning an award, given being nominated?

The stem-and-leaf display given on the next page begins to answer that question. The display summarizes the percentage of Tony nominations in the major categories (Best Musical and Best Play (revival and nonrevival), Best Director (musical and play), Leading Actor and Actress (musical and play), and Featured Actor and Actress (musical and play)) that resulted in Tony wins, for the 61 shows that opened during the 1996–1997 through 1998–1999 seasons that were nominated for at least one major award (Simonoff and Ma, 2003). Note that these percentages are based on different numbers of nominations, ranging from one to six, so they are not precisely equivalent; the analysis that follows is based on the actual counts x_i of number of winners out of n_i nominations for each show.

```
 0 : 0000000000000000000000000000000000000000000
 1 : 7
 2 : 05
 3 : 33
 4 :
 5 : 000000
 6 : 7777
 7 :
 8 : 0
 9 :
10 : 00000
```

The percentages give information about the Tony award process. If each award category for each show had the same probability that a nomination would lead to an award, the total number of awards would follow a binomial distribution. On the other hand, if, for example, the probability of a nomination leading to an award was random, varying from show to show according to some probability distribution, the observed data would be overdispersed relative to the binomial. A binomial distribution (with $\hat{p} = .255$, the overall proportion of nominations that led to awards) fits these data poorly, with $X^2 = 89.4$ ($p = .01$) and $G^2 = 104.1$ ($p < .001$), each on 61 degrees of freedom. The test for overdispersion (4.21) $Z = 5.24$, which is highly statistically significant ($p < .00001$).

A beta-binomial fit to the data yields $\hat{\alpha} = .48$ and $\hat{\beta} = 1.68$. This model fits very well ($X^2 = 59.5$ on 60 degrees of freedom, $p = .49$), and is consistent with an average probability of a nomination leading to an award of $.48/(.48+1.68) = .22$, which is not very different from the binomial estimate of a fixed probability. The beta-binomial implies an estimated overdispersion factor of $\hat{\theta} = (.48 + 1.68 + 1)^{-1} = .32$. The model also provides an interesting perspective on the awards process. Figure 4.6 gives the density function of the beta distribution with parameters (.48, 1.68), which (under the beta-binomial model) is the distribution from which each show's probability of a nomination turning into an award is drawn. This highly asymmetric density implies that a great many shows have virtually no chance of a nomination turning into an award; indeed, half have probability less than .135 (the median of the distribution). On the other hand, more than a quarter of the shows have probability greater than one-third that a nomination will be a winner. This leads to the observed pattern in the stem-and-leaf display, where more than two-thirds of the shows with nominations won nothing, while almost one-quarter of the shows were victorious in at least two-thirds of their nominations (including the big winners, *A Doll's House* (4/4), *Cabaret* (4/6), *Chicago* (4/5), *Death of a Salesman* (4/6), and *The Beauty Queen of Leelane* (4/6)). An alternative, perhaps equally plausible, explanation for the observed overdispersion is that within a show, winning in one category is correlated with winning in other categories, with that correlation estimated to be around .3.

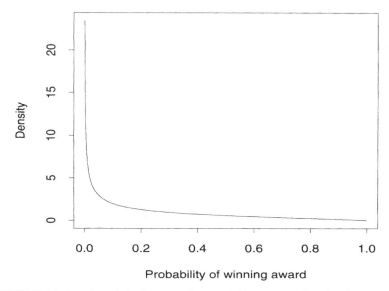

Probability of winning award

FIGURE 4.6. Density of the beta random variable estimated to be the source of variation in the probability of a Tony-nominated show winning the award.

The χ^2 approximations to the test statistics are somewhat suspect here, since many of the n_i values are small. The results don't change in any important way if only those shows with at least three nominations are examined, however, so it is reasonable to believe that the observed over-dispersion is real.

4.6 Underdispersion

Underdispersed count data is much more unusual than overdispersed data. As was noted in the previous section, underdispersion occurs relative to the binomial random variable if the underlying trials are negatively correlated, and the test statistic (4.18) can be used to identify it.

A simple violation of assumed models for count data that leads to underdispersion is truncation, when the range of the distribution of observed counts is restricted. A very common example is truncation from below at zero, since for many count mechanisms nothing is observed unless a single event occurs. For example, if respondents only appear in a sample if they take an action (buy a particular product, say, or attend a certain type of sporting event), the random variable that counts the number of times they take that action in a month is necessarily strictly positive.

Consider, for example, the Poisson random variable truncated at zero. Let $g(x; \mu)$ be the probability density function of a Poisson random variable X with mean μ. By the definition of conditional probability, the probability function of the truncated Poisson random variable Y satisfies

$$P(Y = k) = P(X = k|X > 0) = \frac{P(X = k)}{1 - P(X = 0)},$$

The probability density function is thus

$$f(y; \mu) = \frac{e^{-\mu}\mu^y}{y!(1 - e^{-\mu})}. \tag{4.22}$$

The mean of the random variable is

$$E(Y) = \frac{\mu}{1 - e^{-\mu}},$$

while the variance is

$$V(Y) = \left(\frac{\mu}{1 - e^{-\mu}}\right)\left(1 - \frac{\mu e^{-\mu}}{1 - e^{-\mu}}\right).$$

Note that $V(Y) < E(Y)$, implying underdispersion relative to the Poisson random variable. The MLE of μ for the truncated Poisson random variable is not available in closed form, and must be determined iteratively.

How Many Shoes Do Runners Need?

The World Wide Web contains various sites of use to runners, including web-based running logs where runners can record and share information about their training. New members complete a profile that includes, among other things, a question about how many pairs of running shoes the runner owns. According to the Athletic Footwear Association, the U.S. running shoe market exceeds \$2.4 billion, and presumably marketers would be particularly interested in the sort of dedicated runner that registers a running log.

Table 4.4 gives the distribution of number of pairs of shoes for a sample of 60 members of one of these sites taken in December 1999 (I am indebted to Greg Wilmore for these data). The average number of pairs is 2.45, but a Poisson distribution with $\hat{\mu} = 2.45$ (with expected counts in the second column) does not fit the data very well ($X^2 = 8.1$, $p = .09$, $G^2 = 13.0$, $p = .01$, on 4 degrees of freedom). Since one would expect that any registered member would have to own at least one pair of running shoes, considering a truncated Poisson model seems reasonable. The third column

Number of pairs	Observed counts	Poisson expected counts	Truncated Poisson expected counts
0	0	5.18	0
1	18	12.69	16.78
2	18	15.54	18.21
3	12	12.69	13.17
4	7	7.77	7.15
5 or more	5	6.13	4.70

TABLE 4.4. Counts of pairs of shoes owned by a sample of registered running log members. Observed counts (second column), expected counts under a Poisson model (third column), and expected counts under a truncated Poisson model (fourth column) are given.

gives the expected counts for a truncated Poisson random variable based on the MLE $\hat{\mu} = 2.17$. The truncated Poisson fits the data very well, with both X^2 and G^2 equaling .2 on 2 degrees of freedom ($p = .9$). Thus, it appears that an underlying Poisson distribution with a mean of roughly 2.2 is reasonable for the number of pairs of running shoes among members, given that each member has at least one pair. Presumably how this mean varies as a function of other factors would be of more interest to market researchers. Such regression models for count data are the subject of the next chapter.

4.7   Robust Estimation Using Hellinger Distance

It is very well known that maximum likelihood estimators can be very sensitive to unusual observations in the data (outliers). Estimators and tests that are resistant to outliers are said to be *robust*, which is an obviously highly desirable trait. There has been a great deal of research into robust estimation, although the bulk of it has been for continuous data.

Typically the data analyst must choose between efficiency (precision when the assumed model is true) and robustness (resistance when the assumed model is violated), but a simple approach to robustness avoids this tradeoff, and is easy to implement for categorical data: Hellinger estimation. Consider a vector of counts $\mathbf{n} = \{n_1, \ldots, n_K\}$, and assumed probability function $f(x; \boldsymbol{\theta}), x = 1, \ldots$. The Hellinger estimator $\hat{\boldsymbol{\theta}}$ is defined as the minimizer of the Hellinger distance,

$$H = \sum_{i=1}^{K} \left\{ \left(\frac{n_i}{n} \right)^{1/2} - [f(i; \boldsymbol{\theta})]^{1/2} \right\}^2 .$$

Number of cases	Observed counts	MLE expected counts	Cleaned MLE expected counts	Hellinger expected counts
0	21	13.61	16.83	20.08
1	22	20.19	21.40	21.98
2	5	14.98	13.60	12.03
3	6	7.40	5.77	4.39
4	1	2.75	1.83	1.20
5	3	0.81	0.47	0.26
6	1	0.20	0.10	0.05
7–13	0	0.05	0.02	0.01
14	1	0	0	0
15	0	0	0	0

TABLE 4.5. Counts of cases of polio by month for 1970–1974. Observed counts (second column), expected counts under a Poisson model based on the MLE (third column), expected counts based on the MLE omitting one outlier (fourth column), and expected counts based on the Hellinger estimate (fifth column) are given.

The Hellinger estimator is fully efficient if the assumed model is correct, but has a breakdown point that depends on the distribution (the breakdown point is the smallest proportion of observations in the sample that can lead to infinite bias in the estimator if those observations are allowed to become infinite). For the Poisson distribution the Hellinger estimator has the maximum possible breakdown of 50%.

Monthly Polio Incidence

As part of a larger set of data, Lindsey (1995, page 179) gave the monthly incidence of poliomyelitis in the United States for 1970–1974. Table 4.5 gives counts of the number of months with zero cases, one case, and so on. The highly unusual month with 14 cases was November 1972.

The Poisson MLE of μ, the polio incidence rate per month, is the sample mean, $\hat{\mu} = 1.483$. The third column gives fitted counts based on this estimate, and it is clear that this does not lead to a good fit to the data, with too few expected months were zero cases, and too many with two cases. The problem is that $\hat{\mu}$ has been drawn upwards by the outlying month. Omitting this case improves matters (this "cleaned MLE" equals 1.271), but the number of months with zero cases is still noticeably underestimated.

The final column of the table gives the expected counts based on the Hellinger estimate for these data, $\tilde{\mu} = 1.094$. It is apparent that this estimated rate is the best representation for the bulk of these data, although it does imply the fewest estimated number of months with five or more cases (of which there were five). There is some evidence of overdispersion here (the observed variance is almost twice the mean, even after omitting November 1972), and any kind of seasonal effect would lead to unobserved heterogeneity, so considering a negative binomial model for these data would be a reasonable alternative.

4.8   Background Material

Santner and Duffy (1989, Chapter 2) provides extensive discussion of inference for the binomial, multinomial, and Poisson distributions. Agresti and Coull (1998a), Brown et al. (2001), and Santner (1998) examined the properties of the Wald, score, and exact binomial confidence intervals, and agreed that the score interval should be used routinely as an alternative to the standard interval. Chernick and Liu (2002) discussed the nonmonotonicity of binomial hypothesis tests as a function of sample size in detail.

Chatterjee et al. (1995, pages 37–40) discusses sensitivity and specificity, and the connection of these measures to false positive rates in random drug and disease testing.

Garwood (1936) derived the exact confidence interval for the mean of the Poisson random variable. García-Perez (1999a) gave an algorithm for calculating multinomial probabilities in the fewest number of operations that is possible. García-Perez (1999b) derived and discussed an efficient algorithm for calculating exact p-values and confidence regions for goodness-of-fit statistics for multinomial data. Lancaster (1949) proposed the use of the mid-p-value as a way to combat the conservativeness of exact tests.

Read and Cressie (1988) provides extensive discussion of the power divergence family of goodness-of-fit statistics, pointing out the advantages of the Cressie–Read statistic $2nI^{2/3}$. Rayner and Best (2001, Section 2.6) discusses partitioning of X^2.

Boswell and Patil (1970) describes many different stochastic mechanisms that lead to the negative binomial distribution. Karlis and Xekalaki (2000) investigated various tests of the Poisson assumption using computer simulations, and found that the distribution of the test based on the variance/mean ratio is well-approximated by a standard normal, and that the test has good power to identify over- and underdispersed alternatives.

Kupper and Haseman (1978) proposed and studied the properties of the correlated binomial model. Tarone (1979) gave the asymptotically optimal test for the binomial versus correlated binomial and beta-binomial alternatives, based on the $C(\alpha)$ method, as given in (4.21); see also Smith (1951).

Simpson (1987) investigated the properties of Hellinger estimation for count data. Simpson (1989) proposed using a function of Hellinger distance to test hypotheses about unknown parameters, and showed that the combination of efficiency and robustness that characterizes Hellinger estimation carries over to the testing context.

4.9 Exercises

1. For X distributed $\text{Bin}(n, p)$, verify that $\hat{p} = X/n$ is the maximum likelihood estimator, and that its variance is $p(1-p)/n$.

2. Do women receive equal pay for equal work? A survey of 725 women conducted between July 30 and August 3, 1997, by the AFL–CIO's Working Women's Department found that 450 said that their current job provides equal pay for equal work. Construct a 90% confidence interval for the true probability that a randomly chosen woman feels that her current job provides equal pay for equal work. What assumptions are you making in constructing this interval? Do they seem reasonable here?

3. Agresti and Coull (1998a) discussed a simple adjustment to the ordinary binomial confidence interval that leads to coverage probabilities that are similar to those of the score interval (4.2). This interval uses the standard Wald form, adding "fake" successes and failures. Letting $\tilde{p} = (x + z_{\alpha/2}^2/2)/(n + z_{\alpha/2}^2)$ and $\tilde{n} = n + z_{\alpha/2}^2$, a $100 \times (1 - \alpha)\%$ confidence interval has the form

$$\tilde{p} \pm (z_{\alpha/2})\sqrt{\tilde{p}(1 - \tilde{p})/\tilde{n}}. \tag{4.23}$$

A 95% confidence interval adds $1.96^2/2 \approx 2$ fake successes and 2 fake failures to the data, so $\tilde{n} = n + 4$ and $\tilde{p} = (x + 2)/\tilde{n}$.

(a) Apply this method to the four Lyme disease testing sensitivity and specificity data sets (page 66). How do these adjusted 95% confidence intervals compare to the Wald intervals and score intervals?

(b) Recall that the reproducibility investigation for two-tiered serologic testing for Lyme disease yielded a reproducibility estimate of $\hat{p} = 10/10 = 1$. What is the adjusted 95% confidence interval for the reproducibility?

(c) Consider a situation where zero successes are observed in n trials (that is, $x = 0$). Show that for any n, (4.23) leads to a 95% interval that begins below 0. Similarly, show that if n successes are observed in n trials, the 95% interval based on (4.23) must end above 1. What about if $x = 1$ or $x = n - 1$?

(d) Pan (2002) proposed a further adjustment to the interval (4.23), replacing the Gaussian-based critical value $z_{\alpha/2}$ with a t-based value $t^r_{\alpha/2}$, with the degrees of freedom r being

$$
\begin{aligned}
r = \{&2[\tilde{p}(1-\tilde{p})/\tilde{n}]^2\}/\{\tilde{p}(1-\tilde{p})/\tilde{n}^3 \\
&+ [\tilde{p} + (6\tilde{n} - 7)\tilde{p}^2 + 4(\tilde{n} - 1)(\tilde{n} - 3)\tilde{p}^2 \\
&- 2(\tilde{n} - 1)(2\tilde{n} - 3)\tilde{p}^3]/\tilde{n}^5 \\
&- 2[\tilde{p} + (2\tilde{n} - 3)\tilde{p}^2 - 2(\tilde{n} - 1)\tilde{p}^3]/\tilde{n}^4\}.
\end{aligned}
$$

Construct the intervals of parts (a) and (b) using this method. Are the results very different from those using the Gaussian-based critical value?

4. In 1991, *Fortune* magazine published an article entitled "Why Women Still Don't Hit the Top." This article was a major instigator of the idea of the "glass ceiling," where it is hypothesized that women can advance to middle management levels in corporations, but no further up the corporate ladder. In 1995, the nonprofit research group Catalyst conducted a survey of 461 executive women (all at the vice president level or above). In the study, 240 of the women stated that stereotypical views of women (such as that women have a lower career commitment and won't stick with it for the long haul) prevent them from advancing. Construct a 95% confidence interval for the true proportion of executive women who feel that stereotypical views of women prevent them from advancing.

5. The "rule of three" says that an approximate (one-sided) 95% confidence bound for a binomial probability p based on a sample of size n where no successes are observed is $3/n$ (Louis, 1981).

(a) Derive the rule of three by determining the values of p for which the probability of observing zero successes in n trials is at least .05. (*Hint*: Write down the probability condition, and take natural logs of both sides of the inequality. Then, use the approximation $\log(1 - p) \approx -p$.)

(b) How does the rule of three interval compare to the limit of a one-sided 95% Agresti–Coull interval (equation (4.23))? Show that the Agresti–Coull limit is smaller than $3/n$ for $n \le 35$, and larger than $3/n$ for $n > 35$.

(c) How does the rule of three interval compare to the limit of a one-sided 95% score interval? Show that the score interval upper limit is always smaller than $3/n$.

(d) How does the rule of three interval compare to the limit of a one-sided 95% exact interval when $n = 10$? What does this imply for a rule of three interval estimate for reproducibility in the two-tiered serologic testing for Lyme disease (page 67)?

(e) Browne (2002) suggested using $3/(n+1.7)$ as a better approximate 95% confidence bound for p when no successes are observed. How do the answers to parts (b)–(d) change if this modified rule of three is used?

6. An article in the *New England Journal of Medicine* examined the possible relationship between the results of trials reported in journals and whether a researcher's research was supported by a drug company (Stelfox et al., 1998). The authors of the study examined 70 reports from 1995 and 1996 related to whether the use of calcium-channel blockers to treat hypertension led to an increased risk of heart disease. They classified 30 of the reports as favorable to the drugs, 17 as neutral, and 23 as critical of the drugs. They further determined that 29 of the favorable reports had authors who had received money from manufacturers of calcium-channel blockers, 10 of the neutral reports had such authors, and 8 of the critical reports had such authors.

(a) Construct a 90% confidence interval for the true proportion of all favorable reports that have authors who received money from manufacturers of calcium-channel blockers.

(b) Construct a 90% confidence interval for the true proportion of all neutral reports that have authors who received money from manufacturers of calcium-channel blockers.

(c) Construct a 90% confidence interval for the true proportion of all critical reports that have authors who received money from manufacturers of calcium-channel blockers.

(d) Do the confidence intervals change appreciably depending on whether you construct Wald, score, Agresti–Coull, or exact intervals?

Based on these results, do you think that there is a relationship between the nature of the report and whether the authors received financial support from the drug company?

7. The 64 teams selected to the NCAA Men's Basketball tournament are separated into four regions. The teams are seeded within their region from 1 through 16 based on a consensus of team strength. Thus, in the first four rounds of the tournament, each game involves a higher-seeded team playing a lower-seeded team. During the first ten years of the 64-team format, there were 116 games between teams that were seeded within one level of each other (for example, first versus second seed, fourth versus fifth seed, etc.). The higher-seeded team won 63 of these games. Test the hypothesis that teams that are seeded one level apart are, in truth, exactly evenly matched.

8. Of the 311 movies released in the United States during 1998, 13 were sequels to previously released films. Considering 1998 as a random occurrence from a stable process of movie releases, construct a 99% confidence interval for the true proportion of movies released in a typical year that are sequels.

9. Prove that if the number of successes in n trials is binomially distributed, with the number of trials $n \to \infty$ and the probability of success $p \to 0$ such that $np = \mu$, the distribution of the number of successes is roughly Poisson with mean μ. (*Hint*: Rewrite the binomial probability function in terms of μ using $p = \mu/n$, and use the functional approximations $(1 + x/n)^n \approx e^x$ for fixed x as $n \to \infty$ and $(1 + x/n)^j \approx 1$ for fixed x and fixed j as $n \to \infty$.)

10. Different states (and the District of Columbia) mandate testing for certain disorders in infants immediately after birth, but the number and nature of those tests varies from state to state. The following table summarizes the number of states that were testing for specific numbers of disorders in 2000 (Nelson, 2000). Assume that we can consider these 2000 values as realizations from an ongoing stable testing process.

Number of disorders	Number of states	Number of disorders	Number of states
0	0	6	7
1	0	7	8
2	0	8	6
3	2	9	2
4	13	10	0
5	11	11	1

(a) Does a Poisson distribution (with constant mean number of disorders tested for) fit these data?

(b) The focus of the newspaper article was on the disparities in testing between states. We might hypothesize that such disparities would result in overdispersion relative to a Poisson distribution. Is there any evidence of that here?

(c) The value for Mississippi was not given in the article. Give two estimates for the probability that Mississippi tests for at least ten disorders. Which of the two estimates do you prefer? Can you think of any reason not to trust either estimate very much?

11. Consider again the data on earthquakes in Irpinia (page 26). Construct a 95% confidence interval for μ, the per-decade rate of earthquakes of this severity in the region, based on a Poisson model for

earthquakes. Is a large-sample interval adequate here? How does this interval compare to the one based on a Gaussian assumption?

12. The following table gives the number of Broadway shows that debuted in each month of the year during the 1996–1997 through 1998–1999 seasons (see page 96). Test the hypothesis that the number of shows that open in a month are Poisson distributed with a constant mean number of shows per month. How might you explain the observed pattern of openings, keeping in mind tourism patterns in New York City and the Tony award eligibility criteria?

Month	Shows	Month	Shows
January	3	July	2
February	6	August	2
March	15	September	2
April	28	October	8
May	1	November	15
June	2	December	7

13. The file gviolence contains data on violence in the 74 G-rated animated features released in movie theaters between 1937 and 1999 (Yokota and Thompson, 2000). For each film the number of injuries and number of fatal injuries to good or neutral characters is given, as is the number of injuries and number of fatal injuries to bad characters. Assume that these data can be viewed as a sample from a stable ongoing process of the creation of G-rated animated features.

   (a) Assuming that the number of injuries to good or neutral characters is Poisson distributed, construct a 95% confidence interval for the true average number of such injuries in future G-rated animated features. Be sure to construct your interval in an appropriate way. Does a Poisson assumption seem reasonable here?

   (b) Construct corresponding intervals and tests for injuries to bad characters, and fatal injuries to good or neutral and bad characters, respectively.

   (c) Barker (2002) compared the performance of various confidence intervals for the Poisson mean, ultimately recommending the exact interval, the score interval (if a closed-form interval that held its size well was desired), or an interval based on a Gaussian approximation to the distribution of $\sqrt{X}$,

$$\overline{X} + z_{\alpha/2}^2/(4n) \pm z_{\alpha/2}\sqrt{\overline{X}/n}$$

(with the exact interval used if $\overline{X} = 0$), if a narrow closed-form interval with true coverage potentially further from the nominal coverage was acceptable (this interval is motivated by the approximately constant variance of the square root of a Poisson random variable). Note that this interval is identical to the standard (Wald) interval, other than an additive adjustment to the center of the interval. Construct these variance-stabilized confidence intervals for the injuries data in parts (a) and (b). Are they very different from the intervals you constructed in parts (a) and (b)?

14. (a) Consider two realizations x_1 and x_2 from independent Poisson random variables X_1 and X_2 with means μ_1 and μ_2, respectively. Show that a confidence interval for the ratio of the two Poisson means, μ_1/μ_2, that is appropriate when x_1 and x_2 are large, has the form

$$\left(\frac{\bar{p} - z_{\alpha/2}\sqrt{\bar{p}(1-\bar{p})/n}}{1 - \bar{p} + z_{\alpha/2}\sqrt{\bar{p}(1-\bar{p})/n}}, \frac{\bar{p} + z_{\alpha/2}\sqrt{\bar{p}(1-\bar{p})/n}}{1 - \bar{p} - z_{\alpha/2}\sqrt{\bar{p}(1-\bar{p})/n}} \right),$$

where $\bar{p} = x_1/(x_1 + x_2)$ and $n = x_1 + x_2$. (*Hint*: Use the distribution of $X_1|(X_1 + X_2)$.)

(b) How would you adapt this form to a situation where n is small?

15. Consider again Richardson's data on armed conflicts (page 26). The hypothesis that wars are independent, random, events, with constant probability of occurrence, implies a Poisson distribution for the number of wars in a given year. Is that model consistent with these data? What is the estimated rate of magnitude 4 conflict starts per year?

16. Show that the necessary sample size to detect a true Poisson mean μ_a that is δ away from a hypothesized value μ_0 with probability β based on a one-sided test of size α is roughly

$$n = \left(\frac{z_\alpha\sqrt{\mu_0} - z_\beta\sqrt{\mu_a}}{\delta} \right)^2.$$

(*Hint*: Consider (4.5) on page 60.)

17. (a) Consider again the data on stock returns used to investigate Benford's Law on page 78. Examine the lack of fit of Benford's distribution to these data by partitioning X^2 into location and dispersion components. Is there evidence that the lack of fit comes from such effects? Can you characterize the lack of fit of Benford's distribution in the low absolute return data as being from location or dispersion effects?

(b) The file **benford** gives daily S&P returns for a larger set of trading days (July 3, 1962, through December 29, 1995). Do the same patterns seen for the smaller set of days hold for the larger set?

18. The characterization of Benford's Law as the appropriate distribution for numbers "without known relationship" suggests a practical application of the distribution in auditing and fraud detection, since data that are not consistent with Benford's Law (when they should be) might have been fraudulently altered. An actual fraud case illustrates this possibility. *State of Arizona v. Wayne James Nelson* (CV92–18841) was a 1993 case where the defendant was convicted of attempting to defraud the state of almost $2 million. It was shown during trial that the defendant created a fake vendor, and had 23 checks issued to the vendor for nonexistent services (Nigrini, 2000, pages 82–84).

(a) The following table summarizes the distribution of first digits of the dollar amounts for the 23 checks to the nonexistent vendor. Are these data significantly deviant from the expected Benford's Law distribution? Does your answer change if you use the exact distribution of your goodness-of-fit test, rather than the asymptotic one?

Digit	Count	Digit	Count
1	1	6	0
2	1	7	3
3	0	8	9
4	0	9	9
5	0		

(b) Benford's Law can be generalized to more than just the first digit of numbers. It can be shown that the Benford's distribution for the first two digits $D_1 D_2$ of a number is

$$P(D_1 D_2 = d_1 d_2) = \log_{10}[1 + 1/(d_1 d_2)], d_1 d_2 \in \{10, 11, \ldots, 99\}.$$

This formula can be used to generate the Benford's distribution for the second digit in a number $P(D_2 = d_2)$ by summing the probabilities $P(D_1 D_2 = d_1 d_2)$ over $d_1 = 1, \ldots, 9$. The following table summarizes the distribution of second digits for the Arizona fraud data. Do the second digits show evidence of an unusual distribution of digits?

Digit	Count	Digit	Count
0	1	5	0
1	3	6	3
2	2	7	4
3	3	8	2
4	2	9	3

19. The file **uklottery** gives the number of times each of the 49 distinct numbered balls were chosen in the 416 United Kingdom National Lottery draws between November 19, 1994, and December 18, 1999. Does a uniform distribution (which would be consistent with a fair game) fit these data?

20. The following table gives the expected and observed number of horses winning races in 3785 Hong Kong horse races on the basis of the probabilities implied by the odds of winning of the horse at posttime (Stern, 1998). If the public is adept at picking winners, the observed and expected numbers should be close to each other. Test the fit of the observed data to the model implied by posttime odds. Can any observed lack of fit be ascribed to location or dispersion effects? How might you interpret any such effects?

Probability implied by the odds	Observed number of winners	Expected number of winners
[.000, .010)	15	12.7
[.010, .025)	78	87.0
[.025, .050)	278	242.9
[.050, .100)	695	713.9
[.100, .150)	781	792.6
[.150, .200)	657	643.4
[.200, .250)	451	480.7
[.250, .300)	325	334.6
[.300, .400)	303	290.4
[.400, .500)	121	117.7
[.500, 1.00)	81	69.0

21. The following table gives the number of births in the United Kingdom in 1978, separated by day of week (see Berresford, 1980).

Day of week	Number of births
Sunday	387,057
Monday	447,956
Tuesday	463,201
Wednesday	452,213
Thursday	451,918
Friday	449,074
Saturday	387,929

(a) Is the distribution of births uniform over days of the week? How might you explain the observed pattern of births?

(b) Is the distribution of births on weekdays uniform over days of the week?

22. In 1997, the United Kingdom's Ministry of Defence released documents that showed that germ warfare simulation trials had been carried out along the south coast of England between 1961 and 1977 (Stein et al., 2001). As a result of these trials, up to twice a year clouds of bacteria were sprayed from naval vessels along the Dorset coast. Families in East Lulworth, a coastal village in the area, believed that exposure to the bacteria led to high rates of different types of birth problems and defects, based on a survey of current residents. Ultimately members of Parliament and the media pressed the Dorset Health Authority to conduct a full investigation as to whether a disease cluster existed in East Lulworth. The following table summarizes the results of the villagers' survey and Health Authority survey. Information was available on 163 pregnancies in East Lulworth during the time in question, separated into the categories Miscarriage, Stillbirth, Congenital malformation, Neurodevelopmental disability, and Normal birth. The table gives expected proportion of births in each category, based on published medical literature. One birth was a congenital malformation that led to a stillbirth; this birth is listed as a stillbirth in the table.

Type of birth	Expected proportion	Observed number
Miscarriages	24.0%	19
Stillbirths	1.8%	7
Congenital malformations	2.5%	4
Neurodevelopmental disorders	13.0%	12
Normal births	59.7%	121

Is there any evidence that the observed pattern of births is inconsistent with the expected pattern? That is, does it appear that East Lulworth constituted a disease cluster? Be sure to take into account the small observed counts in several birth categories.

23. Basu and Sarkar (1994) proposed an alternative family of χ^2 goodness-of-fit test statistics based on disparity measures. They suggested the statistic $BWHD_{1/9}$ as an alternative to the Cressie–Read $2nI^{2/3}$ statistic, where

$$BWHD_{1/9} = 81 \sum_{i=1}^{K} \left(\frac{n_i - e_i}{\sqrt{n_i} + 8\sqrt{e_i}} \right)^2.$$

Does this statistic lead to different assessments of goodness-of-fit for the Benford's Law data (page 78)? What about for the data for each of the previous six exercises?

24. Assume that a zero-inflated Poisson model is fit to a random variable X. Show that the fitted value for $P(X = 0)$ is exactly n_0, the observed number of zero observations. (*Hint*: Use that the score equation for p, the probability of coming from the Poisson portion of the distribution, is $p = \frac{1-n_0/n}{1-\exp(-\mu)}$.)

25. Consider again the data on violence in G-rated animated features (page 107).

    (a) Fit a Poisson distribution for the total number of injuries in a feature. Now, fit a zero-inflated Poisson model. Which model do you prefer? How can you interpret the results of the ZIP fit in terms of the moviemaking process?

    (b) Does a ZIP model provide a better description of the distribution of fatalities in a feature than does the Poisson?

26. The file **sexpartners** contains data on the number of (opposite sex) sex partners since the respondent was 18 years of age reported by a sample of respondents (1682 men and 1850 women) to the General Social Survey from 1989 through 1991 (Chatterjee et al., 1995, page 303).

    (a) Fit a Poisson distribution to the number of partners for the sample of women. Does the Poisson fit the data? Is there evidence of overdispersion?

    (b) Fit a zero-inflated Poisson distribution to these data. Does this provide an improved fit? How would you interpret the parameters of the model?

    (c) Extend the model by allowing for inflation of the one-partner cell (corresponding to monogamy). Does this improve the fit?

    (d) There is evidence in these data that responses of more than 10 lifetime partners might not be trustworthy (see Chatterjee et al., 1995, pages 121–122). Exclude these observations and refit the Poisson, zero-inflated Poisson, and zero and one-inflated Poisson models. Does this change your answers to parts (a)–(c)?

    (e) Repeat parts (a)–(d) for the sample of male respondents.

27. (a) Consider a mixture of Poisson distributions, where each observation $X_i|\mu_i$ is distributed $\text{Pois}(\mu_i)$, μ_i a random quantity. Show that as long as $V(\mu_i) > 0$ (that is, μ_i does not follow a degenerate distribution) $V(X_i) > E(X_i)$, and thus the mixture provides a model for overdispersion relative to the Poisson.

(b) Consider a mixture of binomial distributions, where each observation $X_i|(n, p_i)$ is distributed $Bin(n, p_i)$, p_i a random quantity. Show that as long as $V(p_i) > 0$ and $n > 1$, $V(X_i) > E(X_i)(1 - E(X_i)/n)$, and thus the mixture provides a model for overdispersion relative to the binomial.

(c) Show that if $n = 1$, $V(X_i) = E(X_i)(1 - E(X_i)/n)$, and thus Bernoulli trials with random p do not exhibit overdispersion.

(*Hint*: Use the following relationships between conditional and unconditional moments: for random parameter θ, $E(X) = E_\theta[E_X(X|\theta)]$ and $V(X) = V_\theta[E_X(X|\theta)] + E_\theta[V_X(X|\theta)]$.)

28. Show that the maximum likelihood estimator of the mean of the negative binomial variable is the sample mean. (*Hint*: Show that the sample mean satisfies the score equation for μ.)

29. (a) An alternative motivation for the negative binomial distribution that is appropriate when ν is an integer is that it represents the distribution of the number of failures in independent Bernoulli trials X before ν successes are obtained, where the probability p of success in any one trial is $p = \nu/(\mu + \nu)$ (the distribution is sometimes called the *Pascal* distribution in this context). The probability function in this context can be written

$$f(x; p, \nu) = \binom{\nu + x - 1}{x} p^\nu (1 - p)^x, \qquad (4.24)$$

using the fact that $\Gamma(k) = (k - 1)!$ for integral k. Based on this parameterization, $E(X) = \nu(1 - p)/p$ and $V(X) = \nu(1 - p)/p^2$. Use the properties of the binomial random variable to demonstrate that this binomial waiting-time random variable does, in fact, have a negative binomial distribution.

(b) The negative binomial distribution with $\nu = 1$ is called the *geometric* distribution. Using (4.24), show that the maximum likelihood estimate of p, the probability of success of the underlying Bernoulli trial, when all that is observed is a sample $\{x_i\}$ of the number of failures between successes, is $\hat{p} = 1/(\overline{X} + 1)$. Show that this is simply the observed proportion of successes.

(c) The Major League All-Star game annually pits teams of players from the American League against players from the National League. The following table gives the number of National League wins between American League wins for all of the games through 2000 (one tie game has been omitted).

NL wins	Count
0	18
1	6
2	2
3	1
4	2
8	1
11	1

Does a geometric distribution based on evenly matched teams ($p = .5$) fit these data? Does a geometric distribution with estimated p fit better? Now fit a negative binomial distribution to the data. If the geometric distribution is appropriate, $\hat{\nu}$ should be roughly one, but it is considerably less than that here. What are the implications of this result? How might you explain this violation of the geometric distribution assumption?

30. Consider again the data on goals scored and given up by the New Jersey Devils (Table 4.1). Does a negative binomial distribution fit these data better than the Poisson distribution? If so, how does this affect the confidence intervals for true average number of goals scored by high-quality teams?

31. The following table summarizes passing movements in 42 English First Division Football League (soccer) matches in 1957–1958 (Norman, 1998). A passing movement is defined as the number of passes from player to player on the same team until the play ends, whether by loss of possession, stoppage of play due to rule violation, the ball going out of play, or a goal being scored. The values given are the observed counts of the number of passes in each passing movement. Does a Poisson distribution provide an adequate fit to these data? Does a negative binomial distribution fit better? Why might it be that the negative binomial would be a reasonable alternative to the Poisson in this context?

Number of passes	Count	Number of passes	Count
0	10187	5	280
1	6923	6	107
2	3611	7	33
3	1592	8	9
4	608	9 or more	11

32. Two of the earliest applications illustrating overdispersion in real count data were by W.S. Gosset ("Student") in 1907 (see Zelterman,

1999, page 31) and M. Greenwood and G.U. Yule in 1920 (see Santner and Duffy, 1989, page 19). Gosset examined counts of the number of yeast cells in a dilute culture measured using a hemocytometer, while Greenwood and Yule examined the number of accidents that occurred in three months to each of 414 workers. Interestingly, in the first case it is likely stickiness of the yeast cells inducing lack of independence that is causing overdispersion, while in the second it is different underlying accident rates for the workers causing it. The following table summarizes these two data sets. For each, verify that the negative binomial model provides significantly improved fit over the Poisson.

Student data		Greenwood and Yule data	
Number of cells	Count	Number of accidents	Count
0	213	0	296
1	128	1	74
2	37	2	26
3	18	3	8
4 or more	4	4	4
		5	6

33. The discussion of the weekly shampoo use data (page 91) showed that overdispersion can come from both zero-inflation (and inflation of other values) and heterogeneity of underlying means (the negative binomial).

   (a) Show that the mean of the zero-inflated negative binomial random variable (ZINB) is

$$E(X) = p\mu,$$

   and the variance is

$$V(X) = p\mu[1 + (1 - p + \alpha)\mu],$$

   where p is the probability that an observation comes from the underlying negative binomial distribution with parameters μ and $\alpha = 1/\nu$. Show that this implies that

$$\frac{V(X)}{E(X)} = 1 + \left(\frac{1 - p + \alpha}{p}\right) E(X),$$

   implying overdispersion from both the zero inflation and negative binomial sources (Greene, 1994).

(b) One way to try to decide if a zero-inflated Poisson model also exhibits unobserved mean heterogeneity is to construct a test for the ZINB versus the ZIP. The score test comparing the null ZIP distribution to the alternative ZINB distribution is

$$T = \frac{\sum_i [(x_i - \hat{\mu})^2 - x_i] - n\hat{\mu}^2\hat{p}}{\hat{\mu}\sqrt{n(1 - \hat{p})\left(2 - \frac{\hat{\mu}^2}{e^{\hat{\mu}}-1-\hat{\mu}}\right)}},$$

where μ and p are estimated assuming the ZIP model, and T is compared to a χ_1^2 critical value (Ridout, Hinde, and Demétrio, 2001). The following table gives the number of movements made by a fetal lamb in each of 240 consecutive five-second intervals (Ridout, Hinde, and Demétrio, 2001). Does the score test suggest evidence against the ZIP model in favor of the ZINB? Is the result sensitive to the presence of the one unusual interval containing seven movements?

Number of movements	Number of intervals	Number of movements	Number of intervals
0	182	4	2
1	41	5	0
2	12	6	0
3	2	7	1

34. Böhning (1998) gave data originally from Dieckmann (1981) on the distribution of the number of criminal acts of 4301 people with deviating behavior, given in the following table.

Number of criminal acts	Number of people	Number of criminal acts	Number of people
0	4037	3	9
1	219	4	5
2	29	5	2

(a) Fit a Poisson and a ZIP distribution to these data. Which model do you prefer? Does either fit well?

(b) Böhning (1998) fit a zero-inflated mixture model to these data. Does a zero-inflated negative binomial model provide an improved fit over a ZIP model? How might you justify the use of a ZINB model here?

35. Consider a sample $\{x_i\}, i = 1, \ldots, N$, that is hypothesized to come from a single binomial distribution, with each x_i based on the same number of trials n. Define

$$S = \sum_{i=1}^{N} \frac{(x_i - n\hat{p}_0)^2}{\hat{p}_0(1 - \hat{p}_0)}.$$

(a) Show that the statistic Z from (4.21) satisfies

$$Z = \frac{S - nN}{\sqrt{2n(n-1)N}}$$

in this situation.

(b) Show that a standardized version of the binomial version of X^2 (4.19), obtained by subtracting the mean and dividing by the standard deviation of X^2, is roughly

$$Z^* = \frac{S - n(N-1)}{\sqrt{2n^2(N-1)}}.$$

(*Hint*: Use that the mean of a χ^2 random variable equals its degrees of freedom and the variance is twice the degrees of freedom.) Note that the similarity of Z and Z^* implies that Z only provides noticeably better power than X^2 when the number of trials for each x_i varies.

36. The following table gives the number of cases of functional damage that resulted as a side effect from a vaccination, along with the number of vaccinations, for seven years of vaccination (Aitkin, 1992).

Year	Number of cases	Number of vaccinations
1	4	300,293
2	2	247,363
3	0	271,222
4	1	277,163
5	1	283,314
6	0	284,550
7	1	283,116

(a) Fit a binomial distribution to these data, and construct a 95% confidence interval for p, the probability that the vaccination leads to functional damage. Be sure to use a confidence interval that is appropriate for these data.

(b) The internationally accepted damage rate for this vaccination is 1 in 310,000. Is that consistent with your answer to (a)?

(c) Does the binomial distribution with estimated p provide an adequate fit to these data? What about the binomial using the internationally accepted damage rate?

(d) The impetus for this investigation was a court case where the plaintiff alleged that the four damage cases in year 1 was evidence of an increased risk of damage from the batch of vaccine produced in that year. Is 4 cases of damage in 300,293 vaccinations a surprisingly large number? How might you quantify this? (*Hint*: Consider the discussion on page 71 on the connection between the binomial and Poisson random variables.)

37. Show that the beta-binomial probability function approaches that of the binomial distribution if $\alpha \to \infty$ and $\beta \to \infty$ such that $\alpha/(\alpha+\beta)$ is constant. (*Hint*: Use Stirling's approximation, that for large x, $\Gamma(x) \approx \sqrt{2\pi} x^{x-1/2} e^{-x}$.)

38. Many political, civic, and business organizations lobby state and federal legislators to try to influence their votes on key legislative matters. These organizations often track the voting patterns of legislators in the form of "report cards," which summarize the votes of legislators on key pieces of legislation. The file **neavote** summarizes the 1999 U.S. Senate legislative report card produced by the National Education Association (NEA, the nation's oldest and largest organization committed to advancing the cause of public education), based on 10 Senate votes during the year, and was obtained from the NEA's web site (**www.nea.org**).

(a) Considering the 1999 data as a typical year for the NEA, does a binomial distribution with constant probability of votes favorable to the NEA for each senator fit these data?

(b) Is there evidence of overdispersion relative to the binomial? Does the beta-binomial provide a better fit to the data? How would you justify the use of the beta-binomial here?

(c) Fit separate binomial distributions to the Democrats and Republicans in the sample. Does this adequately represent the data? Do separate beta-binomial distributions provide significantly better fit? How would you interpret the parameter estimates in your choice of best model(s)?

39. The study of the human sex ratio has a long history. A. Geissler collected data on the distributions of the sexes of children in families in Saxony during 1876–1885, a data set that has often been used to investigate sex ratio patterns. Two basic questions in the investigation of the sex ratio is whether the probability that a child is a girl (say) changes over time within a family, and whether it differs from

family to family. Data that have been aggregated over families cannot distinguish between these two possibilities, with both being reflected in overdispersion relative to the binomial.

(a) The following table gives the subset of Geissler's data corresponding to families with 10 children (Lindsey and Altham, 1998). Use the beta-binomial distribution to decide if overdispersion relative to the binomial is apparent in these data, which would imply varying sex ratios either within families or between them.

Number of girls	Count	Number of girls	Count
0	30	6	3072
1	287	7	1783
2	1027	8	722
3	2309	9	151
4	3470	10	30
5	3878		

(b) The famous statistician R.A. Fisher also used data from Geissler to explore overdispersion relative to the binomial distribution, with the following table referring to families with eight children (Fisher, 1925, page 69).

Number of boys	Count	Number of boys	Count
0	215	5	11929
1	1485	6	6678
2	5331	7	2092
3	10649	8	342
4	14959		

Does the beta-binomial distribution provide a better fit for these data than the binomial? Is there still evidence of meaningful lack of fit? Fisher claimed that the data show "an apparent bias in favour of even values." Do you agree with this claim?

40. The Katz family of distributions (Winkelmann, 2000, Section 2.5.4) is defined by the recursive relationship

$$\frac{P(X = k)}{P(X = k - 1)} = \frac{\omega + \gamma(k - 1)}{k}, \qquad k = 1, 2, \ldots,$$

where $\omega > 0$ and $k < |\omega/\gamma| + 1$ for $\gamma < 0$. Show that the Katz family is a generalization of many of the distributions described in this chapter, in that

(a) the Poisson distribution is obtained from $\omega = \mu$ and $\gamma = 0$;

(b) the negative binomial distribution is obtained from $\omega = \nu\mu/(\nu + \mu)$ and $\gamma = \mu/(\nu + \mu)$ (*Hint:* Use that $\Gamma(x) = (x - 1)\Gamma(x - 1)$);

(c) the binomial distribution is obtained from $\omega = np/(1 - p)$ and $\gamma = -p/(1 - p)$.

(d) It can be shown that the mean of a random variable with the Katz distribution is $E(X) = \omega/(1 - \gamma)$, while the variance is $V(X) = \omega/(1 - \gamma)^2$. Show that the Katz family exhibits overdispersion if $0 < \gamma < 1$, underdispersion if $\gamma < 0$, and equidispersion if $\gamma = 0$.

(e) Show that the method of moments estimators of ω and γ are $\hat{\omega} = \overline{X}^2/s^2$ and $\hat{\gamma} = 1 - \overline{X}/s^2$, respectively (the method of moments estimators are derived by equating the sample mean and variance to the population mean and variance, respectively).

(f) Investigations of disputed authorship are often based on notions of an author's style, the characteristics of writing that do not depend on subject matter or content. One commonly used measure of style is the frequency of a group of function words (such as prepositions, articles, etc.). The following table gives the combined frequencies of the articles "the," "a," and "an" in samples from Lord Macaulay's "Essay on Milton" (Bailey, 1990). Nonoverlapping samples were drawn from the opening 5 or 10 words of two randomly chosen lines from each of the 50 pages of the Oxford edition of Macauley's literary essays.

Number of occurrences	Five-word sample observed count	Number of occurrences	Ten-word sample observed count
0	45	0	27
1	49	1	44
2	6	2	26
3 or more	0	3	3
		4 or more	0

If one of the articles occurred independently of any others with constant probability, the number of occurrences would be binomially distributed. Does a binomial distribution fit the 5-word sample data? What about the 10-word sample data?

(g) Clearly, in ordinary English grammar, an article cannot be followed in the same sentence by another article. Thus, the occurrence of an article in a sentence reduces the probability of another article in the sentence; that is, the occurrences exhibit negative autocorrelation. We might expect this to result in observed underdispersion. Is that the case here? Fit Katz distributions to each of the data sets using the method of moments estimators. Do the Katz distributions fit the data better than the binomial distributions do? Why is the closeness of the fit to the 5-word sample data less impressive than it might appear at first glance?

41. The following table gives the number of occupants in homes, based on a postal survey (Lindsey, 1995, page 135). Since it is impossible for a house to have zero occupants and be included in the survey, a truncated distribution is appropriate for these data. Fit a zero-truncated Poisson distribution to these data. Does the model fit the data well?

Number of occupants	Number of houses	Number of occupants	Number of houses
1	436	5	1
2	133	6	0
3	19	7	1
4	2		

42. The ZIP model is a special case of the *zero-modified Poisson* (ZMP) model (Dietz and Böhning, 2000). Zero deflation (that is, too few zero observations) can be addressed in this model by allowing $p > 1$ in the ZIP formulation.

(a) Show that in order for the ZMP to be a proper distribution, p must satisfy

$$0 \leq p \leq \frac{e^{\mu}}{e^{\mu} - 1}.$$

(b) The parameter μ can be estimated by fitting a zero-truncated Poisson distribution to the positive observations in the sample. Show that once μ is estimated this way,

$$\hat{p} = \frac{1 - n_0/n}{1 - \exp(-\hat{\mu})},$$

where n_0 is the number of observations with zero counts in the sample. (*Hint*: Write down the score equation for p.)

(c) The following table refers to data from a survey of 457 homeless men and women in Los Angeles County (Gelberg et al., 1995; see also Holcomb, 2002, page 4). The respondents were asked the number of free sources of food that they had visited during the previous week, with the results given in the following table.

Number of free sources	Number of people	Number of free sources	Number of people
0	27	3	96
1	137	4	50
2	128	5 or 6	19

Does a Poisson distribution fit these data well? Is there evidence of underdispersion in these data relative to a Poisson distribution? (Note that this would be consistent with zero deflation.) Are the results different depending on how you treat the response group "5 or 6"?

(d) Fit a zero-modified Poisson distribution to the data. Is there evidence of zero deflation?

(e) The survey was conducted at sites where homeless people tend to congregate, including emergency shelters, a parking lot, parks, a shopping mall, a large beach area, soup kitchens, food distribution centers, and social service assistance areas. Does this suggest why there might be fewer respondents who report not visiting any free sources of food in the past week than a Poisson distribution would predict?

43. Consider again the monthly polio incidence data examined earlier (page 101). Fit a negative binomial distribution to these data using maximum likelihood, and after omitting the November 1972 observation. Does this improve on a Poisson fit? Now, determine the minimum Hellinger negative binomial estimates for these data. How do these estimates compare with the MLEs? (*Hint*: A bivariate grid search over potential values of μ and ν can be used here to find an approximate minimizer of the Hellinger distance.)

44. The following table gives a set of counts of the number of read-write errors discovered in a computer hard disk manufacturing process (Xie et al., 2001).

Number of errors	Count	Number of errors	Count
0	180	6	2
1	11	9	2
2	5	11	1
3	2	15	1
4	1	75	2
5	1		

(a) Fit a ZIP model to these data using maximum likelihood. Does the model fit the data?

(b) The two observations that correspond to disks with 75 errors are obviously very unusual. Omit these observations and refit the ZIP model. The results are very different from those in part (a). Does the model fit better?

(c) Fit a ZIP model to the entire data set using the minimum Hellinger estimate. How do these results compare to those in parts (a) and (b)? (*Hint*: To avoid a bivariate minimization, obtain an approximate Hellinger estimate for μ using a univariate search based on a truncated Poisson model, and then estimate p so as to make the observed and expected zero counts equal.)

45. Hellinger estimation is known to be less effective when there are many cells with zero counts in the data. Harris and Basu (1994), Basu and Basu (1998), and Basu et al. (1996) proposed and investigated penalized Hellinger distance, where empty cells are penalized, resulting in a better-behaved minimizer. One version of the penalized Hellinger distance takes the form

$$H' = \sum_{n_i \neq 0} \left\{ \left(\frac{n_i}{n} \right)^{1/2} - [f(i; \boldsymbol{\theta})]^{1/2} \right\}^2 + \frac{1}{2} \sum_{n_i = 0} f(i; \boldsymbol{\theta}).$$

Construct penalized Hellinger estimates for the polio incidence data and computer hard disk data of the previous two exercises. Are the results very different from the ordinary Hellinger estimates?

5

Regression Models for Count Data

In Chapters 2 and 3 we explored the use of least squares regression to analyze and understand the relationship between a target variable and at least one predicting variable. While it was possible to model the number of deaths monthly from the number of killer tornadoes, there were clear problems in that model fitting, including negative estimated tornado-related deaths, an apparent nonlinear relationship between the target and the predictor, and heteroscedasticity in the residuals from the model. As was noted on page 25, the problem is that least squares is based on a "signal-plus-noise" model using an underlying Gaussian distribution, when the correct analysis uses distributions appropriate for categorical data, such as those discussed in Chapter 4. Before discussing specific models, we present results for a very general regression model, the *generalized linear model*. In later sections and chapters we will see how these general results apply to specific models for distributions such as the Poisson and binomial.

5.1 The Generalized Linear Model

5.1.1 The Form of the Generalized Linear Model

A regression model must specify several things: the distribution of the value of the target variable y_i (the so-called *random component*), the way that the predicting variables combine to relate to the level of y_i (the *systematic component*), and the connection between the random and systematic components (the *link function*). A generalized linear model is defined by the choices of these two components and their link. The random component requires that the distribution of y_i comes from the exponential family,

$$f(y; \theta, \phi) = \exp\left[\frac{y\theta - b(\theta)}{a(\phi)} + c(y, \phi)\right], \tag{5.1}$$

for specified functions $a(\cdot)$, $b(\cdot)$, and $c(\cdot)$. The functions $a(\cdot)$ and $b(\cdot)$ satisfy

$$\mu = E(y) = b'(\theta)$$

and

$$V(y) = a(\phi)b''(\theta),$$

respectively, where $b'(\theta)$ and $b''(\theta)$ are the first and second derivatives of $b(\theta)$ with respect to θ, respectively. The systematic component specifies that the predictor variables relate to the level of y as a linear combination of the predictor values,

$$\eta_i = \beta_0 + \beta_1 x_{1i} + \cdots + \beta_p x_{pi} \tag{5.2}$$

(a *linear predictor*). The link function then relates η to the mean of y, μ, being the function g such that

$$g(\mu) = \eta.$$

The log-likelihood for the entire sample is thus

$$L = \sum_{i=1}^{n} \left[\frac{y_i \theta_i - b(\theta_i)}{a_i(\phi)} + c(y_i, \phi)\right]. \tag{5.3}$$

The Gaussian linear model (2.8) is an example of a generalized linear model. The Gaussian model is

$$f(y; \mu, \sigma^2) = \frac{1}{\sqrt{2\pi\sigma^2}} \exp\left[-\frac{(y - \mu)^2}{2\sigma^2}\right]$$

$$= \exp\left\{\frac{y\mu - \mu^2/2}{\sigma^2} - \frac{1}{2}\left[\frac{y^2}{\sigma^2} + \log(2\pi\sigma^2)\right]\right\}.$$

Thus, the Gaussian is a member of the exponential family, with $\theta = \mu$, $\phi = \sigma^2$, $a(\phi) = \phi$, $b(\theta) = \theta^2/2$, and $c(y, \phi) = -[y^2/\sigma^2 + \log(2\pi\sigma^2)]/2$ in (5.1). The link function is the identity link, $g(\mu) = \mu$, consistent with a linear model.

A particularly desirable form of the generalized linear model occurs when the link function satisfies $g(\mu) = \theta$. This link is called the *canonical link*. When the canonical link is used the sufficient statistic for $\boldsymbol{\eta}$ is a linear function of the data, $X'\mathbf{y}$, the observed and fitted target values have the same marginal totals ($X'\mathbf{y} = X'\boldsymbol{\mu}$), and the observed and expected Fisher information coincide. The canonical link for the Gaussian regression model is the identity link.

5.1.2 Estimation in the Generalized Linear Model

Maximum likelihood is typically used to estimate the parameters of a generalized linear model. The score equations take the form

$$X'W\mathbf{r} = \mathbf{0},$$

where

$$W = \text{diag}\left[\left(\frac{\partial \mu_i}{\partial \eta_i}\right)^2 / V(y_i)\right],$$

using the notation $\text{diag}(c_i)$ to represent the matrix with ith diagonal element equal to c_i and zero elsewhere, and

$$r_i = (y_i - \mu_i)\frac{\partial \eta_i}{\partial \mu_i}.$$

Adding $X'WX\boldsymbol{\beta}$ to both sides of the score equation gives

$$X'WX\boldsymbol{\beta} = X'WX\boldsymbol{\beta} + X'W\mathbf{r}, \tag{5.4}$$

or

$$X'WX\boldsymbol{\beta} = X'W\mathbf{z},$$

where $\mathbf{z} = X\boldsymbol{\beta} + \mathbf{r}$. This ultimately yields

$$\hat{\boldsymbol{\beta}} = (X'WX)^{-1}X'W\mathbf{z}. \tag{5.5}$$

Comparing (5.5) to (3.16) makes it apparent that the MLE takes the form of a weighted least squares estimate, with W taking the place of V^{-1} and $\mathbf{z}$ taking the place of $\mathbf{y}$. Equation (5.5) cannot be used directly to estimate $\boldsymbol{\beta}$, since W and $\mathbf{z}$ are functions of $\boldsymbol{\beta}$, but asymptotic inference for $\boldsymbol{\beta}$ follows from the standard linear weighted least squares formulation. As was discussed in Section 2.1.2, maximum likelihood estimates are consistent and asymptotically Gaussian, with variance equal to the inverse of the expected Fisher information. For the generalized linear model the information matrix is $X'WX$, so for large samples $\hat{\boldsymbol{\beta}} \sim N(\boldsymbol{\beta}, (X'WX)^{-1})$.

The weighted least squares form (5.5) of the score equation suggests a general iterative algorithm to estimate $\boldsymbol{\beta}$. Recall that W and $\mathbf{z}$ are functions of $\boldsymbol{\beta}$. Let $\hat{\boldsymbol{\beta}}^{(k)}$ be the estimate of $\boldsymbol{\beta}$ at the kth iteration (with resultant $W^{(k)}$ and $\mathbf{z}^{(k)}$). Then the $(k+1)$st iteration is based on

$$\hat{\boldsymbol{\beta}}^{(k+1)} = (X'W^{(k)}X)^{-1}X'W^{(k)}\mathbf{z}^{(k)}.$$

That is, $\hat{\boldsymbol{\beta}}^{(k+1)}$ is the weighted least squares estimate of $\boldsymbol{\beta}$ in a regression of $\mathbf{z}$ on X. The weights change at each iteration with the new estimate of β, so iterating the estimation results in an *iteratively reweighted least squares* algorithm. Premultiplying both sides of (5.4) by $(X'WX)^{-1}$ and collecting terms demonstrates that this algorithm corresponds to the Fisher scoring maximum likelihood estimation algorithm, which is equivalent to the Newton–Raphson algorithm when the canonical link function is used.

The different inferential questions that arise in least squares regression, as discussed in Chapters 2 and 3, have analogous forms when analyzing generalized linear models. In the next sections we summarize these inference methods.

5.1.3   Hypothesis Tests and Confidence Intervals for β

F-tests (such as (2.12) and (3.1)) and t-tests (such as (2.13)), the basic tools of inference in least squares regression, are generalized to likelihood ratio and Wald statistics in the generalized linear model framework. The likelihood ratio statistic tests the null hypothesis of a restricted (simpler) model versus the alternative of a more general model, and equals twice the difference between the maximized log-likelihood under the more general (full) hypothesis and the maximized log-likelihood under the simpler (subset) model,

$$LR = 2(L_{\text{general}} - L_{\text{simpler}}), \tag{5.6}$$

where L is determined from (5.3). This is compared to a χ^2_d critical value, where d is the difference in the number of parameters fit under the two models.

The test of the overall significance of the regression tests

$$H_0 : \beta_1 = \cdots = \beta_p = 0$$

versus

$$H_a : \text{some } \beta_j \neq 0, \qquad j = 1, \ldots, p.$$

The likelihood ratio test is (5.6) with L_{general} based on the fitted $\hat{\beta}$ and L_{simpler} based on a model with only the intercept, and is compared to a χ^2_p critical value.

The significance of any individual regression coefficient β_j tests

$$H_0 : \beta_j = 0$$

versus

$$H_a : \beta_j \neq 0,$$

and can be tested using the likelihood ratio form (5.6). This is not the test that typically appears in output from statistical packages, however.

Rather, the asymptotically equivalent Wald test, which corresponds to a t-test, is typically used. Recall that asymptotically $\hat{\boldsymbol{\beta}} \sim N(\boldsymbol{\beta}, (X'WX)^{-1})$. An appropriate asymptotic test of H_0 is the Wald test

$$z_j = \frac{\hat{\beta}_j}{\widehat{s.e.}(\hat{\beta}_j)},$$

where $\widehat{s.e.}(\hat{\beta}_j)$ is the square root of the jth diagonal element of $(X'WX)^{-1}$. The correspondence to (2.13) is obvious, with the exact t-distribution in the least squares case replaced by an approximate standard normal distribution. The Wald test generalizes in the straightforward way when testing the more general hypotheses

$$H_0 : \beta_j = \beta_{j0}$$

versus

$$H_a : \beta_j \neq \beta_{j0}$$

based on the statistic

$$z_j = \frac{\hat{\beta}_j - \beta_{j0}}{\widehat{s.e.}(\hat{\beta}_j)}. \tag{5.7}$$

An asymptotic $100 \times (1 - \alpha)\%$ confidence interval for β_j has the form $\hat{\beta}_j \pm z_{\alpha/2}\widehat{s.e.}(\hat{\beta}_j)$.

Wald, likelihood ratio, and score tests of hypotheses are asymptotically equivalent, and tend to be similar when the null hypothesis is true (or almost true) even for smaller samples, but can be appreciably different under the alternative hypothesis. This is not necessarily a problem, however, since if all three tests give small p-values, the inferential implications are effectively the same, whichever test is used.

5.1.4   The Deviance and Lack of Fit

The likelihood ratio statistic (5.6) can be used to assess overall lack of fit of a model by taking the more general model to be the full (saturated) model that has a parameter for each observation, and $\hat{\boldsymbol{\mu}} = \mathbf{y}$. Let $\hat{\boldsymbol{\theta}}^S$ be the value of $\boldsymbol{\theta}$ under the saturated model, and let $\hat{\boldsymbol{\theta}}^M$ be the value under a model M. Then the likelihood ratio statistic is

$$D(\mathbf{y}, \hat{\boldsymbol{\mu}}^{(M)}) = 2\sum_{i=1}^{n} [y_i(\hat{\theta}_i^S - \hat{\theta}_i^M) - b(\hat{\theta}_i^S) + b(\hat{\theta}_i^M)]/a_i(\phi). \tag{5.8}$$

$$\equiv \sum_{i=1}^{n} d_i \tag{5.9}$$

This statistic is known as the *deviance*, as it measures the discrepancy between the observed and fitted values (that is, the overall lack of fit of the model). The deviance for the Gaussian regression model is the residual sum of squares divided by σ^2, although some authors and statistical packages define the deviance in this case to simply be the residual sum of squares. Some authors refer to (5.8) as the *scaled deviance*, using deviance for just the numerator term (by this definition the deviance for the Gaussian regression model is, in fact, the residual sum of squares). D can be used to formally test lack of fit. For the regression models discussed in this book, when the entries in $\boldsymbol{\mu}$ are large enough, D is roughly χ^2 distributed on $n - p - 1$ degrees of freedom.

The other common measure of lack of fit for generalized linear models is the generalized Pearson statistic,

$$X^2 = \sum_{i=1}^{n} \frac{(y_i - \hat{\mu}_i)^2}{\hat{V}(y_i)},$$

where $V(y_i)$ is estimated under the model being assessed for lack of fit (we continue to use the standard notation X^2 to represent the Pearson statistic, but this should not be confused with X, the matrix of predictor values). As was true for the deviance, for the regression models discussed in this book, when the entries in $\boldsymbol{\mu}$ are large enough, X^2 is roughly χ^2 distributed on $n - p - 1$ degrees of freedom.

A standard measure of lack of fit in least squares regression is the R^2. R^2 measures are difficult to define in this context; since sums of squares are not being used to determine closeness of fit, the usual definition of R^2 isn't appropriate. The common approach is to use ordinary least squares as a template to derive R^2 measures (a *pseudo-R^2*) from the available generalized linear model tests. One version is to argue that since the deviance corresponds to the sums of squares in the Gaussian regression model, the "proportion of variability accounted for" in a generalized linear model should be the proportion of the deviance of the intercept-only model M_0 accounted for by the fitted regression model,

$$R_D^2 = 1 - \frac{D(\mathbf{y}, \hat{\boldsymbol{\mu}}^{(M)})}{D(\mathbf{y}, \hat{\boldsymbol{\mu}}^{(M_0)})}. \tag{5.10}$$

5.1.5   Model Selection

Section 3.4 discussed the issues that arise when trying to choose among candidate least squares regression models. The same issues arise when fitting generalized linear models, and analogous methods are available to the data analyst.

Equations (5.6) and (5.8) together provide the framework for a systematic way of investigating certain model selection issues. Consider two nested models $M_1 \subset M_2$ (that is, M_1 is a special case of M_2). By (5.6) and (5.8), the likelihood ratio test of the adequacy of M_1 given M_2 is

$$\begin{aligned} LR(M_1|M_2) &= 2(L_{M_2} - L_{M_1}) \\ &= 2(L_S - L_{M_1}) - 2(L_S - L_{M_2}) \\ &= D(M_1) - D(M_2), \end{aligned} \qquad (5.11)$$

where S refers to the saturated model. That is, the likelihood ratio test is based on the difference in deviance measures for the two models. This can be generalized to a series of nested models $M_1 \subset M_2 \subset \cdots \subset M_k$,

$$D(M_1) = LR(M_1|M_2) + LR(M_2|M_3) + \cdots + LR(M_{k-1}|M_k) + D(M_k).$$

The lack of fit of the simplest model (which might be the one with all slopes equal to zero, for example) is thus partitioned into components assessing the adequacy of simpler models given more general counterparts, and the lack of fit of the most general model. This can be termed an analysis of deviance by analogy with the analysis of variance for Gaussian regression models, with the likelihood ratio tests taking the role of (partial) F-tests. These tests tend to follow the assumed χ^2 distributions reasonably well, even if the underlying deviances do not.

As noted in Section 3.4, however, there are clear limitations to using such tests to choose a set of candidate models. These include the dependence of p-values on sample size (and the associated lack of interpretability of p-values as assessments of the practical importance of the improvement in fit of one model over another) and the inability to compare non-nested models. Once again, information measures can be useful in choosing among models. Recall that the Akaike Information Criterion AIC has the form

$$AIC^* = -2L + 2\nu,$$

where L is the maximized log-likelihood function and ν is the number of estimated parameters in the model (the reason for the * superscript will be apparent in a moment). This is equal to

$$AIC^* = D - 2L_S + 2\nu,$$

where D is the deviance for the model and L_S is the log-likelihood under the saturated model. Since it is only differences in AIC that matter in model selection, the term $2L_S$ can be ignored, implying ordering models on the basis of minimizing the sum of the deviance and twice the number of parameters in the model,

$$AIC = D + 2\nu, \qquad (5.12)$$

and it is this version that we will refer to as AIC. (Note that this is only true if the saturated model is the same for all models being considered. This is true if all of the models are based on the same family of distributions for the random component, but not if different families are being considered. In the latter situation, the AIC^* form, based on the log-likelihood rather than the deviance, should be used.) As AIC tends to lead to overfitting, the corrected AIC criterion, which more strongly penalizes complex models, also can be adapted to generalized linear models, equaling

$$AIC_C = D + 2\nu \left(\frac{n}{n - \nu - 1} \right) = AIC + \frac{2\nu(\nu + 1)}{n - \nu - 1}. \tag{5.13}$$

Although the theory justifying AIC_C is based on Gaussian models, and does not generalize directly to generalized linear models, (5.13) is still intuitively reasonable as an adjustment to AIC that addresses particularly highly parameterized models, although some care must be taken in defining the sample size n in the formula.

5.1.6   Model Checking and Regression Diagnostics

Sections 2.4 and 3.3 described how the assumptions underlying least squares regression are checked. Analogous checks for generalized linear models are clearly also appropriate, and we sketch some of them in this section.

The weighted least squares form for the MLE given in (5.5) is the building block for much of the generalized linear model checking. Broadly speaking, $\mathbf{y}$ and $\hat{\mathbf{y}}$ in least squares diagnostics and plots are replaced with $\mathbf{z}$ and $\hat{\boldsymbol{\eta}}$, respectively, $\hat{\sigma}^2$ is replaced with $\hat{\phi}$, and the hat matrix (defined by analogy with least squares regression) takes the weighted least squares form

$$H = W^{1/2} X (X' W X)^{-1} X' W^{1/2}, \tag{5.14}$$

with the ith diagonal element of H, h_{ii}, providing a measure of the potential of an observation to have a large effect on the fitted regression (that is, the leverage).

Residuals are defined based on χ^2 goodness-of-fit statistics. The *Pearson residuals* are defined as

$$r_i^P = \frac{y_i - \hat{\mu}_i}{\sqrt{\hat{V}(y_i)}},$$

and are simply the raw residuals scaled by the estimated standard deviation of y. Clearly $X^2 = \sum_i (r_i^P)^2$. This relationship suggests a corresponding form of residuals based on the deviance, the *deviance residuals*

$$r_i^D = \text{sign}(y_i - \hat{\mu}_i) \sqrt{d_i},$$

where d_i is defined as in (5.9). The Pearson and deviance residuals do not account for the dependence of the precision of $\hat{\mu}$ as an estimate of μ, so standardized versions analogous to (3.8) can be constructed, as

$$\tilde{r}_i^P = \frac{r_i^P}{\sqrt{\hat{\phi}(1 - h_{ii})}}$$

and

$$\tilde{r}_i^D = \frac{r_i^D}{\sqrt{\hat{\phi}(1 - h_{ii})}},$$

respectively (these are often called *scaled* residuals). When the entries in $\hat{\mu}$ are large (say $\mu_i \geq 3$) the Pearson and deviance residuals will be similar, but they can differ for small values in $\hat{\mu}$. The Pearson residuals are probably the most commonly used residuals, but the deviance residuals (or standardized deviance residuals) are actually preferred, since their distribution is closer to that of least squares residuals. Either version of the standardized residuals, and the leverage values, can be combined to construct an influence measure analogous to Cook's distance (3.9),

$$CD_i = \frac{\tilde{r}_i^2 h_{ii}}{(p + 1)(1 - h_{ii})}.$$

Model checking for a generalized linear model involves checking the three parts of the model (the random component, the systematic component, and the link function), as well as the possibility of unusual and influential observations. It should be standard practice to construct plots of residuals, including plots of residuals versus the estimated linear predictor $\hat{\eta}$ and residuals versus each predictor, with the desired result being the lack of a discernable pattern. An inappropriate mean function (from an incorrect link function, or the need for transformations of the variables) can result in a pattern in the plots. A plot of absolute residuals versus the estimated linear predictor is sometimes useful in highlighting a possibly incorrect variance function. Another useful plot is one of $\mathbf{z}$ versus $\hat{\eta}$, which should look like a straight line (based on the weighted least squares form of the maximum likelihood estimate). We will reserve discussion of addressing errors in the specification of the random component to Section 5.3.

5.2   Poisson Regression

Consider a regression problem based on data $\{x_{1i}, x_{2i}, \ldots, x_{pi}, y_i\}$, where the target variable $\mathbf{y}$ is a count variable. A Poisson regression model takes as the random component a Poisson distribution, $y_i \sim \text{Pois}(\mu_i)$. The Poisson model is

$$f(y; \mu) = \exp[-\mu + y \log \mu - \log y!],$$

(5.1) with $\theta = \log \mu$, $\phi = a(\phi) = 1$, $b(\theta) = \exp(\theta)$, and $c(y) = \log y!$. The systematic component is the usual linear predictor,

$$\eta_i = \beta_0 + \beta_1 x_{1i} + \cdots + \beta_p x_{pi}.$$

The link function could be the identity, $\mu_i = \eta_i$, but then negative values of μ_i would be possible, which is of course not appropriate for a Poisson mean. The usual link used is the logarithm, $\eta_i = \log(\mu_i)$, which is the canonical link for the generalized linear model, and results in a loglinear model. Thus, the Poisson regression model is ultimately

$$y_i \sim \text{Pois}(\exp(\beta_0 + \beta_1 x_{1i} + \cdots + \beta_p x_{pi})). \tag{5.15}$$

Figure 5.1 illustrates the Poisson regression model with one predictor, and can be compared to Figure 2.2, the corresponding figure for the least squares regression model. The shaded bars represent the Poisson probability distributions at three different values of x (only probabilities greater than .001 are plotted). The figure reinforces the fundamental deficiencies in using least squares regression when the Poisson regression model is appropriate. The relationship between $E(y)$ and x is nonlinear, although it can be roughly linear over parts of the range of x. When $E(y)$ is small the distribution of $y|x$ is noticeably asymmetric. For large $E(y)$ the distribution is reasonably symmetric, but the variability around $E(y)$ is also larger. These properties are, of course, immediate implications of the underlying Poisson distribution for $y_i|x_i$.

All of the statistics and diagnostics derived in the last section for generalized linear models apply to Poisson regression models using the appropriate values of θ, ϕ, $b(\cdot)$, and $c(\cdot)$, and after substituting μ_i for $V(y_i)$. Equation (5.15) can be used to estimate the probability distribution of y given a set of values for the predictors, by substituting $\exp(\hat{\eta}_i)$ for μ in the Poisson probability function (4.9). The deviance for a Poisson regression model is

$$D = 2 \sum_{i=1}^{n} \left[y_i \log \left(\frac{y_i}{\hat{\mu}_i} \right) - (y_i - \hat{\mu}_i) \right].$$

If the model includes an intercept term $\sum y_i = \sum \hat{\mu}_i$, and this is simply the likelihood ratio goodness-of-fit statistic G^2 from (4.14).

The loglinear nature of the (canonical) Poisson regression model means that the interpretation of the estimated regression coefficients is more complicated than for the (linear) least squares model, although the basic structure is the same. The intercept term has the following interpretation.

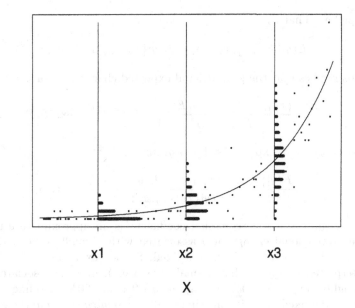

FIGURE 5.1. The Poisson regression model

$\hat{\beta}_0$: The estimate of the logarithm of the expected value of y when the predictors all equal zero.

As was true for least squares regression, this might not have any physical interpretation, since a zero value for the predictors might be meaningless. Note that $\exp(\hat{\beta}_0)$ is the maximum likelihood estimate of the expected value of y when all of the predictors equal zero.

The estimated coefficient for the jth predictor $(j = 1, \ldots, p)$ is interpreted in the following way.

$\hat{\beta}_j$: $\exp(\hat{\beta}_j)$ is the estimated expected multiplicative change in y associated with a one unit change in the jth predicting variable, holding all else in the model fixed.

Since the model for μ is loglinear, the regression coefficients must be exponentiated to assess the associative effect they summarize, which will be multiplicative, rather than additive. Since the absolute magnitude of the effect differs based on the level of y, the absolute change in y corresponding to a one unit change in x_j for a few x_j values of particular interest are sometimes reported (substituting the mean values of the other predictors into the estimated model). For small changes in a predictor variable, an approximation to the multiplicative effect takes a particularly intuitive form. Say $x_j \to x_j + \delta$ for small δ. By the definition of the regression coefficient above, this is associated with an expected multiplicative change in y of

$\exp(\beta_j \delta)$. That is,

$$E(y|\{x_k\}_{k \neq j}, x_j + \delta) = E(y|\{x_k\}_{k \neq j}, x_j) \times \exp(\beta_j \delta).$$

This implies that the proportional expected change in y satisfies

$$\frac{E(y|\{x_k\}_{k \neq j}, x_j + \delta) - E(y|\{x_k\}_{k \neq j}, x_j)}{E(y|\{x_k\}_{k \neq j}, x_j)} = \exp(\beta_j \delta) - 1.$$

But for small δ, $\exp(\beta_j \delta) \approx 1 + \beta_j \delta$, implying

$$\frac{E(y|\{x_k\}_{k \neq j}, x_j + \delta) - E(y|\{x_k\}_{k \neq j}, x_j)}{E(y|\{x_k\}_{k \neq j}, x_j)} \approx \beta_j \delta.$$

Thus, the estimated regression coefficient is an approximate estimate of the proportional change in y associated with a small additive change in x_j, holding all other predictors fixed. So, for example, $\hat{\beta}_1 = .2$ can be interpreted as saying that a small increase, δ, in x_1 is associated with a (roughly) expected $20\delta\%$ change in y (.2 being 20%), holding all else in the model fixed. This is sometimes called a *semielasticity* in the economics literature. If the predictor x_j is itself a logged variable (say $x_j = \log z$, where log is the natural logarithm), the relationship is even simpler. For small δ, $x_j \to x_j + \delta$ corresponds to

$$\log z \to \log z + \delta = \log[\exp(\delta)z] \approx \log[(1 + \delta)z],$$

the logarithm of a proportional change in z of δ. So, for example, if $x_1 = \log z$, $\hat{\beta}_1 = 2$ can be interpreted as saying that a 1% increase in z is associated with a (roughly) expected 2% change in y, holding all else in the model fixed. This proportional/proportional relationship is called an *elasticity*.

One important advantage of using a Poisson regression model, rather than a least squares model (even if the latter model is a weighted least squares loglinear model) is that a prediction from the model is not merely of the expected y, but of the entire Poisson distribution with that mean. This is a discrete distribution, of course, meaning that it allows direct estimation of probabilities that might be of interest, such as $P(y = 0|\mathbf{x} = \mathbf{x}_0)$ for some $\mathbf{x}_0$, based on the observed $\hat{\boldsymbol{\beta}}$. It should be noted that such a probability estimate ignores the variability associated with estimating $\boldsymbol{\beta}$, but this is relatively small for reasonable sample sizes.

Monthly Tornado Mortality Statistics (continued)

We will now examine the monthly tornado statistics data previously analyzed using least squares regression in Chapters 2 and 3 using a more appropriate Poisson regression model. Recall that we are attempting to

Predictors	AIC_C difference
(a) Models using only main effects	
None	650.91
Killer tornadoes	117.98
Tornadoes	486.11
Year	574.45
Month	237.74
Killer tornadoes, Tornadoes	109.88
Killer tornadoes, Year	119.72
Killer tornadoes, Month	5.83
Tornadoes, Year	443.21
Tornadoes, Month	92.70
Year, Month	161.73
Killer tornadoes, Tornadoes, Year	107.87
Killer tornadoes, Tornadoes, Month	1.29
Killer tornadoes, Year, Month	5.53
Tornadoes, Year, Month	63.60
Killer tornadoes, Tornadoes, Year, Month	0.00
(b) Different slopes by month	
Killer tornadoes, Month	63.69
Killer tornadoes × Month	0.00

TABLE 5.1. AIC_C differences of models for tornado mortality data.

model the number of tornado-related deaths nationally from the number of killer tornadoes, tornadoes, year, and month (possibly dichotomized into months during tornado season and not during tornado season), allowing for different slopes by time of year. The month variable is a categorical predictor, and is handled the same way such predictors are used in least squares regression models (see Section 3.2).

Table 5.1 gives AIC_C differences of Poisson regression models for these data, and is directly comparable to Table 3.2 (on page 137). We don't bother to include AIC values in this table, based on the better performance of AIC_C when the number of parameters in the model is large relative to the number of observations. Note that in this context n is the total number of tornado-related deaths, not the number of months.

It shouldn't be a surprise that the model comparisons are considerably different based on Poisson regression than when based on least squares regression. Consider first models using only main effects. It is clear that the number of killer tornadoes and the month are necessary in any model for tornado-related deaths. The addition of the number of tornadoes and year are weakly indicated, with AIC_C choosing as the best model the one including both. Although these variables are statistically significant predictors (the Wald statistic for number of tornadoes is 2.51, while the p-value for the likelihood ratio statistic for year is .05), a model for tornado-related deaths that depends on the number of tornadoes that did not lead to deaths seems counterintuitive. Similarly, from a predictive point of view using year as a predictor is problematic. Since the gain (according to AIC_C) is not very large, we will focus on the model based on only the number of killer tornadoes and month. Output for this model follows.

Predictor	Coef	s.e.	Wald	p
Constant	0.206	0.149	1.38	0.167
Killer tornadoes	0.254	0.018	13.89	0.000
Month				
January	0.212	0.248	0.85	0.393
February	1.730	0.194	8.91	0.000
March	0.979	0.195	5.02	0.000
April	0.630	0.217	2.90	0.004
May	1.344	0.183	7.36	0.000
June	-0.020	0.372	-0.05	0.958
July	-0.398	0.431	-0.92	0.356
August	-1.035	0.662	-1.56	0.118
September	-0.708	0.546	-1.30	0.195
October	-1.053	0.661	-1.59	0.111
November	-1.053	0.661	-1.59	0.111
December	-0.629	0.547	-1.15	0.250

The deviance for this model is 140.07 on 35 degrees of freedom. Comparing this to the null deviance 810.34 on 47 degrees of freedom gives an overall likelihood ratio test equaling 670.27 on 12 degrees of freedom, obviously overwhelmingly statistically significant (and a deviance-based "R^2" of $1 - 140.07/810.34 = 82.7\%$). The coefficient for the number of killer tornadoes is highly significant, and implies that, given the month, each additional killer tornado is associated with an increase in the expected number of tornado-related deaths of 28.9% ($= \exp(.254) - 1$). The regression coefficients clearly identify the "tornado season" effect discussed in Chapter 3, with January through May associated with increased tornado-related deaths (given the number of killer tornadoes) and June through December associated with decreased deaths; February, March, April, and May all have coefficients significantly greater than zero. Figure 5.2 represents the model graphically, by showing the fitted relationship between deaths and killer tornadoes for each month. The lines for January through May are identified on the plot, and the multiplicative pattern inherent in the loglinear model is apparent. In order to avoid misleading impressions,

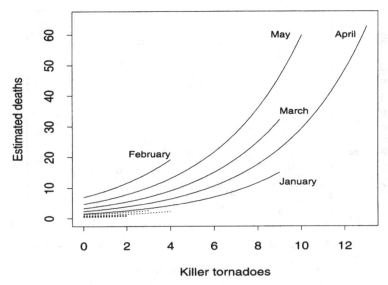

FIGURE 5.2. Estimated tornado-related deaths by month based on the Poisson regression model using the number of killer tornadoes and the month.

we have only given lines out to the maximum observed number of killer tornadoes for each month, since there is a strong seasonal pattern in that variable.

While this model is appealing in its simplicity, there is evidence that it is not an adequate representation of the data. Since many of the estimated fitted values are small (more than half of the observations have fitted values less than two), the deviance should not be used as a formal χ^2 goodness-of-fit test, but the fact that it equals four times its degrees of freedom is still suggestive of lack of fit (the Pearson statistic X^2 is similar, as it equals 132.1). In fact, the ratio X^2/df is often used as a multiplicative correction factor for apparent overdispersion in Poisson regression data (see Section 5.3). Figure 5.3 gives a plot of the absolute scaled deviance residuals versus the estimated linear predictor for this model, and strongly suggests increasing variance with the mean (beyond the expected direct mean/variance relationship for the Poisson model).

Two possible approaches suggest themselves when faced with this lack of fit. First, we could hypothesize that the lack of fit comes from using a model that is missing important predictors, and we could try to determine and include those predictors. This is what we will, in fact, do here, and is almost invariably the approach taken when fitting loglinear (Poisson) models to contingency tables (as will be discussed in Chapters 6–8). A second approach is to hypothesize that the linear predictor is a reasonable repre-

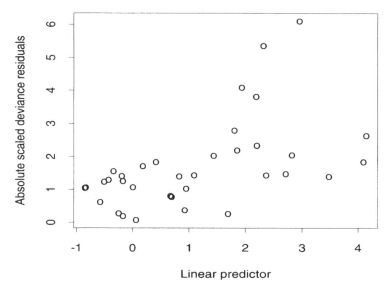

FIGURE 5.3. Plot of absolute scaled deviance residuals versus estimated linear predictor for the Poisson regression model using the number of killer tornadoes and the month.

sentation of the systematic component, but that the random component is incorrect (because of an incorrect probability family, for example, or an incorrect variance function). We will defer discussion of this approach to Section 5.3.

Our earlier discussion of these data suggests how we might enrich this model. As was noted on page 49, a reasonable generalization of the purely additive model is to include an interaction between the month and the number of killer tornadoes, allowing for different slopes for each month (in addition to the different intercepts in the noninteraction model). Table 5.1 shows that this is a good idea, as the value of AIC_C for the model including the interaction is 63.7 less than that for the model without it, based on a deviance of 51.46 on 24 degrees of freedom. The likelihood ratio test for the significance of the Killer tornadoes $\times$ Month interaction is 88.61 on 11 degrees of freedom, which is highly statistically significant.

Figure 5.4 graphs the fitted model, and can be compared to Figure 5.2. The plots are surprisingly similar (given the apparent superiority of the different slopes model), with the only major difference being the much steeper increase in expected deaths in February. Figure 5.5 provides a clue as to what is going on. The regression diagnostics show that in a sense the data are being pushed about as far as they can be. The different slopes model estimates a separate Poisson regression of deaths on killer tornadoes for

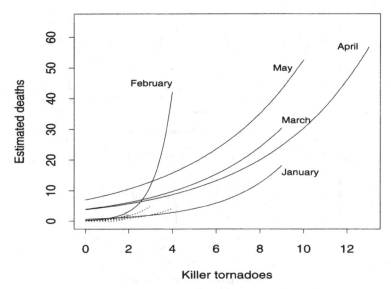

FIGURE 5.4. Estimated tornado-related deaths by month based on the Poisson regression model using different slopes and intercepts for the number of killer tornadoes by month.

each month on the basis of only four data points, which means that a good deal of faith is being put in a few observations. Many of the observations have leverage values that are very close to one, reflecting the paucity of data (one observation has leverage exactly equal to one, which is why the scaled deviance residual, which equals 0/0, is not given).

Two observations show up as being particularly influential — observations 26 (February 1998) and 37 (January 1999). There is also one apparent outlier (observation 17 — May 1997) that has a smaller Cook's distance. Omitting these observations has a strong effect on model choice. Consider, for example, February 1998, the very unusual month that was discussed on page 42. If this observation is omitted, the evidence for the need for different slopes largely disappears, as the AIC_C for the model without the interaction term is larger than that for the model including the interaction by only 6.1 (the intercept term for the February line also gets smaller, as would be expected). That is, the curve for February in Figure 5.4 is being driven to a great extent by a single month (indeed, a single day), and should not be trusted very much. Omitting the other observations decreases the intercept for January and May, respectively. All of the models have similar slopes for the number of killer tornadoes, implying a 25–30% increase in the number of deaths associated with an additional killer tornado, given month. Even the model with the three unusual observations omitted is not

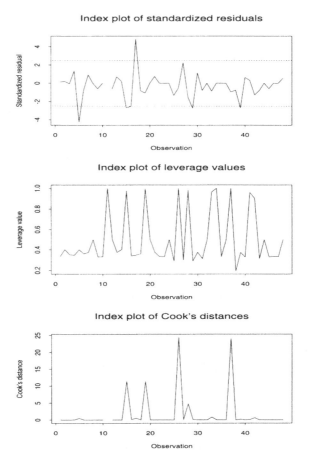

FIGURE 5.5. Index plots of scaled deviance residuals (top plot), leverage values (middle plot), and Cook's distances (bottom plot) for different slopes Poisson regression model.

perfect, as there is still some evidence of overdispersion (see Figure 5.6). Still, the basic pattern seems clear. March, April, and May are the highest risk months for tornado-related deaths, with the occasional January and February also potentially problematic. Further, when meteorological conditions are favorable for multiple killer tornadoes, the death toll rises quickly, and multiplicatively.

After all this, there is an obvious problem with all of these models that we've ignored. If there are no killer tornadoes in a month, there cannot be any tornado-related deaths; that is, the estimated number of deaths should be zero when the number of killer tornadoes is zero. Figures 5.2 and 5.4 clearly demonstrate that the models considered so far do not satisfy this condition. This is not a straightforward condition to impose on

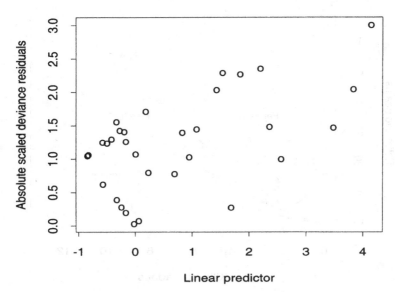

FIGURE 5.6. Plot of absolute scaled deviance residuals versus estimated linear predictor for the Poisson regression model using the number of killer tornadoes and the month and omitting three unusual observations.

the model, since the loglinear form makes any estimated mean number of deaths positive. One somewhat *ad hoc* approach is as follows. First, omit all of the months with zero killer tornadoes, since they do not provide any useful information. Now, fit the Poisson regression model using log(Killer tornadoes) instead of Killer tornadoes as a predictor. As the number of killer tornadoes approaches zero, its log approaches $-\infty$, and (since the estimated slope is positive) the estimated number of deaths approaches $\exp(-\infty)$, or zero.

Figure 5.7 gives the resultant model using all of the months with nonzero killer tornadoes, and can be compared to Figure 5.2. The estimated expected number of deaths when the number of killer tornadoes is zero is effectively zero for all months, as is required, but otherwise the patterns in the two plots are similar. Since the predictor is in the logged scale, the common slope for all of the months, 1.35, is an elasticity, implying that given the month, a 1% increase in the number of killer tornadoes is associated with roughly a 1.35% increase in the number of tornado-related deaths.

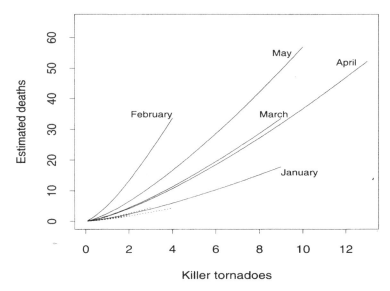

FIGURE 5.7. Estimated tornado-related deaths by month based on the Poisson regression model using the logged number of killer tornadoes and the month.

A common complication when faced with count data regression is that the actual measure of interest is the rate of occurrence of some event (rather than the number of occurrences of the event), where the rate is appropriately standardized. For example, when analyzing marriages by state, it is marriage rate (marriages per 100,000 population, say) that is comparable across states, not the actual number of marriages, because of differing state population sizes. That is, the appropriate model is

$$y_i \sim \text{Pois}(k_i \times \exp(\beta_0 + \beta_1 x_{1i} + \cdots + \beta_p x_{pi})), \qquad (5.16)$$

where k_i is the standardizing value (such as population) for the ith observation, and

$$\exp(\beta_0 + \beta_1 x_{1i} + \cdots + \beta_p x_{pi})$$

now represents the mean rate of occurrence, rather than the mean occurrence. An equivalent form for (5.16) is

$$y_i \sim Pois\{\exp[\beta_0 + \beta_1 x_{1i} + \cdots + \beta_p x_{pi} + \log(k_i)]\}.$$

That is, a Poisson rate model is fit by including $\log(k_i)$ (the so-called *offset*) as a predictor in the model, and forcing its coefficient to equal one.

Offsets can, in fact, be used more generally in generalized linear modeling. For example, consider a hypothesis that a slope coefficient β_j equals a specified value β_{j0}. This condition can be imposed on a model by including $\beta_{j0}x_j$ as an offset. The difference between the deviance values with and without the offset is then the likelihood ratio test of the hypothesis.

Fatal Train Accidents in Great Britain

It is ironic that while fatal accidents involving private automobiles are relatively common, and fatal accidents involving trains or airplanes are rare, it is the latter accidents that attract a great deal of public attention, almost invariably resulting in public inquiries into the causes of the accident. One aspect of such inquiries is to propose potential safety measures that could be taken to prevent future accidents (or reduce the seriousness of their consequences). This requires estimates of the number of injuries and deaths that could be prevented using such measures, which requires projections of these numbers into the future, based on past experience.

Fatal train accidents on the national rail system of Great Britain during the period 1967–1997 provide an example of this type of modeling (Evans, 2000, with data given in the file uktrainacc). There were a total of 75 fatal train accidents during this 31-year period. We could attempt to model the number of accidents each year, but this ignores the scale of railroad activity in a given year; more activity should naturally lead to a higher chance of an accident. This can be accounted for by examining the rate of accidents per billion train-kilometers, where a train-kilometer corresponds to one train covering one kilometer during the year, making log(Train-kilometers) an offset variable.

Figure 5.8 gives the observed accidents per billion train-kilometers by year. Table 5.2 summarizes three different Poisson fits of a time trend for these data. The first model is a Poisson regression of the total accident rate on year, and is given in the plot by the solid line (the regressions actually use the time index as a predictor, which is equivalent to years since 1966). The regression is highly significant (the likelihood ratio statistic $LR = 10.05$, with $p < .001$), and the deviance is 27.6 on 29 degrees of freedom, suggesting no lack of fit (with the expected number of accidents being small for many years, the χ^2 approximation for the deviance is suspect here, so we will not report a p-value for this deviance). The slope coefficient implies that each passing year is associated with a proportional change of $\exp(-.041) - 1 = -.04$ in the accident rate, or a 4% decrease.

Accidents actually can be classified into two broad categories, those that would have been preventable using automatic train protection (ATP, modern computer-based systems intended to prevent human error in passing signals at danger, overspeeding, or overrunning buffers), and those that would not have been prevented using such systems. Preventable accidents

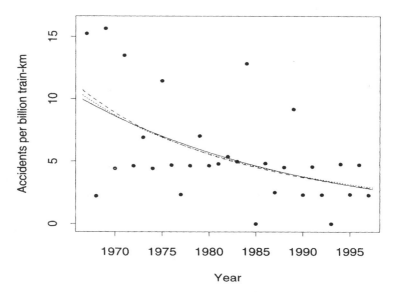

FIGURE 5.8. Plot of fatal train accident rate per billion train-kilometers in Great Britain by year. Solid line is the estimated accident rate based on the annual total accidents Poisson regression; the dotted line is the estimated accident rate based on separate annual Poisson regressions based on type of accident; the dashed line is the estimated accident rate based on separate grouped-year Poisson regressions based on type of accident.

are split into ones caused by signals passed at danger (SPADs) and other ATP-preventable causes. It is reasonable to expect that the time trend in accident rates could be different for these three different types of accidents, so in the second model separate Poisson regressions were fit to each of these accident rates, with the expected total accident rate then being the sum of the separate accident rates. This leads to the dotted curve in Figure 5.8. While SPAD-preventable and nonpreventable accidents are both apparently decreasing at a slower annual rate (4.3% and 2.9%, respectively), other preventable accidents are decreasing at a much faster rate (11.6% annually). Indeed, the last eight accidents in the data set come from the two former classes, while the last accident due to a preventable cause other than SPAD was in early 1991, almost seven years before the end of the study. In any event, the estimated total accident rate based on the three separate models is obviously similar to that from the single model.

Evans (2000) noted that the number of accidents in each year is often small, and therefore aggregated the data to increase the counts. This is the third model fit here. The data are combined into six time periods, 1967–1971, 1972–1976, 1977–1981, 1982–1986, 1987–1991, and 1992–1997, and

Target variable	Constant	Year	(s.e.)	LR	p
(a)					
Total accidents	2.321	−.041	(.013)	10.05	< .001
(b)					
SPAD preventable accidents	1.293	−.044	(.023)	3.90	.048
Other preventable accidents	0.806	−.112	(.051)	6.19	.013
Nonpreventable accidents	1.553	−.029	(.018)	2.72	.099
(c)					
SPAD preventable accidents	0.820	−.021	(.024)	0.75	.386
Other preventable accidents	0.911	−.123	(.055)	6.89	.009
Nonpreventable accidents	1.841	−.044	(.018)	6.50	.011

TABLE 5.2. Summaries of three models fit to British railway accident rate data. (a) Single regression for total accident rate on annual data. (b) Separate regressions for SPAD preventable, other preventable, and nonpreventable accident rates on annual data. (c) Separate regressions for SPAD preventable, other preventable, and nonpreventable accident rates on grouped year data.

separate Poisson regressions are fit to these data for the three types of accidents. The fitted regressions are similar to the ones on the annual data, although now there is no significant evidence of a declining trend in the SPAD-related accident rate. The estimated total accident rate implied by these models is given by the dashed line in Figure 5.8. Again, the implications of this model are similar to those of the other models, particularly at the right side of the plot (which is where projections would be made). Having said this, there are differences. Figure 5.9 gives forecasted expected total accident rates using the three models. The single total accident rate model is more optimistic than the other two models, ultimately predicting a 25% lower accident rate in 2027. Of course, putting much faith in *any* prediction 30 years ahead is probably not a very good thing to do!

5.3   Overdispersion

The Poisson regression model imposes the strong assumption of equality of means and variances, an assumption that is often violated. Just as was true in the univariate problems discussed in Section 4.5, the most common situation is overdispersion, often as a result of unobserved heterogeneity. Recall that a plot of absolute (scaled deviance) residuals versus the linear predictor can help identify the presence of overdispersion. Score tests are easy to construct to test for it as well, with the form of the test depending on

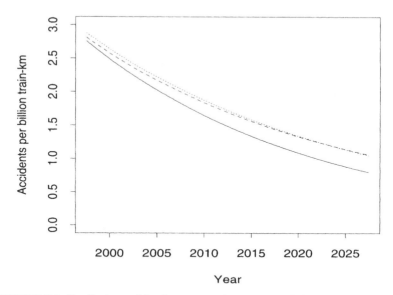

FIGURE 5.9. Predictions of fatal train accident rate per billion train-kilometers in Great Britain by year based on three Poisson regression models.

the form of overdispersion hypothesized. One possibility is to hypothesize that the actual variance satisfies

$$V(y_i) = \mu_i(1 + \alpha), \tag{5.17}$$

which is a constant multiplicative inflation of the Poisson variance. Let

$$u_i = \frac{(y_i - \hat{\mu}_i)^2 - y_i}{\hat{\mu}_i}.$$

The score test z_1 for $\alpha = 0$ based on a mixed Poisson model with this variance structure is the ordinary t-test of $E(u) = 0$, or equivalently the t-test for the intercept in a regression of $\mathbf{u}$ on only a column of ones. An alternative form for the variance is

$$V(y_i) = \mu_i(1 + \alpha\mu_i), \tag{5.18}$$

where the variance inflation is a quadratic function of the mean. The score test z_2 for $\alpha = 0$ based on a mixed Poisson model with this variance structure is the t-test for the significance of the slope in a regression without the constant term of $\mathbf{u}$ on $\hat{\boldsymbol{\mu}}$.

Assuming the specification of the regression relationship is correct (that is, the loglinear model using the chosen predictors is appropriate), the

Poisson-based coefficients $\hat{\beta}$ are consistent even if the variance is misspecified, and are usually not very inefficient. Not accounting for overdispersion leads to misleading inference, however, since the estimated variance of the estimated coefficients is too small, making confidence intervals too narrow and significance tests anticonservative. There are several approaches to addressing this problem.

5.3.1   The Robust Sandwich Covariance Estimator

As noted above, Poisson regression-based coefficients $\hat{\beta}$ are consistent even if the variance is misspecified, assuming the systematic component is correct. More specifically, the estimator is asymptotically normally distributed with mean β and variance

$$V(\hat{\beta}) = (X'\text{diag}(\mu_i)X)^{-1}(X'\text{diag}[V(y_i)]X)(X'\text{diag}(\mu_i)X)^{-1}.$$

A consistent estimator of this variance is the so-called *robust sandwich estimator*, which in this case has the form

$$\hat{V}(\hat{\beta}) = (X'\text{diag}(\hat{\mu}_i)X)^{-1}(X'\text{diag}[(y_i - \hat{\mu}_i)^2]X)(X'\text{diag}(\hat{\mu}_i)X)^{-1}. \quad (5.19)$$

This variance estimate then can be used to construct intervals and tests for the regression coefficients that will be asymptotically correct.

Since (5.19) provides a consistent estimate of variance in broad circumstances, it is tempting to suggest that it be used as the primary estimate in all situations. This would be a mistake. The sandwich estimate can be very inefficient relative to estimates based on the correct parametric model. These highly variable estimates of the variance, when used in inference methods, can lead to confidence intervals and tests that are anticonservative and not trustworthy. If appropriate parametric forms for $V(y)$ are available, these are preferable to the use of the sandwich estimate.

5.3.2   Quasi-Likelihood Estimation

If a specific form for $V(y_i)$ is hypothesized, but a distributional form for y_i is not, methods similar to maximum likelihood can be used to derive parameter estimates that are reasonably efficient. This approach is termed *quasi-likelihood*. Say $E(\mathbf{y}) = \boldsymbol{\mu}$ and $V(\mathbf{y}) = \sigma^2 V(\boldsymbol{\mu})$, where σ^2 may be unknown and $V(\boldsymbol{\mu})$ is a matrix of known functions. For independent observations this is a diagonal matrix. The estimator solves the quasi-score equations,

$$\sum_{i=1}^{n} \left[\frac{y_i - \mu_i}{\sigma^2 V(\mu_i)} \right] \left(\frac{\partial \mu_i}{\partial \eta_i} \right) \mathbf{x}_i = \mathbf{0}.$$

These equations have the same form as the score equations for maximum likelihood estimation in the generalized linear model, but the maximum quasi-likelihood estimator is not an MLE unless the true distribution of y_i is a member of the exponential family (since in that case the variance function characterizes the distribution).

The variance form (5.17), a simple multiplicative inflation over the equality of mean and variance, corresponds to taking $\sigma^2 = 1 + \alpha$ and $V(\mu_i) = \mu_i$. The quasi-score equations in this case are identical to the Poisson score equations, so the quasi-likelihood parameter estimates are identical to the Poisson MLEs. Since

$$\frac{V(y_i)}{\mu_i} = E\left[\frac{(y_i - \mu_i)^2}{\mu_i}\right] = 1 + \alpha,$$

a moment-based estimate of $1 + \alpha$ is

$$\widehat{1 + \alpha} = \frac{1}{n - p - 1} \sum_{i=1}^{n} \frac{(y_i - \hat{\mu}_i)^2}{\hat{\mu}_i},$$

which is simply the Pearson statistic X^2 divided by its degrees of freedom. Thus, quasi-likelihood in this case translates into the particularly simple technique of dividing the Wald statistics from the standard Poisson regression output by $\sqrt{\widehat{1 + \alpha}} = \sqrt{X^2/(n - p - 1)}$ (and extending the upper and lower limits of the standard confidence intervals by that same factor). The significance of a categorical predictor with $d + 1$ categories can be tested by forming an approximate F-statistic

$$F = \frac{\text{Change in deviance}/d}{\widehat{1 + \alpha}}$$

referenced to an $F_{d, n-p-1}$ distribution, where the numerator is the change in the deviance from omitting the predictor divided by the number of free parameters that define it. Tests based on these F-statistics can be used to form an analysis of deviance table, by analogy with the usual analysis of variance in least squares regression.

Even though the quasi-likelihood approach is not based on maximizing an explicit likelihood (although the quasi-likelihood that is maximized has properties similar to a likelihood function), model selection criteria can be easily adapted to the (5.17) situation. The *quasi-AIC* criterion $QAIC$ takes the form

$$QAIC = \frac{-2L}{1 + \hat{\alpha}} + 2\nu,$$

while the corresponding bias-corrected version is

$$QAIC_C = QAIC + \frac{2\nu(\nu + 1)}{n^* - \nu - 1}, \tag{5.20}$$

where n^* is the total number of counts (recall that ν is the total number of parameters estimated in the model, which would include α here). Some care is needed in applying these criteria for model selection. All of the $QAIC$ (or $QAIC_C$) values should be calculated using the same value of $\hat{\alpha}$, or else the values are not comparable from model to model. A reasonable strategy is to determine $\hat{\alpha}$ from the most complex model available, and then calculate the model selection criterion for each model using that value. Alternatively, $\hat{\alpha}$ could be chosen based on the "best" Poisson regression model according to AIC (or AIC_C).

Quasi-likelihood also can be applied assuming quadratic overdispersion, as in (5.18). This does not result in the same parameter estimates as Poisson maximum likelihood. This variance assumption corresponds to the variance of the negative binomial distribution, so we postpone treatment of it to the discussion of negative binomial regression in Section 5.4.1.

Wildcat Strikes in a Coal Mining Company

A wildcat strike is a strike that originates spontaneously, without a formal vote or without the sanction of a union. Being able to estimate the distribution of wildcat strikes as a function of known characteristics of the workplace is clearly useful for management. Marsh and Mukhopadhyay (1999) provide data from a company's coal mining operations relating the number of wildcat strikes in different mines to various factors, including whether or not the mine is unionized, the size of the workforce, the number of grievances filed, and whether or not shift rotation is permitted at the mine, which are given in the file wildcat.

Table 5.3 summarizes the results of a Poisson regression fit to these data (the workforce data are quite long right-tailed, so logged workforce is used as a predictor). Given that all else in the model is held fixed, one additional grievance is associated with an expected 1% increase in wildcat strikes, the availability of shift rotation is associated with a 58.2% decrease in wildcat strikes, a unionized mine is associated with expected wildcat strikes more than seven times higher than those in a nonunionized mine, and a 1% increase in the workforce is associated with a .19% increase in expected wildcat strikes. Given that in a nonunion mine without a general contract any strike is considered a wildcat strike, we might have expected a decrease in expected wildcat strikes with unionization, but that is not the case (perhaps the presence of a union increases the general level of labor activism in a mine); otherwise, the patterns are as would be expected.

Each of the coefficients is highly statistically significant based on the Poisson regression, and the model with all four predictors has the lowest AIC_C value. Unfortunately, the Poisson model does not fit the data very well. The deviance $D = 491.0$ on 158 degrees of freedom; since more than 60% of the observations have fitted counts less than three it is not appropriate to use this for a formal test of goodness-of-fit, but it is still a worrisome indicator. Figure 5.10 strongly suggests overdispersion, with

	Intercept	Grievances	Rotate	Union	Logged workforce
Coefficient	−1.483	0.010	−0.872	1.962	0.185
(a) Poisson maximum likelihood					
Wald	−4.48	6.27	−8.04	8.03	2.80
p	< .001	< .001	< .001	< .001	.005
(b) Robust sandwich estimator					
Wald	−2.31	2.91	−4.24	4.30	1.13
p	.021	.004	< .001	< .001	.258
(c) Quasi-likelihood using (5.17)					
Wald	−2.58	3.61	−4.63	4.63	1.61
p	.010	< .001	< .001	< .001	.107

TABLE 5.3. Regression estimation for wildcat strike data. Coefficients are based on a Poisson regression model, with Wald statistics based on Poisson maximum likelihood, the robust sandwich estimator, and quasi-likelihood using (5.17).

variability increasing with expected number of strikes. The score tests for overdispersion support the presence of overdispersion, with $z_1 = 4.67$ and $z_2 = 6.38$, respectively.

Table 5.3 describes the changes in inference if the overdispersion is addressed. The robust sandwich estimates of $\widehat{s.e.}(\hat{\beta})$ are all larger than the Poisson-based estimates, as would be expected, resulting in smaller (and less significant) Wald statistics. Logged workforce is now considered an unnecessary predictor. The results based on (5.17), where the Poisson Wald statistics are divided by $\sqrt{X^2/(n - p - 1)} = 1.74$, are similar, although the significance is assessed as stronger (with logged workforce almost significant at a .10 level).

We might very well leave the analysis there, but careful examination of residuals shows that there's more to the story. Figure 5.11 is a boxplot of the scaled deviance residuals from the Poisson regression fit, separated by whether or not the mine was unionized. It is apparent that the dispersion is very different in these two types of mines, with unionized mines far more variable than nonunionized ones. This suggests splitting the mines into these two groups and analyzing them separately (fitting an interaction model that allows for different slopes in the two groups won't necessarily address this problem, since the Poisson model doesn't allow for different variance structures in the two groups).

Table 5.4 summarizes the fits to the two types of mines. Consider first the nonunion mines. There are no grievances in nonunion mines (since grievances are filed with the union), so that variable does not appear in the model (this might well have suggested analyzing the two types of mines separately from the start). Given workforce size, shift rotation availability is associated with 88% lower expected wildcat strikes, while a 1% increase

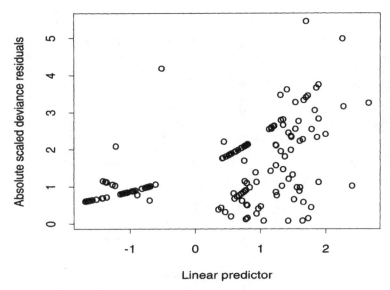

FIGURE 5.10. Plot of absolute scaled deviance residuals versus linear predictor for Poisson regression fit to wildcat strikes data.

in workforce is associated with a 2% increase in wildcat strikes given the availability of shift rotation. There is no evidence of overdispersion in the data ($z_1 = -1.82$ and $z_2 = -0.82$), and no evidence of lack of fit ($D = 29.7$ and $X^2 = 49.5$, both on 55 degrees of freedom). Thus, it would seem a prudent strategy for the mining company to allow shift rotation in nonunion mines, and to be aware that larger mines are more prone to wildcat strikes. Overall, nonunion mines are not very prone to wildcat strikes, as the largest estimated expected number of strikes is only 3.3.

The situation in unionized mines is very different. Given the other predictors, each additional grievance is associated with a 1.2% increase in expected wildcat strikes, and shift rotation availability is associated with a 58% decrease in expected strikes. Logged workforce size has no predictive power for the number of wildcat strikes in any of the models. Overdispersion is apparent here ($D = 423.5$ on 101 degrees of freedom, $z_1 = 5.88$, and $z_2 = 7.56$), and both of the overdispersion-based tests lead to similar inferences (the quasi-likelihood tests are based on a scale factor of $\sqrt{X^2/(n - p - 1)} = 1.91$). Figure 5.12 indicates a potential problem, however. This is a plot of the absolute scaled Pearson residuals based on the model that drops logged workforce, accounting for (5.17), versus the linear predictor, and should have no structure (with the values centering on one in the vertical scale). There is some evidence of widening from left to right, but this is to a large extent because of one unusual observation. This mine

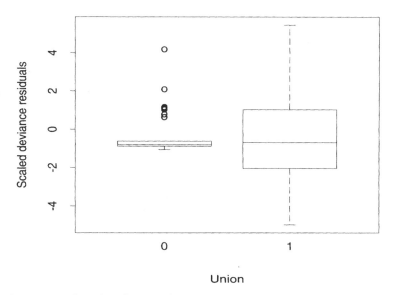

FIGURE 5.11. Boxplot of scaled deviance residuals separated by unionization of mine for Poisson regression fit to wildcat strikes data.

had the largest number of wildcat strikes (22), despite only a moderate number of grievances. If this observation is omitted, logged workforce size is now a statistically significant predictor in the Poisson regression, but the tests corrected for overdispersion are relatively unchanged from before (results not given here). Thus, shift rotation in unionized mines seems to be a good idea, and grievances can possibly be viewed as an "early warning sign" of labor unrest. On the other hand, the size of the workforce seems to have little relationship to wildcat strikes. Higher rates of unrest occurred at unionized mines, as the largest estimated expected number of strikes is roughly 13.

5.4   Non-Poisson Parametric Regression Models

The corrections for non-Poisson behavior discussed in the previous section suffer from several weaknesses. While the Poisson MLE $\hat{\beta}$ is consistent, it is inefficient. This means that fitting a correct parametric model yields estimates that are more precise than those based on only moment assumptions (and could lead to appreciably different interpretations of the relationship

	Intercept	Grievances	Rotate	Logged workforce
Nonunionized mines				
Coefficient	−9.433		−2.109	2.028
Wald	−5.09		−3.37	5.07
p	< .001		< .001	< .001
Unionized mines				
Coefficient	1.039	0.012	−0.878	0.075
(a) Poisson maximum likelihood				
Wald	3.07	7.12	−7.85	1.11
p	.002	< .001	< .001	.267
(b) Robust sandwich estimator				
Wald	1.25	3.36	−4.11	0.46
p	.211	< .001	< .001	.646
(c) Quasi-likelihood using (5.17)				
Wald	1.60	3.72	−4.11	0.58
p	.110	< .001	< .001	.562

TABLE 5.4. Regression estimation for wildcat strike data, separating nonunionized mines from unionized ones.

between the target and a predictor, and appreciably different predictions). A parametric regression model also allows predictive probability statements of the form, "What is the estimated probability that the target exceeds y_0 for a given set of predictor values?". In this section we generalize the negative binomial model of Section 4.5.2, the zero-inflated Poisson model of Section 4.5.1, and the zero-truncated Poisson model of Section 4.6 to the regression situation.

5.4.1 Negative Binomial Regression

The most common parametric model for overdispersion, the negative binomial distribution, generalizes in a straightforward way to the regression situation. The probability distribution for the negative binomial is

$$f(y_i; \mu_i, \nu) = \frac{\Gamma(y_i + \nu)}{y_i!\Gamma(\nu)} \left(\frac{\nu}{\nu + \mu_i} \right)^\nu \left(\frac{\mu_i}{\nu + \mu_i} \right)^{y_i},$$

which is (4.16), allowing for different means μ_i for each y_i. The means are based on the logarithmic link, $\mu = \exp(X\beta)$. The model leads to overdispersion of the quadratic form (5.18), where $\alpha = 1/\nu$. The negative binomial parameters β and α can be estimated using maximum likelihood. The as-

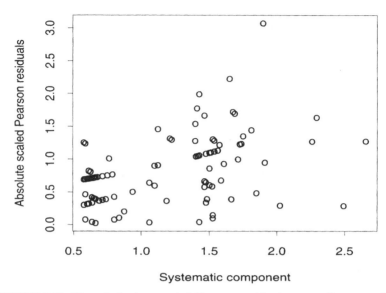

FIGURE 5.12. Plot of absolute scaled Pearson residuals versus linear predictor for quasi-likelihood regression fit based on (5.17) to unionized mines portion of wildcat strikes data.

ymptotic variance of $\hat{\boldsymbol{\beta}}$ can be estimated using

$$\hat{V}(\hat{\boldsymbol{\beta}}) = \left(X' \text{diag} \left[\frac{\hat{\mu}_i}{1 + \hat{\alpha}\hat{\mu}_i} \right] X \right)^{-1}.$$

The score test z_2 described in Section 5.3 is an appropriate test for the negative binomial model versus the Poisson model, since (5.18) corresponds to the negative binomial variance function. Parametric tests are also possible. The likelihood ratio test of $\alpha = 0$ is twice the difference between the log-likelihood for the negative binomial model and the log-likelihood for the Poisson model, and is compared to a reference χ_1^2 distribution. Since $\alpha \geq 0$, the test is one-sided, so the actual tail probability is one-half the value from the χ_1^2 distribution (equivalently, the critical value for the test at a .05 level (for example), would be the value corresponding to the 90% point in the χ_1^2 distribution, or 2.71). Similarly, the Wald test for $\alpha = 0$ is a one-sided test, with a .05 level test based on the 95% point of the standard normal distribution, or 1.645. Finally, AIC or AIC_C can be used to assess the benefit of the negative binomial model over the Poisson, as well as to help choose among candidate models. Recall that when comparing models based on different distributional families, all of the constants in the log-likelihood normally ignored must be included in construction of AIC or AIC_C to make these measures meaningful. For the same reason, AIC

and AIC_C must be based on the actual log-likelihood, not the deviance, because the saturated model for two different distributional families can be different.

Voting Irregularities in the 2000 Presidential Election

The 2000 Presidential election, which pitted Republican Texas Governor George W. Bush against Democratic Vice President Al Gore was almost certainly the closest and most contentious in history. Not only were the popular votes of the two candidates almost equal to each other (with election winner Governor Bush actually having fewer popular votes), but the election ultimately came down to whichever candidate won the 25 electoral votes of Florida. Allegations of voting irregularities in the state were made almost immediately, and a recount of ballots was eventually begun by order of the Florida Supreme Court. Vice President Gore did not concede the election until after a U.S. Supreme Court ruling effectively ended recount efforts, five weeks after Election Day.

One alleged irregularity that gained a great deal of national attention occurred in Palm Beach County. Palm Beach is heavily Democratic and politically liberal. Despite this, extremely conservative Reform candidate Pat Buchanan received 3407 votes, almost 20% of his Florida total. Attention quickly focused on the now-infamous *butterfly ballot*, the ballot used in Palm Beach County. Voters cast their ballots by punching out the appropriate hole down the center of the ballot. Governor Bush was listed first on the left side, Vice President Gore was listed second, and Mr. Buchanan was listed first on the right side. Voters who wished to vote for Vice President Gore were required to punch out the third hole down, and many voters claimed after voting that they mistakenly punched out the second hole, which corresponded to voting for Mr. Buchanan (many also claimed that after mistakenly punching the second hole, they then also punched the third hole, which would make their ballot invalid).

Many analyses of the election appeared on the Internet, with many claiming that the butterfly ballot was a major cause of the Vice President losing the election; see, for example, Wand et al. (2001) and Elms and Brady (2001). Before irregularities can be blamed on the butterfly ballot, however, there first has to be evidence that the Buchanan vote was actually too high in Palm Beach County. One picture that quickly popped up on various Internet sites was similar to Figure 5.13. This is a plot of the Buchanan vote versus the Bush vote for the 67 Florida counties. A positive correlation would be expected here (reflecting political conservativeness of the county), and that is apparent. What is also apparent, of course, is that the Buchanan vote for Palm Beach County is far larger than would be implied by the relationship in the rest of the counties.

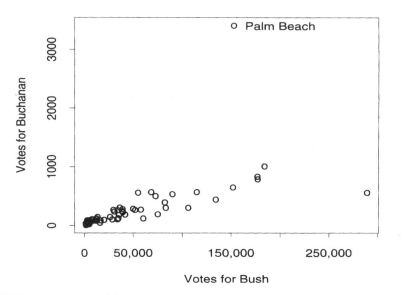

FIGURE 5.13. Plot of Buchanan vote versus Bush vote for 67 Florida counties in the 2000 presidential election.

There are several different ways that one might imagine using regression models to analyze the voting data (the data are given in the file election2000). One possibility is to use total votes for different candidates to predict Buchanan votes, corresponding to Figure 5.13. One difficulty that arises from this approach is that potentially informative votes from other elections must be addressed somehow, since they are based on different total turnout figures. For this reason, we will model Buchanan votes as the target variable, but include logged total votes as an offset (effectively modeling Buchanan's percentage of the total county vote as the target), and use other candidates' percentages as predictors. A similar approach would be to condition on the total vote in the county and model the Buchanan vote as the number of "successes" out of that total; models of that sort are the subject of Chapter 9. All of the models examine the Palm Beach County votes relative to the pattern in the rest of Florida, and are based on the 66 Florida counties other than Palm Beach.

A simple place to start is using a model based on only the Bush voting percentage. The following output refers to a Poisson regression using this predictor (the number of Buchanan votes can be viewed as binomially distributed with a small probability of success, justifying a Poisson regression approach).

Predictor	Coef	s.e.	Wald	p
Constant	-7.443	0.047	-158.57	0.000
Bush percentage	2.853	0.088	32.38	0.000

Substituting Bush's 35.36% of the vote in Palm Beach into this model gives an estimated expected Buchanan vote of 694.3, more than 2700 votes lower than Buchanan's actual vote total. There is strong evidence of overdispersion here, with score tests $z_1 = 4.80$ and $z_2 = 4.89$. The Poisson model's fit is overwhelmingly rejected, as the deviance $D = 2984.7$ on 64 degrees of freedom (the fitted Buchanan counts range from 7 to 1370, so the χ^2 approximation to D is reasonable here).

A negative binomial regression model is a viable alternative to consider here, since we would expect heterogeneity related to differences between counties in minority population, income level, education level, and so on, that are not accounted for in the regression. In fact, using a negative binomial model leads to a remarkable improvement in fit (standard errors are based on the expected Fisher information).

Predictor	Coef	s.e.	Wald	p
Constant	-7.096	0.422	-16.82	0.000
Bush percentage	3.011	0.755	3.99	0.000

The deviance for the model is $D = 70.26$, indicating a reasonable fit to the data ($p = .28$). The estimated expected Buchanan vote is 1038.7 votes, a noticeable difference from the Poisson fit, but still considerably less than the observed vote. The variance inflation parameter $\hat{\alpha}$ equals 0.287, implying an estimated standard deviation of 557.4 for the Palm Beach vote, more than 20 times the Poisson regression-implied value.

It is reasonable to wonder if a richer model might be an even better representation of the voting pattern. In particular, adding the percentage of voters who voted for Green Party candidate Ralph Nader in 2000 and Reform Party candidate Ross Perot in 1996 could potentially help account for the Buchanan Palm Beach effect, since they could reflect voters that are more extreme politically, or simply have reservations about the two-party system. A Poisson regression using these predictors fits far better than just using the Bush voting percentage, although it still fits poorly ($D = 1774.8$ on 62 degrees of freedom).

Predictor	Coef	s.e.	Wald	p
Constant	-8.129	0.060	-136.61	0.000
Bush percentage	2.296	0.096	23.80	0.000
Nader percentage	-4.146	1.665	-2.49	0.013
Perot percentage	10.671	0.327	32.64	0.000

The estimated expected Palm Beach Buchanan count is even smaller, being 621.5.

Predictors	AIC_C	Expected vote	$\hat{P}$(Buchanan) $\geq$ 3407	95% limit	99% limit
Bush percentage	39.1	1038.7	.001791	2092	2750
Bush percentage Nader percentage Perot percentage	0.0	948.3	.000011	1630	2018
Nader percentage Perot percentage	2.7	1253.0	.001119	2197	2740

TABLE 5.5. Properties of negative binomial regression fits to Florida election data.

The negative binomial model using these predictors is very different, and fits very well ($D = 67.8$, $p = .26$).

Predictor	Coef	s.e.	Wald	p
Constant	-7.005	0.410	-17.10	0.000
Bush percentage	1.347	0.605	2.23	0.026
Nader percentage	-33.201	9.244	-3.59	0.000
Perot percentage	10.773	1.562	6.90	0.000

The deviance values for the two negative binomial models cannot be compared, because the dispersion parameter α is different (it is $\hat{\alpha} = .152$ for the latter model). Table 5.5 gives the AIC_C values for the two models, indicating that the AIC_C value for the second model is 39.1 lower than that for the first (as in previous chapters, we report differences in AIC_C from the smallest value, since these are the values that matter). This suggests a strong preference for the latter model. This model implies an estimated expected Buchanan vote of 948.3, more than 2450 votes less than the observed vote. The simpler model omitting the least significant predictor (Bush percentage) has similar fit ($D = 67.1$, $p = .31$), and very close (but slightly higher) AIC_C, and is also a viable choice, although one might feel less comfortable with a model that uses only small party voting.

Predictor	Coef	s.e.	Wald	p
Constant	-6.176	0.227	-27.16	0.000
Nader percentage	-41.623	9.010	-4.62	0.000
Perot percentage	11.225	1.589	7.06	0.000

So, then, just how "irregular" was the Palm Beach Buchanan vote? Table 5.5 presents several ways of answering that question, based on each of the three negative binomial regression models. One answer is to estimate the probability of Buchanan receiving as many as 3407 votes (or more), given the fitted negative binomial distribution for Buchanan Palm Beach votes. These values are given in the fourth column of the table. All three

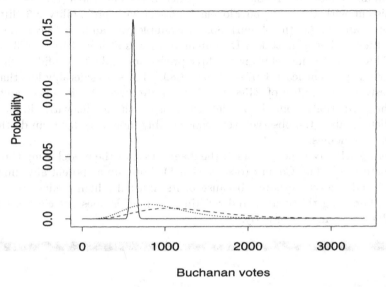

FIGURE 5.14. Estimated probability functions of Palm Beach Buchanan vote based on Poisson regression (solid line) and negative binomial regression (dotted line) using Bush, Nader, and Perot votes, and negative binomial regression using Nader and Perot votes (dashed line).

models estimate that probability as less than 1 in 500, so it seems clear that the Palm Beach Buchanan vote was surprising. Note, however, that the "best" model (at least according to AIC_C) estimates this probability as roughly 1 in 100,000, a very different assessment from that of the other two models. This highlights an important point — different models with comparable fits to the data can lead to very different implications about the tails of a distribution. The probability functions in Figure 5.14 reinforce this. The Poisson fit is very far from the others, and the predictive distribution is much less variable. The overdispersion relative to the Poisson of the negative binomial distribution is apparent from the dotted curve (model with all three predictors) and the dashed curve (model using Nader and Perot percentages only), but the two negative binomial fits clearly result in very different inferences about the likelihood of a Buchanan total between 2000 and 3000 votes.

Given, therefore, considerable evidence that the Buchanan Palm Beach vote was "too high," and given the belief that this was because of voting errors from the butterfly ballot, how could we estimate the number of mistaken votes? The difference between the observed 3407 votes and the expected number doesn't take into account the randomness in the votes, so some sort of probability-related measure seems more informative. The last two columns in Table 5.5 give such a measure. The entries are the estimated maximum number of "regular" Palm Beach Buchanan votes (that is, ones consistent with the other 66 Florida counties) with probability .95 (fifth column) and .99 (sixth column). So, for example, we can be 95% confident that the number of mistaken Buchanan votes was at least $3407 - 1630 = 1777$ based on the model using all three predictors, and 99% confident that the number was at least $3407 - 2018 = 1389$. This is considerably less than the estimated number of 2458, but it accurately reflects the uncertainty in the predictions from the model. Once again the model with all three predictors labels the observed vote considerably more extreme than do the other two models.

Figure 5.15 gives index plots of the diagnostics for the model using three predictors. Alachua County (observation 1) shows up as potentially influential and a leverage point, because of its unusually high Nader vote of 3.7%. Omitting this county and refitting the models does not change the results of the analysis greatly.

5.4.2 Zero-Inflated Count Regression

The univariate zero-inflation models discussed in Sections 4.5.1 and 4.5.2 can be extended to the regression situation as well. The zero-inflated Poisson model satisfies

$$P(Y_i = y_i | \mathbf{x}_i) = \begin{cases} \psi_i + (1 - \psi_i)e^{-\mu_i} & \text{if } y_i = 0 \\ (1 - \psi_i)e^{-\mu_i}\mu_i^{y_i}/y_i! & \text{if } y_i \neq 0. \end{cases}$$

Counts are generated from one of two processes: the usual Poisson random variable with mean $\mu_i = \exp(\mathbf{x}_i'\boldsymbol{\beta})$ with probability $1 - \psi_i$, and zero with probability ψ_i. The regression model generalizes the univariate ZIP model by allowing ψ_i to vary from observation to observation. The probability is modeled to satisfy

$$\psi_i = F(\mathbf{z}_i'\boldsymbol{\gamma}),$$

where $F(\cdot)$ is a cumulative distribution function, $\mathbf{z}_i$ is a set of predictors for the probability of coming from the constant zero distribution, and $\boldsymbol{\gamma}$ is a set of parameters corresponding to the predictors. The distribution function F is typically taken to be either the standard normal distribution (a *probit* model) or the logistic distribution

$$F(x) = \frac{\exp(x)}{1 + \exp(x)}$$

Index plot of standardized residuals

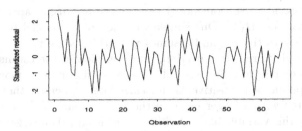

Index plot of leverage values

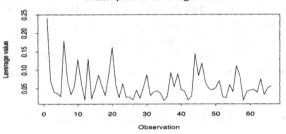

Index plot of Cook's distances

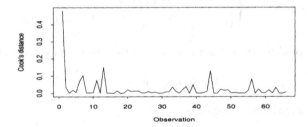

FIGURE 5.15. Index plots of scaled deviance residuals (top plot), leverage values (middle plot) and Cook's distances (bottom plot) for negative binomial regression model.

(a *logit* model). The logit and probit models will be discussed extensively in Chapter 9, but for our purposes at this point it is enough to recognize that for either model, a positive coefficient $\hat{\gamma}_j$ implies that, given the other z variables, a larger value of z_j is associated with a greater estimated probability that the observation comes from the zero point mass distribution. The z variables can include some (or all) of the x's, or can be different variables. The parameters β and γ can be estimated using maximum likelihood. The conditional mean of the target variable is

$$E(y_i|\mathbf{x}_i, \mathbf{z}_i) = \mu_i(1 - \psi_i),$$

while the conditional variance is

$$V(y_i|\mathbf{x}_i, \mathbf{z}_i) = \mu_i(1 - \psi_i)(1 + \mu_i\psi_i),$$

implying overdispersion if $\psi_i > 0$. The need for the ZIP regression model over the Poisson regression model can be assessed using an information criterion, such as AIC_C. One simple variation of the model takes the z's to be identical to the x's, and assumes $\boldsymbol{\gamma} = \tau \boldsymbol{\beta}$ for some τ.

If the $\mathbf{z}$ and $\mathbf{x}$ predictors overlap, interpretation of the estimated parameters $\hat{\boldsymbol{\beta}}$ requires care. For example, if a β coefficient corresponding to a particular predictor is positive, then clearly larger values of that predictor are associated with a larger mean in the Poisson part of the distribution. If, however, this variable has a positive γ coefficient, then larger values are associated with a higher probability of coming from the constant zero distribution, lowering the overall expected value of the target variable. These two tendencies can combine in complex ways. If the ZIP formulation of two distinct (nonoverlapping) processes for y is a reasonable representation of reality, things are easier, as then the $\boldsymbol{\beta}$ coefficients have the usual loglinear interpretation, only now given that the constant zero distribution is not being observed.

Negative binomial regression also can be modified to allow for zero-count inflation. The zero-inflated negative binomial (ZINB) model satisfies

$$P(Y_i = y_i | \mathbf{x}_i) = \begin{cases} \psi_i + (1 - \psi_i) \left(\frac{\nu}{\nu + \mu_i} \right)^{\nu} & \text{if } y_i = 0 \\ (1 - \psi_i) \frac{\Gamma(y_i + \nu)}{y_i! \Gamma(\nu)} \left(\frac{\nu}{\nu + \mu_i} \right)^{\nu} \left(\frac{\mu_i}{\nu + \mu_i} \right)^{y_i} & \text{if } y_i \neq 0. \end{cases}$$

The conditional mean of the target is identical to that for the ZIP model,

$$E(y_i | \mathbf{x}_i, \mathbf{z}_i) = \mu_i (1 - \psi_i),$$

while the conditional variance is

$$V(y_i | \mathbf{x}_i, \mathbf{z}_i) = \mu_i (1 - \psi_i)[1 + \mu_i(\psi_i + 1/\nu)].$$

If $\psi_i = 0$ this is the standard negative binomial regression model variance, but overdispersion relative to the negative binomial occurs if $\psi_i > 0$.

The ZIP model is a special case of the ZINB model (with $\alpha = 1/\nu = 0$), so the usual tests (Wald, score, and likelihood ratio) can be constructed to test against the one-sided alternative $\alpha > 0$. The likelihood ratio and Wald tests have the usual forms ((5.6), where the simpler model takes $\alpha = 0$, and (5.7), based on α rather than β_j, respectively), while Ridout, Hinde, and Demétrio (2001) derived the score test for these hypotheses (see also Hall and Berenhaut, 2002). It should be noted that misspecification of the nondegenerate portion of the model (that is, Poisson or negative binomial) is more serious in the zero-inflated context than in the standard situation, since (unlike in the standard case) parameter estimators assuming a Poisson distribution are not consistent if the actual distribution is negative binomial.

Congressional Responses to Supreme Court Decisions

When the U.S. Supreme Court reverses an action of the Congress, it is often controversial and a major news event. A much rarer occurrence, but one with similar political importance, is when the Congress responds to a decision of the court, through action of the appropriate House or Senate committee. We will attempt to stochastically model such action using the 4052 Supreme Court decisions handed down between 1953 and 1988 that fell under Judiciary Committee jurisdiction during the 96th through 101st Congress, 1979–1988 (Zorn, 1998, with data given in the file **supreme**). I am indebted to Chris Zorn for making these data available here.

The vast majority of Supreme Court decisions elicit no action from Congress (95.8% for these data). This should not be surprising. It is reasonable to assume that without specific pressure to examine a Supreme Court decision, Congress will not act to modify or overturn it. Thus, a sensible representation of this process is based on a two-part procedure: first, the probability that a group or groups will lobby Congress for redress from a decision, and if they do, the number of actions that Congress will take in response. If the latter number is Poisson distributed (allowing, for example, for no Congressional action even if the Congress is lobbied), this is consistent with a ZIP model.

The marginal distribution of the number of actions taken by Congress in response to Supreme Court decisions suggests an overabundance of zero values, compared to a Poisson distribution with mean $\hat{\mu} = .11$ (see Table 5.6). There are 6.8% too many zeroes in the observed sample given this Poisson mean, and the Poisson fits poorly at other values as well. A univariate ZIP model fits the data much more closely (its AIC_C value is more than 460 smaller than that of the Poisson model). According to this model, 95.4% of Supreme Court decisions are never considered for Congressional action, while those that are result in an average of 2.35 committee actions per decision.

We now extend this analysis by allowing for predictors. It would be expected that the political environment and the character of the cases could be useful in modeling the number of Congressional actions taken. We consider here the variables discussed in Zorn (1998). The political environment is incorporated using an indicator variable of whether the decision was a liberal one. Several aspects of the nature of the decision are included: whether the Court made a declaration of unconstitutionality as part of the decision (expected to be negatively associated with Congressional action, since it would involve a constitutional ammendment), whether the decision altered previous precedent (expected to be positively associated with action), and whether the decision was unanimous (expected to be negatively associated with action). If there was lower court disagreement on the case, this could be expected to encourage action. Finally, older decisions would be expected

Number of actions	Observed count	Poisson fitted count	ZIP fitted count
0	3882	3634.2	3882.0
1	63	395.5	42.2
2	38	21.5	49.5
3	32	0.8	38.7
4	8	0.0	22.7
5	12	0.0	10.7
6	12	0.0	4.3
7	3	0.0	1.4
8 or more	2	0.0	0.5

TABLE 5.6. Observed and fitted counts of numbers of Congressional actions taken for univariate Poisson and ZIP models.

to be associated with fewer actions (so the year of the decision would be directly related to actions). Since some cases were handed down after 1979, and in some cases Congress overturned a decision before 1988, not all cases were "available" for action for the entire 10-year period, so the logarithm of the number of years of "exposure" is also included (expected, therefore, to be directly related to the number of actions taken).

Table 5.7 summarizes the fits of two competing models (the ZIP model was fit using the statistical package LIMDEP, which uses the outer-product variance estimator in all of its count regression calculations; see page 10). The Poisson model is attractive in that several of the predictors are highly statistically significant, and those imply the expected relationships. For example, given the other covariates, a declaration of unconstitutionality is associated with a reduction in the estimated average number of actions by almost 85%, and a unanimous decision is associated with a one-third reduction. Unfortunately, there is clear overdispersion in the data, as the score tests show ($z_1 = 5.6$ and $z_2 = 4.7$, respectively; note that χ^2 goodness-of-fit statistics are not useful here, given the many observations with small fitted counts).

The argument given earlier suggests the ZIP model as an alternative model worth considering. Table 5.7 includes two columns related to this model: the estimated coefficients for a logit fit to estimate ψ (the probability of coming from the zero point mass, which corresponds here to a decision that elicited no pressure on Congress for action), and the Poisson fit for μ given the case did elicit pressure on Congress. According to AIC_C the ZIP model is an overwhelming improvement over the Poisson model, with a value almost 1300 lower than that of the Poisson model. The probability of the absence of Congressional pressure is significantly related to several variables, including a declaration of unconstitutionality (associated with higher probability of no pressure), a unanimous decision (associated with higher probability of no pressure), and the two time-related variables (with

Variable	Poisson model	ZIP model ψ	ZIP model μ	Reduced ZIP model ψ	Reduced ZIP model μ
Intercept	−160.125 (−9.91)	153.580 (6.35)	−8.793 (−0.63)	153.113 (6.65)	0.747 (13.82)
Liberal	0.296 (3.02)	−0.091 (−0.54)	0.190 (2.08)		0.212 (2.53)
Declaration of unconstitutionality	−1.838 (−4.78)	1.590 (2.42)	−0.421 (−0.69)	1.673 (2.77)	
Precedent altering	−0.254 (−0.67)	−0.401 (−0.65)	−0.582 (−1.08)		
Unanimous decision	−0.407 (−3.74)	0.499 (2.58)	0.098 (0.96)	0.458 (2.42)	
Lower court disagreement	−0.212 (−1.79)	0.043 (0.22)	−0.138 (−1.30)		
Year of decision	0.079 (9.82)	−0.076 (−6.24)	0.005 (0.68)	−0.076 (−6.54)	
Log(exposure)	0.544 (4.77)	−0.487 (−2.64)	0.089 (0.65)	−0.508 (−2.82)	

TABLE 5.7. Results for Poisson and ZIP regression models for U.S. Supreme Court data. Wald statistics for each coefficient are given in parentheses.

more recent decisions having higher probability of pressure, and decisions exposed to Congressional action longer having higher probability of pressure). These patterns are all intuitively appealing, and apparently account for most of the relationship between Congressional action and the predictors. Given pressure on Congress, the only predictor that is even weakly related to the expected number of Congressional actions is whether the decision is a liberal one, with liberal decisions likely to elicit more responses. If we simplify the model to include only the four significant predictors for ψ and the one significant predictor for μ (which decreases AIC_C further by 15), we are left with the very simple result that, given the existence of pressure on Congress, the estimated expected number of Congressional actions from a liberal decision is 2.61, while it is 2.11 for a nonliberal decision. Not surprisingly, this is not very different from the univariate analysis, but the ZIP regression does have the advantage of also incorporating inference about what factors are related to the existence of pressure on Congress in the logit regression for ψ. If we substitute the average values of the predictors for ψ into the logit equation, we obtain a "typical" probability of no pressure on Congress of $\hat{\psi} = .962$, reinforcing the result that the vast majority of Supreme Court decisions are truly final.

5.4.3 Zero-Truncated Poisson Regression

Section 4.6 described count variables that are truncated from below from a univariate point of view, but such variables also can arise as the target variable in a regression framework. The standard count regression models generalize to this situation. Zero-truncated Poisson regression is based on the distribution

$$f(y_i; \mu_i) = \frac{e^{-\mu_i} \mu_i^{y_i}}{y_i!(1 - e^{-\mu_i})}, \qquad y_i = 1, 2, \ldots,$$

with μ_i following the usual logarithmic link with predictors $x_1, \ldots, x_p$. As in the univariate case, the mean of the target variable is

$$E(y_i) = \frac{\mu_i}{1 - e^{-\mu_i}}, \tag{5.21}$$

while the variance is

$$V(y_i) = \left[\frac{\mu_i}{1 - e^{-\mu_i}} \right] \left[1 - \frac{\mu_i e^{-\mu_i}}{1 - e^{-\mu_i}} \right].$$

In contrast to the nontruncated situation, the estimates of μ_i (or equivalently, of β) are not consistent if the Poisson assumption is violated (even if the link function and linear predictor are correct), because the expected value depends on correct specification of $P(y_i = 0)$.

The interpretation of regression coefficients for this model is more complicated than that for nontruncated data. If the population of interest is actually the underlying Poisson random variable without truncation (if, for example, values of the target equal to zero are possible, but the given sampling scheme does not allow them to be observed), the coefficients have the usual interpretation consistent with the loglinear model (since in that case $E(y) = \exp(\mathbf{x}'\beta)$). If, on the other hand, observed zeroes are truly impossible, β_j no longer has a simple connection to the expected multiplicative change in y. Rather, the instantaneous expected change in y given a small change in x_j, holding all else fixed, is

$$\frac{\partial E(y_i | x_i; y_i > 0)}{\partial x_{ij}} = \beta_j \mu_i \left\{ \frac{1 - \exp(-\mu_i)(1 + \mu_i)}{[1 - \exp(-\mu_i)]^2} \right\}. \tag{5.22}$$

For nontruncated data the righthand side of (5.22) is simply $\beta_j \mu_i$ (reflecting the interpretation of β_j as a semielasticity), so the truncation adds an additional multiplicative term that is a function of the nontruncated mean and the nontruncated probability of an observed zero. The righthand side of (5.22) divided by μ_i represents the semielasticity of the variable for this model.

If the target variable exhibits overdispersion relative to a (truncated) Poisson variable, a truncated negative binomial is a viable alternative model. Since for the negative binomial distribution $P(Y = 0) = (1 + \alpha\mu)^{-1/\alpha}$, the distribution of the zero-truncated negative binomial is

$$f(y_i; \mu_i, \alpha) = \frac{\frac{\Gamma(y_i+1/\alpha)}{y_i!\Gamma(1/\alpha)} \left(\frac{1/\alpha}{\mu_i+1/\alpha}\right)^{1/\alpha} \left(\frac{\mu_i}{\mu_i+1/\alpha}\right)^{y_i}}{1 - (1 + \alpha\mu_i)^{-1/\alpha}}.$$

The mean of this random variable is

$$E(y_i) = \frac{\mu_i}{1 - (1 + \alpha\mu_i)^{-1/\alpha}},$$

while the variance is

$$V(y_i) = \left[\frac{\mu_i}{1 - (1 + \alpha\mu_i)^{-1/\alpha}}\right]\left[1 + \alpha\mu_i - \frac{\mu_i(1 + \alpha\mu_i)^{-1/\alpha}}{1 - (1 + \alpha\mu_i)^{-1/\alpha}}\right].$$

How Many Shoes Do Runners Need? (continued)

In Chapter 4 we studied the number of running shoes for a sample of runners who registered an online running log. A running shoe marketing executive would be interested in knowing how that number relates to other factors, and several are available on the web site, including demographic variables such as gender, marital status, age, education, and income, and running-related variables such as typical number of runs per week, average miles run per week, and the preferred type of running in which the runner participates (given in the file runshoes).

Table 5.8 summarizes the results of Poisson regression fits. For these data the demographic variables did not provide any predictive power (income was not considered, as almost one-fourth of the runners did not give income information), so we focus on regressions based on runs per week, miles run per week, and an indicator variable identifying if preferred running was distance running (half-marathon or marathon). The likelihood ratio statistic for the overall regression is highly significant ($LR = 17.0$, $p < .001$). The Wald statistics for the Poisson regression do not show very much going on, but there is clear evidence of underdispersion ($z_1 = -6.04$ and $z_2 = -4.75$). Wald statistics using the robust sandwich estimator are noticeably larger, implying weakly significant contributions from all three variables. As would be expected, more runs, more mileage, and a preference for distance running are all associated with more running shoes (given the other variables are held fixed).

Since these data are naturally truncated from below, however, this is not the correct model to be fitting. The last two columns summarize a truncated Poisson fit to these data. The coefficients for all three predictors are

Predictor	Poisson MLE	Poisson Robust	Truncated Poisson MLE	Truncated Poisson Robust
Intercept	0.00969		−0.64988	
	(0.03)	(0.04)	(−1.25)	(−1.54)
Runs per week	0.10099		0.16661	
	(1.13)	(1.82)	(1.47)	(1.88)
Miles run per week	0.01072		0.01514	
	(1.35)	(2.14)	(1.59)	(2.04)
Distance	0.41015		0.50679	
	(1.98)	(2.30)	(2.26)	(2.48)

TABLE 5.8. Regression results for ordinary and truncated Poisson regression models for running log data. Wald statistics based on maximum likelihood and the robust sandwich estimator for each coefficient are given in parentheses.

larger, and the likelihood ratio statistic for the overall regression is considerably larger than before ($LR = 36.0$, $p < .001$). There is still evidence of underdispersion ($z_1 = -4.86$ and $z_2 = -3.33$), and the Wald statistics based on robust variances are all marginally significant.

Interpretation of the estimated parameters depends on what is considered the natural population. One possibility is a population of "reasonably active/athletic" adults, where runners constitute a subsample that always own at least one pair of running shoes. In this case the coefficients have the usual interpretation. For example, $\hat{\beta}_{\text{Runs per week}} = 0.1666$ implies that, holding all else fixed, an increase of δ in runs per week is associated with a $16.66\delta\%$ increase in the number of running shoes; equivalently, one additional run per week is associated with an 18.1% increase in average running shoes. Given the level of running (in runs and miles), being a distance runner is associated with 66% more running shoes on average.

A more reasonable approach in this case, however, would seem to be to treat the population as "serious runners," which exhibits an inherently truncated-from-below distribution of running shoes. As noted above, interpretation of regression coefficients in this case is more challenging, since the multiplicative effect changes depending on the expected number of shoes. One way to account for this is to substitute the mean values for numerical variables in the equation. Consider again the coefficient for runs per week. The average number of runs per week in this sample is 4.98, and the average number of miles run per week is 24.71. The estimated mean number of running shoes for nondistance runners with the predictors at their averages (based on the nontruncated distribution) is 1.74, while that for distance runners is 2.88. Substituting into (5.22) gives a semielasticity of .127 for runs per week and a semielasticity of .012 for miles run per week, respectively, for non-distance runners, and .146 for runs per week and .013 for miles run per week, respectively, for distance runners. Direct computation of the estimated expected number of running shoes for the truncated distribution using (5.21), with runs and miles per week at their means, is 2.11 for nondistance runners and 3.05 for distance runners, or 45% higher for distance runners. Thus, the estimated effects related to the numerical predictors are larger in the truncated distribution, compared with the non-truncated distribution, but the estimated effect of the indicator variable predictor is smaller (at the mean values of the numerical predictors). From a market research point of view, the results aren't too surprising: among runners, distance runners are a prime market for running shoes, as they own more shoes on average, and have an ownership pattern more sensitive to the amount of running done than do nondistance runners.

It must be pointed out that the observed underdispersion is a fundamental issue in this analysis. As was pointed out earlier, the estimates of β are inconsistent unless the nontruncated probability of zero is consistently estimated. Unfortunately, the truncated negative binomial cannot address underdispersion, so there is little that can be done using the methods discussed here.

5.5 Nonparametric Count Regression

All of the models discussed so far suffer from the drawback that predictors only relate to the target variable (log-)linearly. Allowing for polynomial relationships extends the applicability of the generalized linear model, but there are still situations where additional flexibility over a linear predictor is necessary. This can be accomplished by replacing the linearity assumption with the weaker assumption that, whatever the precise form of the relation between the target and predictor(s), it can be represented by a smooth curve or surface. There are several ways of incorporating smoothness into the count regression framework (or, more generally, the generalized linear model framework), but we will focus here on local likelihood estimation, an intuitively attractive generalization of (global) maximum likelihood.

5.5.1 Local Likelihood Estimation Based on One Predictor

Consider first the simple situation of a single predicting variable. The generalized linear model imposes a linear systematic component (5.2), which can be too inflexible to adequately represent the relationship between η and x. An alternative is to estimate η at any evaluation point x_0 *locally* based on a linear model, weighting observations closer to x_0 more heavily than ones farther away. That is, the idea is to approximate an unknown smooth function $\eta(x_0)$ as being a straight line in a small neighborhood around x_0, for any value of x_0. This can done using kernel weights. Let W be a symmetric probability density with compact support, and define

$$W_h(z) \equiv \frac{1}{h} W\left(\frac{z}{h}\right).$$

The local linear likelihood estimate of $\eta(x_0)$ is then $\hat{\beta}_0$, the constant term of the maximizer of the local log-likelihood,

$$\sum_{i=1}^{n} [y_i(\beta_0 + \beta_1 x_i) - b(\beta_0 + \beta_1 x_i) + c(y_i)] W_h(x_i - x_0). \tag{5.23}$$

Note that we are using the canonical link, as that results in desirable properties, and are using a one-parameter version of the generalized linear model (with the scale parameter ϕ not needed). The estimated mean function for y is then $\hat{\mu}(x_0) = g^{-1}(\hat{\eta}(x_0))$, where g is the link function.

The value h is called the *bandwidth*, and controls the amount of smoothing done. If the kernel function W has support on $[-1, 1]$, the estimate $\hat{\eta}(x)$ is based on a weighted maximum likelihood estimate over a sliding window of width $2h$ centered at x (and is thus a generalization of the regressygon estimate given in Figure 2.4). If h is very small, the estimated mean function interpolates the observed data (and is therefore low in bias, but high in variance); if h is very large, the local likelihood converges to the ordinary likelihood, and the estimated mean function is based on a linear predictor (and is therefore low in variance, but high in bias if a linear relationship is inappropriate). Thus, an important task is to choose h properly, to balance fidelity to the data with smoothness of the estimated relationship. As the sample size $n \to \infty$, if the bandwidth is chosen so that $h \to 0$ and $nh^3 \to \infty$, and certain regularity conditions hold (including adequate smoothness of η), the squared bias is of order $O(h^4)$, while the variance is of order $O([nh]^{-1})$. Balancing bias and variance thus yields a mean squared error (MSE)-optimal $h = O(n^{-1/5})$, with resultant $MSE = O(n^{-4/5})$. Note that this is slower than the $O(n^{-1})$ squared error rate for standard parametric problems; this slower convergence rate is the price one pays for not specifying a parametric systematic component.

While the choice of kernel function W is not very important, choosing the bandwidth incorrectly can lead to a poor estimator. The bandwidth can be chosen by eye (in order to give a pleasing graphical representation of η or μ), or chosen automatically. Recall that the sum of the leverage values (5.14) in a generalized linear model equals $p + 1$, the number of parameters in the linear predictor. Leverage values also can be defined for the local likelihood fit, based on the local regressions constructed at each observed x_i, with the sum of these leverage values being the effective number of degrees of freedom of the estimated smooth curve, ν^* (say). The bandwidth and degrees of freedom are inversely related, with a rougher curve (smaller h)

corresponding to larger ν^* (approaching a maximum of n), and a smoother curve (larger h) corresponding to smaller ν^* (approaching a minimum of 2). An automatic way of choosing h is then to minimize AIC_C, where the effective degrees of freedom ν^* is substituted for the estimated number of parameters ν. In the smoothing context the number of observations is typically taken to be n, rather than $\sum_i y_i$, by analogy with results for Gaussian-based nonparametric regression.

The local linear estimator can be generalized to higher-order polynomials, such as local quadratic or local cubic. If the true η is sufficiently smooth, higher-order polynomials can achieve faster convergence rates, with odd-order polynomials theoretically preferable (for example, a local cubic likelihood estimate can achieve $MSE = O(n^{-8/9})$), although the practical benefit of using higher-order polynomials is not always clear.

Move Over, Roger Maris

In 1998, the top story in the sports world in general, and in the baseball world in particular, was the attempts of Mark McGwire (of the St. Louis Cardinals) and Sammy Sosa (of the Chicago Cubs) to break Roger Maris' 37-year old all-time season home run record. In fact, the attempt eventually transcended sports entirely, becoming the lead story in newspapers and television news reports. McGwire broke the record with his 62nd home run on September 8, and Sosa hit his 62nd five days later. Ultimately McGwire ended the season with 70 home runs, while Sosa finished with 66.

The dynamics of this home run race can be summarized nicely using individual game statistics for each player (Simonoff, 1998a, given in the file homerun). Figure 5.16 gives the number of home runs hit in each game played by McGwire (top plot) and Sosa (bottom plot) over the season, with the estimated home runs per game based on a Poisson regression on the number of days into the season. It is very difficult to see any pattern in the home run rates for the two players, but the loglinear model is obviously of no help, since there is no apparent relationship between home run rate and time for McGwire, and little relationship for Sosa.

Figure 5.17 gives smooth curves estimating the home run per game rate for each player. The curves are derived as local linear likelihood estimates based on a Poisson likelihood, using a tricube kernel,

$$W(x) = \begin{cases} (70/81)(1 - |x|^3)^3 & \text{if } -1 \le x \le 1, \\ 0 & \text{otherwise.} \end{cases}$$

The bandwidths were chosen to minimize AIC_C. The tick marks at the bottom of the plots identify games in which each player hit at least one home run. The curves show that McGwire was a reasonably consistent home run hitter, with his rate not varying very much through the season.

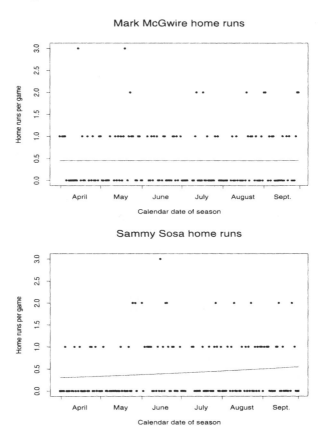

FIGURE 5.16. Plots of home runs per game hit by Mark McGwire (top plot) and Sammy Sosa (bottom plot) during the 1998 season, with estimated home runs per game based on a Poisson loglinear model superimposed.

He started off hitting home runs at a much higher rate than Sosa, with the rate slowly decreasing until late July. Sosa's rate increased rapidly from the beginning of the season, passing McGwire's at the end of May and peaking in mid-June, before he too slowed down until late July. At that point both of their home run rates began to increase again. Sosa's rate remained higher than McGwire's until September, when it decreased. McGwire, on the other hand, finished the season very strongly, hitting 15 home runs in September. Without further information about what home run per game rate functions look like for other players, it's not possible to say whether the observed patterns for Sosa and McGwire are typical or atypical, but they are illuminating from an exploratory point of view.

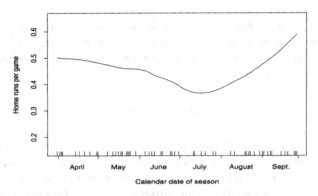

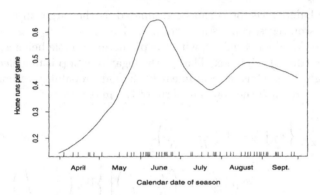

FIGURE 5.17. Estimated home run rates for Mark McGwire (top plot) and Sammy Sosa (bottom plot) during 1998 season, based on local linear likelihood estimation.

5.5.2   Smoothing Multinomial Data

The connection between the Poisson and multinomial random variables described in Section 4.3.3 means that the techniques described in the previous section can be adapted easily to multinomials. Consider a K-cell multinomial random variable, with observed counts $\mathbf{n} = \{n_1, \ldots, n_K\}$, $\sum_i n_i = n$, and underlying probability vector $\mathbf{p}$. The frequency estimates $\hat{p}_i = n_i/n$ are accurate estimates of the true probabilities if np_i is large enough, but if the multinomial is sparse (that is, has many small counts n_i), the frequency estimates can be quite poor estimates, even if the total sample size n is large.

This can be formalized using what are known as *sparse asymptotics*. In this framework, both K and n become larger, but at the same rate, thereby modeling a large sparse table. In this situation, the frequency estimates are not even consistent, in the sense that

$$\sup_i \left| \frac{\hat{p}_i}{p_i} \right| - 1 \not\to 0$$

as n and K get larger. If the categories in the multinomial vector have a natural ordering, however, it is possible to construct sparse asymptotic consistent estimates of $\mathbf{p}$ by borrowing information from nearby cells in the estimation of the probability of any given cell; that is, by smoothing across nearby cells.

The probability vector $\mathbf{p}$ can be estimated as the target in a nonparametric Poisson regression, as in Section 5.5.1 (since each n_i can be treated as a Poisson random variable), with the predictor variable being $x_i = i/K$, the (standardized) cell index. That is, the logarithm of $\mathbf{p}$ is modeled locally as a straight line. Thus, a local linear likelihood probability estimate $\tilde{p}_i$ is $\exp(\hat{\beta}_0)$, where $\hat{\beta}_0$ is the constant term of the maximizer of

$$\sum_{j=1}^{K} \left\{ n_j \left[\beta_0 + \beta_1 \left(\frac{i}{K} - \frac{j}{K} \right) \right] \right.$$
$$\left. - \exp\left[\beta_0 + \beta_1 \left(\frac{i}{K} - \frac{j}{K} \right) \right] \right\} W_h \left(\frac{i}{K} - \frac{j}{K} \right)$$

(the estimator generalizes to higher-order local polynomials in the obvious way). In order to make the estimates be true probabilities, they are then normalized to sum to one. It is useful to model $\mathbf{p}$ as being generated from an unknown smooth density f on $[0,1]$ through the relation

$$p_i = \int_{(i-1)/K}^{i/K} f(u)\, du.$$

If f is sufficiently smooth, the resultant probability estimates are sparse asymptotic consistent, with properties broadly similar to those of $\hat{\mu}$ in Section 5.5.1. Smoothing can improve probability estimation even if all of the cells in a multinomial vector have large counts, but the real benefits appear in sparse tables, reflecting this sparse asymptotic context.

Draws for the Texas Lottery (continued)

On page 82, we examined the draws from a machine used in the Texas lottery, and found that while overall χ^2 goodness-of-fit tests could not reject the hypothesis of equal probabilities for each number drawn, the linear component of X^2 indicated that there was a violation of uniformity. Since the balls were loaded into the machine in sequential order, there is a natural ordering to the categories, and it is reasonable to examine a smoothed nonparametric estimate to try to identify a reasonable alternative to uniformity. Figure 5.18 gives the frequency estimates (bars) and local linear likelihood estimates (circles connected by lines, choosing the bandwidth based on AIC_C) for these data. The smoothed estimates clearly show the nonuniformity of the probabilities. Nonuniformity is relatively small for the lower numbers (which were loaded first into the machine), but is very strong for the higher numbers, with the estimated probability of drawing a 9 less than 1/15 (rather than the null .1). This corresponds to a shift in favor of the lower numbers, which is why the linear component of X^2 identified it, but the smooth estimates are able to characterize the nonuniformity more specifically than the nonsmooth analysis can.

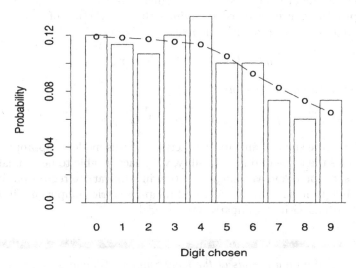

FIGURE 5.18. Probability estimates for Texas lottery data. The bars are frequency estimates, while the circles connected by solid lines are local linear likelihood estimates.

5.5.3 Regression Smoothing with More Than One Predictor

Local polynomial likelihood estimation can be generalized in a straightforward way to regression models with $p > 1$ predictors. Let W be a p-variate kernel function, typically taken to be a p-variate symmetric density function with covariance matrix equal to cI, where I is the identity matrix and c is a constant. Let $W_H(\mathbf{z}) = |H^{-1/2}|W(H^{-1/2}\mathbf{z})$, where H is a $p \times p$ positive definite matrix of bandwidths, typically taken to be diagonal. The p-variate local linear likelihood estimator of $\eta(\mathbf{x})$ at the point $\mathbf{x}_0$ is $\hat{\beta}_0$, the constant term of the maximizer of

$$\sum_{i=1}^{n}[y_i(\beta_0 + \beta_1 x_{1i} + \cdots + \beta_p x_{pi})$$
$$- b(\beta_0 + \beta_1 x_{1i} + \cdots + \beta_p x_{pi}) + c(y_i)]W_h(\mathbf{x}_i - \mathbf{x}_0).$$

Unfortunately, this estimator suffers from the so-called "curse of dimensionality," whereby the performance of the estimator deteriorates as the number of predictors p increases, with the optimal MSE for the p-variate local linear likelihood estimate being $MSE = O(n^{-4/(p+4)})$. Thus, in higher dimensions progressively larger sample sizes are needed to achieve comparable accuracy. The reason for this is that in higher dimensions "local" neighborhoods are almost surely empty, so smoothing must be done over larger neighborhoods, resulting in higher bias.

This problem can be avoided sometimes through the use of *generalized additive models*. In such models the linear predictor

$$\eta_i = \beta_0 + \beta_1 x_{1i} + \cdots + \beta_p x_{pi}$$

is replaced with an additive predictor

$$\eta_i = \beta_0 + f_1(x_{1i}) + \cdots + f_p(x_{pi}),$$

where the f_j are smooth univariate functions. If this model is appropriate, it overcomes the curse of dimensionality, with each f_j able to be estimated with the same precision as a single function in a univariate regression. It is also possible to include both linear and nonparametric components in the predictor, resulting in a semiparametric model.

Winning Medals at the 2000 Summer Olympics

The 2000 Summer Olympic games in Sydney, Australia, brought 11,000 athletes from 199 nations to complete on what is arguably the world's biggest sports stage. While the essence of the games is that any athlete in a particular competition should have the same chance of winning as any other athlete, it is certainly not the case that all nations are created equal.

It seems clear that larger and wealthier nations are more likely to produce medalists, but what precisely are those relationships? We will examine this question for total 2000 medals (gold, silver, and bronze) by using national population, per capita Gross Domestic Product (GDP), and total medals from the 1996 Atlanta Olympics. The analysis is based on the 66 countries that won at least one medal in both 1996 and 2000. The population values range from less than 300,000 (the Bahamas) to more than one billion (China), so we will used logged population as the predictor. Clearly the number of medals won by a country is related to the number of athletes competing for that country, so we will model the rate of medals won per athlete using Poisson regression with the logged number of athletes as an offset.

The following output is from a Poisson regression on the three predictors.

Predictor	Coef	s.e.	Wald	p
Constant	-2.862	0.319	-8.97	.000
Logged population	0.028	0.032	0.87	.384
Per capita GDP	-0.015	0.003	-4.65	.000
Total 1996 medals	0.012	0.002	7.36	.000

Given the other two predictors, logged population adds little to the fit. As might be expected, there is a direct relationship between 1996 and 2000 medals, but (perhaps surprisingly) there is an inverse relationship with per capita GDP, implying that given the performance in the previous Olympic games, wealthier countries had lower rates of medals per athlete. The deviance for this model is 131.63 on 62 degrees of freedom.

Would nonlinear functions of the predictors be more effective here? Figure 5.19 gives the estimated f_j functions based on a local quadratic generalized additive model, with the smoothing parameters for each curve chosen so as to minimize the overall AIC_C. This model has deviance 89.98 on 58.8 effective degrees of freedom, with AIC_C 0.26 lower than for the loglinear model. The component for logged population is linear, so its nonparametric component can be replaced with a linear one. The component for total 1996 medals appears exponential, suggesting replacing its nonparametric component with a linear one on logged total 1996 medals (this is also supported by the long right tail in the 1996 medals variable). The component for per capita GDP is clearly nonlinear, a relationship that could not be picked up in the generalized linear model. Given the other predictors, the medal rate drops steeply with increasing per capita GDP for poorer countries (GDP less than $15,000 per person), but then increases steadily with increasing per capita GDP. That is, given performance in the previous Olympics and population, the best performance in the 2000 Olympics came from the poorest and richest countries. Examples of this include poor countries such as Ethiopia (from three medals in 1996 to eight in 2000) and Georgia (from

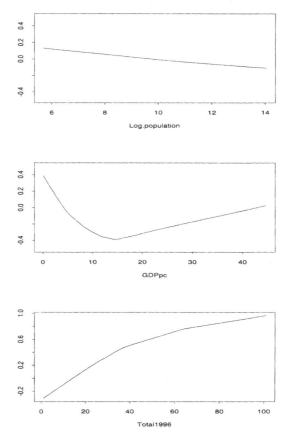

FIGURE 5.19. Estimated additive components of generalized additive model for Olympics data for logged population (top plot), per capita GDP (middle plot), and total 1996 medals (bottom plot).

two to six), and Chinese Taipei, a rich country where the medal count went from one to five.

The semiparametric model based on linear components for logged population and logged 1996 medals and a nonparametric component for per capita GDP increases AIC_C by 0.02. The logged population variable adds little to the fit (with AIC_C value virtually identical the one based on three nonparametric components). The nonparametric component for per capita GDP is virtually identical to that in Figure 5.19, while the coefficient for logged 1996 medals is 0.324. The model does a reasonable job of fitting the observed 2000 medals (Figure 5.20), although there is some evidence of overdispersion ($X^2/(n-\nu^*) = 1.81$). Reassessing the models using $QAIC_C$ (equation (5.20)) does not change the choice of models.

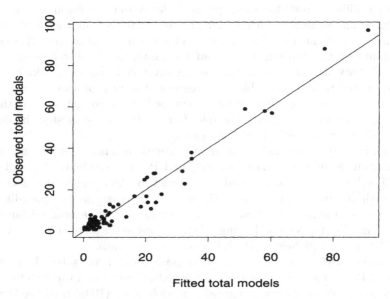

FIGURE 5.20. Observed versus fitted 2000 medals based on a semiparametric model. The line corresponds to equality of observed and fitted values.

5.6 Background Material

Nelder and Wedderburn (1972) introduced and discussed the broad utility and unifying nature of generalized linear models. McCullagh and Nelder (1989) provide extensive discussion of the theory and applications of generalized linear models. Hardin and Hilbe (2001) discuss generalized linear model estimation methods and issues extensively. Cressie and Pardo (2002) investigated hypothesis testing for loglinear models using tests based on the ϕ-divergence family (which includes the power divergence family that leads to the general goodness-of-fit statistic (4.15), and specifically the likelihood ratio test statistic). Long (1997), Cameron and Trevidi (1998), and Winkelmann (2000) give more extensive discussion of count regression models, including Poisson regression and methods designed to address violations of equidispersion.

Cameron and Windmeijer (1997) discussed various pseudo-R^2 measures for count regression models, proposing the R_D^2 measure (5.10).

Cameron and Trevidi (1986), Dean and Lawless (1989), and Dean (1992) explored different tests for overdispersion relative to the Poisson regression model. Kauermann and Carroll (2001) noted that the sandwich variance estimator can be far more inefficient than the usual estimator when there is constant variance, which results in confidence intervals based on it being potentially very anticonservative (that is, the coverage is too low). Nakashima (1997) noted that negative binomial regression estimation can be ineffective if the underlying distribution is not negative binomial, even if the variance satisfies the negative binomial form (5.18), and suggested the use of quasi-likelihood methods in some cases.

An alternative approach to modeling situations where the variance is a function of the mean, based on weighted Pearson residuals, is pseudo-likelihood (see, for example, Hinde and Demétrio, 1998, and Nelder, 2000), although the two terms are sometimes used interchangeably. Quasi-likelihood estimation can be viewed as a generalization of generalized linear modeling, allowing for a wide range of extensions to the standard methods; see McCullagh and Nelder (1989, Chapter 9) for details.

Lambert (1992) and Greene (1994) suggested and investigated the zero-inflated Poisson and negative binomial models. Grogger and Carson (1991) investigated truncated count regression models. Zorn (1998) used the U.S. Supreme Court Congressional review data (pages 165–167) to explore and compare the ZIP regression model to an alternative, the "hurdles" model (in which positive counts are driven by a zero-truncated Poisson distribution), finding strong similarities between the implications provided by those two models for that data set.

Fan et al. (1995) and Fan and Gijbels (1996) investigated the properties of local likelihood regression estimation. Simonoff (1995; 1996, Chapter 6; 1998b) studied different approaches for smoothing multinomial data. Hurvich et al. (1998) discussed how smoothing parameter selection can be put in the form of a model selection problem, and proposed the use of AIC_C for this purpose for a wide range of Gaussian regression problems. The improved accuracy of smoothed probability estimates for sparse multinomials can translate into more effective goodness-of-fit statistics based on them (rather than the observed counts); see Simonoff (1985; 1996, Section 7.2) for further discussion. Hastie and Tibshirani (1990) explore the theory and practice of generalized additive models. Simonoff and Tsai (1999) discussed the use of AIC_C to choose smoothing parameters for additive models.

5.7   Exercises

1. The deviance residual for the Poisson regression model takes the form

$$r_i^D = \text{sign}(y_i - \hat{\mu}_i)\sqrt{2[y_i \log(y_i/\hat{\mu}_i) - y_i + \hat{\mu}_i]}.$$

Pierce and Schafer (1986) investigated different types of residuals for generalized linear models, including an adjustment to the deviance residual (designed to remove bias from it),

$$\tilde{r}_i^D = r_i^D + 1/(6\sqrt{\hat{\mu}_i}),$$

and the so-called *Anscombe residual*,

$$r_i^A = \frac{3(y_i^{2/3} - \hat{\mu}_i^{2/3})}{2\hat{\mu}_i^{1/6}},$$

designed to reduce the skewness of the Pearson residual. Construct these residuals for the Poisson regression models fit in this chapter. Are they very different from the deviance and Pearson residuals? Are they very different from each other?

2. Consider again the geometric random variable (page 113).

   (a) Show that the probability function of the random variable can be written

   $$f(y, \mu) = \frac{\mu^y}{(\mu + 1)^{y+1}}, \qquad y = 0, 1, \ldots,$$

   where $\mu = (1 - p)/p$ is the mean of the random variable.

   (b) Show that the geometric distribution is a member of the exponential family with $\theta = \log[\mu/(\mu + 1)]$, $b(\theta) = -\log[1 - \exp(\theta)]$, $a(\phi) = 1$, and $c(y, \phi) = 0$. Use this representation to confirm that the mean of the geometric random variable is μ, while the variance is $\mu(\mu + 1)$.

   (c) Consider now a regression model based on the linear predictor $\eta_i = \mathbf{x}_i'\boldsymbol{\beta}$. Show that the canonical link does not imply an exponential relationship between the mean of the ith observation μ_i and η_i, but rather the relationship

   $$\mu_i = \frac{\exp(\mathbf{x}_i'\boldsymbol{\beta})}{1 - \exp(\mathbf{x}_i'\boldsymbol{\beta})}.$$

3. Show that the outer-product estimator of variance of $\hat{\boldsymbol{\beta}}$ for Poisson regression is

   $$\hat{V}(\boldsymbol{\beta}) = (X'\text{diag}[(y_i - \hat{\mu}_i)^2]X)^{-1}.$$

4. Recall from Section 3.4.3 that in the least squares analysis of the tornado deaths data treating month as a six-category "tornado season" variable (January, February, March, April, May, June–December) results in improved model fitting. Is that true when using the more correct Poisson regression formulation?

5. Consider again the data on unprovoked shark attacks in Florida (page 27).

   (a) Fit a Poisson regression model of attacks per million residents on year. Is there a significant time trend in attack rate? Are the results very different from those based on the linear (least squares) model constructed in the earlier exercise?

   (b) According to the U.S. Census, the population of Florida in 2000 was 15,982,378. What is the estimated number of unprovoked shark attacks in Florida in 2000? There were, in fact, 34 such attacks. Is this a surprising number, given your estimate?

   (c) The U.S. Census Bureau estimates Florida's population in 2005, 2015, and 2025 to be 16,279,000, 18,497,000, and 20,710,000, respectively. Construct 95% prediction intervals for the number of unprovoked shark attacks in each of those years, based on the regression-based estimated distributions of shark attacks.

   (d) We might hypothesize that these data would exhibit heterogeneity unaccounted for this model, reflecting, for example, changes in tourism through the post-World War II years. Is there evidence of overdispersion in these data? Does a negative binomial regression improve on the Poisson regression?

6. (a) The R^2 measure R_D^2 (5.10) does not depend on the number of predictors in the model, and hence can lead to misleading overassessments of the strength of a regression model when it is based on many predictors. Mittlböck and Waldhör (2000) discussed several adjustments to R_D^2 to correct for this problem, ultimately preferring

$$R_{D1}^2 = 1 - \frac{(n-1)D(\mathbf{y}, \hat{\boldsymbol{\mu}}^{(M)})}{(n-p-1)D(\mathbf{y}, \hat{\boldsymbol{\mu}}^{(M_0)})}$$

when fitted values are generally large, and

$$R_{D2}^2 = 1 - \frac{D(\mathbf{y}, \hat{\boldsymbol{\mu}}^{(M)}) + p/2}{D(\mathbf{y}, \hat{\boldsymbol{\mu}}^{(M_0)})}$$

when fitted values are generally small. Construct each R_D^2 measure for the Poisson regression models fit in this chapter. Are they very different from one another?

   (b) Zheng and Agresti (2000) proposed $\text{corr}(y_i, \hat{y}_i)$ as a measure of the predictive power of a generalized linear model, in analogy with (2.11). Is this measure (squared) similar to the R_D^2 measures for the Poisson regression models fit in this chapter?

7. The file **floridashark** also contains data for the number of shark attack related fatalities in Florida.

   (a) Build a regression model for the fatality rate per million residents as a function of year and the number of attacks. Does the model make sense to you?

   (b) What is the estimated probability of any shark-related fatalities in 2000? Is the fact that there was one such fatality surprising?

   (c) Use the estimated attack numbers from the previous exercise to estimate the number of shark-related fatalities in the years 2005, 2015, and 2025. Do you think these estimates are very trustworthy? Why or why not?

8. Consider again the data on annual homicides in Vermont (page 28).

   (a) Test the fit of a constant annual homicide rate to these data (assume that the population in Vermont has stayed the same over this time period).

   (b) Is there a time trend in homicides apparent in these data? Fit a Poisson regression model on year to assess this question.

   (c) There is no particular reason to think that a time trend (if it existed) would be exponential in time. Refit a Poisson regression model using the (noncanonical) identity link, thereby fitting a linear model in time. Which model fits better? How does the fit of this model compare to that of a linear least squares model?

   (d) Is there evidence of overdispersion in these data? Fit quasi-likelihood versions of the models in parts (b) and (c) to address this. Does this change your feelings about whether there is a time trend in homicides? How do the results compare to those in part (a)?

9. The Poisson regression model can be used to fit univariate exponential family variables via loglinear modeling (Lindsey and Mersch, 1992). We will focus in this question on discrete random variables, although the same method can be used for categorized versions of continuous random variables. Consider a vector $\mathbf{n} = \{n_0, \ldots, n_K\}$ hypothesized to represent frequencies from a discrete random variable Y from the exponential family (that is, n_0 is the number of observations equaling zero, n_1 is the number of observations equaling one, and so on, in a sample of size n), where K is large enough to cover all frequencies with appreciable positive probabilities. Then, if $p_j \equiv P(Y = j)$, we can treat n_j as being Poisson distributed with mean np_j, because of the relationship between the multinomial and Poisson distributions.

(a) Write down $\log(np_j)$ as a function of λ if $Y \sim \text{Pois}(\lambda)$. Show that a simple Poisson regression model with n_j as the target variable values, j as the predictor variable values, and $-\log(j!)$ as values of an offset variable, yields a fitted regression with $\hat{\lambda} = \log(n) - \hat{\beta}_0 = \exp(\hat{\beta}_1)$.

(b) Of course, $\hat{\lambda}$ in part (a) is just the sample mean, so this regression construction isn't necessary. Consider now the truncated Poisson random variable with probability function (4.22). Show that a Poisson regression with the same structure as in part (a) (although now j starting at one, rather than zero) yields a fitted regression with $\hat{\lambda} = \exp(\hat{\beta}_1)$.

(c) Use this method to estimate λ for the postal survey data on page 121. What is the estimated mean number of occupants? Now, add observations corresponding to 8 through 10 occupants, all with zero frequencies. Does this change the estimate of λ appreciably?

This method of fitting distributions using Poisson regression can be generalized to bivariate distributions; see Section 7.1.4.

10. An article in the February 4, 2002 issue of *Sports Illustrated* claimed that being the host country in a Winter Olympics provides an advantage that leads to more medals being won (Cazeneuve, 2002). The file `winterolympic` contains data from that article giving the number of medals won by the host country in each Winter Olympics from 1932 through 1998, along with the number that that country won in the previous Winter Games.

(a) Test the hypothesis that the expected number of medals won by the host country can be well-represented as an additive constant over the number won in the previous Olympics. Does this model fit the data well? (*Hint:* Fit a linear Poisson regression model predicting the number of medals from the previous Olympics' medals, and compare it to a linear Poisson model with previous medals as an offset and no other predictors.)

(b) Answer the questions in part (a) using least squares regression. Do you come to the same conclusion? Are your estimates of the "home country advantage" the same based on the two models?

(c) The 2002 Winter Olympic Games were held in February 2002 in Salt Lake City, Utah. In April 2001, the U.S. Olympic Committee projected that Americans would win 20 medals in the Olympics, and the interim USOC Chief Executive Officer was quoted in the *Sports Illustrated* article as saying "If we end up winning 19, I'll be disappointed." The United States won 13 medals at the 1998 Games in Nagano. Based on your chosen model in part (a), estimate the probability that the U.S. would win at least 20 medals at the 2002 Games.

(d) The United States ultimately won 34 medals at the Salt Lake City Games. Does a performance that good surprise you?

(e) The 2006 Winter Games will be held in Torino (Turin), Italy. Italy won 12 medals in the 2002 games. Using the model derived in part (a), construct a 95% prediction interval for the number of medals Italy will win in the 2006 games. Now, rederive the model, including the 2002 U.S. results as part of the data set. Does the resultant prediction interval for 2006 Italian medals change appreciably?

11. Consider again the data on vineyard production (page 52).

(a) Answer the questions in parts (a)–(c) on page 52, now based on a Poisson regression model. Are the implications different from those based on least squares? Which models do you prefer? Do information measures provide a clear choice between the appropriateness of the Gaussian and Poisson assumptions?

(b) Consider again year and row number as numerical variables. Fit a Poisson (nonparametric) additive model on these two variables. Is this model an improvement over the model in part (b)?

(c) Is there evidence of overdispersion here? Do models based on a negative binomial regression provide a better fit?

12. The home run data file **homerun**, which contains the data related to the Mark McGwire–Sammy Sosa home run race (pages 173–175) also includes game information, such as the location of the game (home or away) and the score of the game.

(a) Construct regression models for the number of runs scored in each game by St. Louis as a function of the location of the game, the number of runs scored by the opposition, and the number of home runs hit by McGwire. Do the same thing for Chicago games and Sosa's home runs. Was scoring related to this home run hitting? Do the models fit the data?

(b) Given the discussion on pages 89–89 concerning fitting a negative binomial distribution to goal scoring in soccer, we might consider negative binomial regression models for these data. Do they provide a better fit to the data? What factors seem to be important in modeling St. Louis and Chicago run scoring, respectively?

(c) Barry Bonds, of the San Francisco Giants, broke McGwire's season home run record in 2001, when he hit 73 home runs. The file **homerun** contains information for his season corresponding to the information for McGwire and Sosa. Construct a smoothed estimate of Bonds' home run rate over the season. Is it qualitatively different from the curves for the other two hitters?

(d) Repeat the analyses in parts (a) and (b) for San Francisco runs as a function of Bonds home runs and game characteristics. Are the results similar to those you obtained for McGwire and Sosa, respectively?

13. The 1978 data on births in the United Kingdom (page 110) actually gives the number of births for each day of the year, and is given in the file **birthday**. Fit a Poisson regression model for the number of births based on the day of the week and the month as categorical predictors. Does the model fit the data? Is the estimated day of the week effect the same as was seen on page 110 now that month is taken into account? Is there a seasonal (monthly) component to births? If you looked at several years' data, do you think the monthly effect would persist? What about the day-of-week effect? What do these results imply regarding an assumption of uniformity of birth rate on a particular date (say, June 1) when looking at data over several years?

14. Consider again the previously examined data on tornadoes in the United States (in the file **tornado**).

(a) Fit a Poisson regression model for the number of killer tornadoes from the total number of tornadoes and month. Is the monthly pattern of killer tornadoes similar to the pattern for deaths from tornadoes found on page 138? Does the model fit the data? Are you comfortable using the deviance or Pearson's X^2 to assess fit for these data? Why or why not?

(b) Fit a nonparametric (smooth) term for the number of tornadoes. Does the estimated component suggest a transformation for the number of tornadoes in the linear model? If so, does this lead to a better fit? Do the results make sense?

15. (a) The United Kingdom train accident data (pages 145–147, in the file **uktrainacc**) includes for each year the percent of the U.K. rolling stock that is of "Mark 1" (pre-1970) design. It would be expected that post-Mark 1 stock would be associated with lower accident rates because of improved safety designs. Construct a Poisson regression modeling accident rate as a function of the percent of rolling stock of Mark 1 design. Is this an improvement over the time trend model?

(b) The percentage of Mark 1 stock is expected to drop by two percentage points per year from 1998 through 2007, when it reaches zero. Based on this projection, construct estimates of the accident rate using the model constructed in part (a). How do these forecasts compare to those from the model based on a time trend?

(c) The data file also includes figures for the number of fatalities annually, separated by type of accident. Construct regression models fitting the total number of fatalities as a function of time, number of accidents, and percentage of Mark 1 stock. Simplify the models if necessary. Do the models fit the data? Construct plots and diagnostics for the models you've constructed. Do they indicate any problems? How might you correct these problems?

16. There are several score tests for overdispersion that are asymptotically equivalent under the Poisson model to z_1 and z_2. Two such tests are

$$z_1^* = \frac{1}{\sqrt{2n}} \sum_{i=1}^{n} \frac{(y_i - \hat{\mu}_i)^2 - y_i}{\hat{\mu}_i}$$

and

$$z_2^* = \frac{\sum_{i=1}^{n}[(y_i - \hat{\mu}_i)^2 - y_i]}{(2\sum_{i=1}^{n} \hat{\mu}_i^2)^{1/2}},$$

respectively. These statistics are score tests against negative binomial overdispersion, and Katz system (page 119) underdispersion, respectively.

(a) Construct these tests for the wildcat strikes in the coal industry data (pages 151–154, in the file wildcat) and Florida election data (pages 157–162, in the file election2000). Do these versions of the tests support the existence of overdispersion? Note that even though z_1 and z_1^* (and z_2 and z_2^*) are asymptotically equivalent under the null, they can be very different if overdispersion exists.

(b) Dean (1992) proposed adjustments to z_1^* and z_2^* designed to improve their small-sample performance. These adjusted statistics are

$$\tilde{z}_1^* = \frac{1}{\sqrt{2n}} \sum_{i=1}^{n} \frac{(y_i - \hat{\mu}_i)^2 - y_i + h_{ii}\hat{\mu}_i}{\hat{\mu}_i}$$

and

$$\tilde{z}_2^* = \frac{\sum_{i=1}^{n}[(y_i - \hat{\mu}_i)^2 - y_i + h_{ii}\hat{\mu}_i]}{(2\sum_{i=1}^{n} \hat{\mu}_i^2)^{1/2}},$$

respectively, where h_{ii} are the leverage values from (5.14). Construct the adjusted tests $\tilde{z}_1^*$ and $\tilde{z}_2^*$ for the wildcat strike and election data. Are they very different from the unadjusted statistics z_1^* and z_2^*, respectively?

17. Consider again the examination of wildcat strikes in the coal industry data.

    (a) Examine the regression diagnostics for the Poisson regression model on nonunionized locations. Is there evidence of problems related to any of the individual mines?

    (b) Fit a negative binomial regression model to the unionized location data. Does this model have different implications than the quasi-likelihood model? Is the mine that appeared unusual in the earlier analysis unusual when fitting a negative binomial model?

18. In August 1991, two beluga whale calves were born at the Aquarium for Wildlife Conservation in Coney Island, Brooklyn, New York. These two calves, subsequently named Hudson and Casey, were the first two calves in the world to be successfully born and raised to maturity in captivity. They were also the first cetaceans born in captivity for which detailed quantitative information about the births and early lives of the animals was gathered. The file whale contains information about the nursing behavior of the calves during the first 55 days of their lives (Chatterjee et al., 1995, pages 256–275; Russell et al., 1997). The observations refer to 6-hour time periods, and give the time period, the number of nursing bouts for that time period (a successful nursing episode where milk was obtained from the mother), the number of lockons for that period (when the calf attached itself to the mother while suckling milk from a mammary gland), and whether the time period was during the day (coded 1) or night (coded 0). It is well known that nursing patterns can be highly informative regarding the health of an infant, so it is important to try to understand nursing behavior as well as possible.

    (a) Construct a Poisson regression model for each calf for the number of lockons in each period as a function of time, number of nursing bouts, and time of day. Do the variables provide predictive power? How would you interpret the coefficient for time period for each calf? Are the models for the two calves similar to each other? Could you use the model for one calf to predict well the number of lockons of the other? Do the models fit the data?

(b) Fit a semiparametric model for Hudson's data allowing a non-linear time trend. Does the time trend appear to be nonlinear? How might you explain the pattern you see?

(c) Fit a negative binomial model for number of lockons for each calf. Does this model have different implications than the Poisson model? Is it a better fit to the data?

(d) These data form a time series, and it is reasonable to suppose that there could be structure in the data related to this. Explore this possibility by considering regression models with lagged values of lockons and bouts as predictors (that is, predict the tth lockon value from the $(t-1)$st lockon and bout values, as well as other predictors). Do such models improve on the ones examined in (a)-(c)?

19. Figure 5.6 suggests the presence of overdispersion in the Poisson regression model for the number of deaths from tornadoes in the United States. Explore alternative formulations for these data that take that overdispersion into account. Do the different alternatives yield results that are similar to each other? Are they similar to the Poisson regression results?

20. The file **supreme** contains the data relating to U.S. Supreme Court Congressional reviews (page 165–167). Investigate whether the overdispersion in these data can be addressed using methods designed for that purpose, such as quasi-likelihood, the sandwich variance estimate, or negative binomial regression. Is there any model that you prefer to the ZIP fit for these data?

21. Consider again the Brazil adventure tourism data (page 53).

(a) Construct a Poisson regression model for the number of trips to the adventure tourism areas. Does the model seem to fit the data?

(b) Construct a negative binomial regression model for the number of trips to the adventure tourism areas. Does the model seem to fit the data?

(c) Construct a ZIP regression model for the number of trips to the adventure tourism areas. Why would a ZIP formulation be sensible for these data? Does the model seem to fit the data? Do the results make sense from an economic point of view? What variables are most important in determining whether or not a respondent would consider visiting the adventure tourism areas?

22. (a) The analysis of the Florida election data focused on voting percentages, rather than the actual counts of votes. Construct a

Poisson regression model for Buchanan votes, rather than Buchanan voting percentage. Is there a big difference in the model fits between the two approaches? Are the estimated Palm Beach Buchanan votes similar?

(b) Now construct a negative binomial regression model for Buchanan votes. Are the implications from this model very different from the one based on voting percentage?

(c) Remove Alachua County from the data set and reconstruct the different regression models. Does anything change in an important way?

23. Consider again the data on Michigan traffic accidents (page 54).

(a) Fit a Poisson regression model for the number of accidents as a function of time, unemployment rate, and season. Do the variables provide significant predictive power? What would you consider the "best" model for these data?

(b) Construct score tests for over/underdispersion. Is there evidence against equidispersion here? In fact, there is considerable evidence for underdispersion in these data. Construct tests and information measures based on a quasi-likelihood fit to these data. Does this change your answers to part (a)?

(c) Fit a semiparametric model, allowing for a nonlinear time trend. Is there evidence that the trend in accidents, given the other predictors, is nonlinear (in the logged scale)?

(d) Replace the season effect with a month effect. Does this change your answers to parts (a)-(c)?

24. The file dmft contains the full data set related to the BELCAP caries prevention study (pages 85–88). In addition to the DMFT index of each participant at the beginning and end of the study, the gender, ethnicity (Dark, White, or Black), and prevention strategy (oral health education, school diet enrichment with ricebran, administration of a fluoride mouthwash, oral hygiene, all four methods, or none) for each participant is given. Fit a ZIP regression model to the ending DMFT index as a function of the other predictors. Simplify the model as seems appropriate. Does the ZIP model provide significant improvement over a Poisson regression model fit?

25. The file gsssexsurvey contains a sample taken from the 1991 General Social Survey with demographic information (age, marital status, years of education, income, and gender), personal information (whether a respondent considers him or herself to be a member of a religion, whether they have used illegal drugs), and information related to sexual practices (whether they have ever had sex with

someone of the same gender). I am indebted to Ron Shokes for these data.

(a) Use these predictors to model the number of sex partners the respondent had had in the previous 5 years. Does a Poisson regression model adequately represent these data?

(b) Explore fitting regression models that allow for count inflation. For example, we might expect to see more zero counts than the Poisson distribution implies (corresponding to people who have never been sexually active), or more one counts (corresponding to monogamous people). Do such models improve the fit?

(c) As was noted on page 112, for data of this type, responses of more than 10 sex partners are often unreliable. Does removing those responses for these data change things very much?

(d) The file also includes information on how many people the respondent knows who have AIDS (acquired immunodeficiency syndrome). Fit a zero-inflated Poisson regression model to these data. What factors appear to be predictive for this variable?

26. The file **creditscore** refers to data from an investigation into consumer behavior and credit scoring (Greene, 1994; Greene, 2000b, page 949). The target variable is the number of major derogatory reports of an individual within a fixed time period (defined as a delinquency of 60 or more days on a credit account). Predictors include age, income per dependent, monthly credit card expenses, and indicators of home ownership and self-employment.

(a) Fit an appropriate Poisson regression model to these data. Now fit a ZIP regression model. Does the latter model provide an improvement over the former? Why do you think a ZIP representation might be more sensible than an ordinary Poisson one here?

(b) Is there evidence that a ZINB model improves on the ZIP model? Why might a negative binomial representation be appropriate? Is a negative binomial regression model just as good as a ZINB model?

27. The file **runshoes** contains the data relating to registrants of a running web site (pages 169–171). The file includes other potential predictors for the number of running shoes, including gender and income. Do these variables provide additional predictive power in a zero-truncated Poisson fit? Do the results change very much if you separate the distance runners from the nondistance runners in the sample?

28. The following table (also given in the file **hurricanes**) gives the number of major North Atlantic hurricanes (wind speed greater than 50 meters per second) annually from 1944 through 2000 (Goldenberg et al., 2001).

Year	Count	Year	Count	Year	Count
1944	3	1963	2	1982	1
1945	2	1964	5	1983	1
1946	1	1965	1	1984	1
1947	2	1966	3	1985	3
1948	4	1967	1	1986	0
1949	3	1968	0	1987	1
1950	7	1969	3	1988	3
1951	2	1970	2	1989	2
1952	3	1971	1	1990	1
1953	3	1972	0	1991	2
1954	2	1973	1	1992	1
1955	5	1974	2	1993	1
1956	2	1975	3	1994	0
1957	2	1976	2	1995	5
1958	4	1977	1	1996	6
1959	2	1978	2	1997	1
1960	2	1979	2	1998	3
1961	6	1980	2	1999	5
1962	0	1981	3	2000	3

(a) Construct a test for constant annual hurricane rate. Do you reject the hypothesis of constant rate?

(b) Combine the years into three year groups, and reconstruct the test in part (a). Does this change the results?

(c) Construct a smooth estimate for the annual average number of major hurricanes. Does this curve lead you to accept or reject a constant annual average number of storms?

(d) The title of the source for these data refers to "the recent increase in Atlantic hurricane activity." Do you agree that the data indicate such an increase? Might it just be random fluctuation?

29. (a) The file **mines** summarizes the time intervals between explosions in mines involving more than 10 men killed in Great Britain from December 8, 1875, to May 29, 1951. The data are given as frequencies of observed intervals for 30-day time periods (0-30 days between explosions, 31-60 days, and so on), as given in Simonoff (1983). Construct a smoothed estimate of the probability distribution of time intervals between explosions.

(b) If explosions occur randomly and independently at a constant rate, the number of explosions in a fixed time period would be Poisson distributed, and the time period between explosions would be exponentially distributed, with density

$$f(t; \lambda) = \lambda e^{-\lambda t}, \qquad t > 0$$

(where $1/\lambda$ is the average time between explosions). Construct an estimated probability distribution based on an underlying exponential distribution with average time between explosions of 241 days. Does the smoothed probability function look similar to the exponential fit?

(c) Nonexponentiality can come from (among other things) a non-constant rate of explosions. The file **mines** also includes data that shed some light on this possibility. It represents the number of disasters in British coal mines annually for the years 1850-1961 (Biller, 2000). Construct a smoothed estimate of the average number of disasters per year. Does this estimate support the possibility of changes in the safety of British mines, and hence nonexponentiality of the times between disastrous explosions?

30. The file **olympic2000** also contains variables giving the number of Gold, Silver, and Bronze medals for the countries in the sample. Construct generalized additive models predicting each of these variables as a function of country characteristics. Simplify the models as you see fit. Are your results similar to those for total medals (pages 178–180)?

31. The file **galapagos** gives information related to plant life on the Galapagos Islands (Andrews and Herzberg, 1985, pages 291–293). Variables given include the observed number of species on each island, the observed number of native species, the area of the island (in square kilometers), the elevation of the island (in meters), the distance to the nearest island (in kilometers), the distance to Santa Cruz Island, the central island of the chain and home to its main port and largest town (in kilometers), and the area of the nearest island (in square kilometers).

(a) Fit an appropriate Poisson regression model for the observed number of species from the properties of the island. Does the model fit the data well? What relationships might you expect between species diversity and the different variables, given the way pollen is spread by air and water? Are the results in line with your expectations?

(b) Construct diagnostics for the model fit. Are there any outliers or leverage points? Does omitting them change the results very much?

(c) Fit a Poisson additive model to these data. Is it an improvement over the (log)linear model? Are there observed nonlinear
relationships? Can you think of reasons why the nonlinearities
would take the form they do? (*Hint*: You might find that the
model is easier to fit based on sensible transformations of the
predictors.)

(d) Repeat parts (a)-(c) with the number of native species on each
island as the target variable. Are the results similar to those for
total observed species?

6

Analyzing Two-Way Tables

The count regression models of Chapter 5 are general enough to be applied in situations where predictors are categorical, which leads to data in the form of tables of counts, or *contingency tables*. Despite this, it is worthwhile to examine such tables as a separate topic, as the tabular structure can lead to useful insights into appropriate models and modeling strategies. In this chapter we focus on two-way tables, starting with the simplest situation, tables with two rows and two columns.

6.1 Two-by-Two Tables

6.1.1 Two-Sample Tests and Comparisons of Proportions

Recall that in Section 3.2.1 it was shown that a two-sample problem could be viewed as a regression analysis on an indicator or effect coding variable. The test of the hypotheses

$$H_0 : \mu_1 = \mu_2$$

versus

$$H_a : \mu_1 \neq \mu_2,$$

where μ_j is the mean target value for group j, is (3.3), which is equivalent to the t-statistic testing whether the slope coefficient in a regression on an indicator variable defining the groups equals zero.

Now consider the scenario where a different sort of comparison between two groups is of interest: that of the probability of occurrence of some event. This is a two-sample problem, but now referring to binomial proportions. The goal is to determine if the observed proportions of successes in two samples provide sufficient evidence to believe that the true underlying probabilities of success in the two groups are different. Let p_1 be the probability of success for a member of the first group, and p_2 be the probability for a member of the second. The hypotheses being tested are

$$H_0 : p_1 = p_2$$

versus

$$H_a : p_1 \neq p_2.$$

The appropriate statistic has a similar construction to the Gaussian-sample t-statistic (3.3). The data consist of two independent binomial samples of size n_1 (of which x_1 are successes) and n_2 (of which x_2 are successes), respectively. Let $\overline{p_1} = x_1/n_1$ and $\overline{p_2} = x_2/n_2$ be the observed frequencies of success for group one and group two, respectively. The natural estimate for the difference in population probabilities is the difference in observed frequencies, $|\overline{p_1} - \overline{p_2}|$. Since under the null hypothesis $p_1 = p_2 = p$ (say), the standard error of this difference is based on a pooled estimate of the common probability,

$$\overline{p} = \frac{x_1 + x_2}{n_1 + n_2}.$$

The test statistic is then

$$z = \frac{|\overline{p_1} - \overline{p_2}|}{\sqrt{\overline{p}(1 - \overline{p})\left(\frac{1}{n_1} + \frac{1}{n_2}\right)}} \tag{6.1}$$

The close conceptual connection between (3.3) and (6.1) is clear. An appeal to the Central Limit Theorem justifies referencing this to a Gaussian random variable.

Just as was true in the single binomial sample situation, the standard (Wald) confidence interval is not equivalent to inverting the corresponding hypothesis test, since the interval is based on the estimated values of p_1 and p_2, rather than a pooled estimate of a common probability. The $100 \times (1 - \alpha)\%$ interval for $p_1 - p_2$ is

$$\overline{p_1} - \overline{p_2} \pm (z_{\alpha/2})\sqrt{\frac{\overline{p_1}(1 - \overline{p_1})}{n_1} + \frac{\overline{p_2}(1 - \overline{p_2})}{n_2}}. \tag{6.2}$$

Predicting the Relapse of Leukemia

Between 3000 and 4000 people are diagnosed with acute lymphoblastic leukemia in the United States each year, two-thirds of them children. The standard criterion for considering a child with leukemia to be in remission is if doctors cannot see any cancerous cells in the child's bone marrow when looking through a microscope. A genetic fingerprinting technique called polymerase chain reaction (PCR), however, can detect as few as 5 cancer cells in every 100,000 cells, a far smaller amount than can be seen through a microscope.

Investigators examined 178 children who appeared to be in remission using the standard criterion after undergoing chemotherapy (Cave et al., 1998). Using the PCR test, traces of cancer were detected in 75 of the children; during three years of followup, 30 of these children suffered a relapse. Of the 103 children who did not show traces of cancer, 8 suffered a relapse. Is there a significant difference in relapse rate in these two groups? The estimated relapse rate for the children with traces of cancer was $\overline{p_1} = 30/75 = .4$, that for the children without apparent traces of cancer was $\overline{p_2} = 8/103 = .078$, while the pooled relapse rate was $\overline{p} = 38/178 = .213$. Thus, the test whether the relapse rates are different in the two groups is

$$z = \frac{|.4 - .078|}{\sqrt{(.213)(.787)\left(\frac{1}{75} + \frac{1}{103}\right)}} = 5.18.$$

This is, of course, highly significant, indicating that even small traces of cancer are a bad sign in children in apparent remission. A 95% confidence interval for the difference in relapse rates between the two groups is

$$.4 - .078 \pm (1.96)\sqrt{\frac{(.4)(.6)}{75} + \frac{(.078)(.922)}{103}} = (.200, .445).$$

6.1.2 Two-by-Two Tables and Tests of Independence

Another way to think about the question of whether success rates are related to membership in one or the other of two groups is as follows. Each sample member has two attributes recorded for it: the group it belongs to, and success or failure. The question of whether success rates differ between the groups is equivalent to the question of whether knowing group membership gives any information about whether the result is a success or failure. If the success rates are the same in the two groups, that means that knowing group membership gives no information about success status; that is, these two attributes are *independent* of each other.

Say two characteristics, X, and Y, are recorded for respondents, each of which has two possible outcomes, 1 and 2. Bivariate data of this type can be put in the form of a 2×2 contingency table:

$$Y$$

X	1	2	Total
1	n_{11}	n_{12}	$n_{1\cdot}$
2	n_{21}	n_{22}	$n_{2\cdot}$
Total	$n_{\cdot 1}$	$n_{\cdot 2}$	n

Here n_{11} is the number of respondents with characteristics $(X = 1, Y = 1)$, n_{12} is the number of respondents with characteristics $(X = 1, Y = 2)$, and so on, and a $\cdot$ in a subscript means summation over the variable that corresponds to that position. A corresponding table of probabilities is

$$Y$$

X	1	2	Total
1	p_{11}	p_{12}	$p_{1\cdot}$
2	p_{21}	p_{22}	$p_{2\cdot}$
Total	$p_{\cdot 1}$	$p_{\cdot 2}$	1

Independence of X and Y is defined by the condition

$$P(X = i \text{ and } Y = j) = P(X = i) \times P(Y = j)$$

for all i and j. Thus, a test as to whether the table is consistent with independence is a test of the hypotheses

$$H_0 : p_{ij} = p_{i\cdot} p_{\cdot j}$$

versus

$$H_a : p_{ij} \neq p_{i\cdot} p_{\cdot j},$$

where $p_{ij} \equiv P(X = i \text{ and } Y = j)$, $p_{i\cdot} \equiv P(X = i)$ (the row marginal probabilities), and $p_{\cdot j} \equiv P(Y = j)$ (the column marginal probabilities).

The closeness of the observed table to independence can be assessed using the goodness-of-fit tests of Section 4.4. Consider the counts in the table as having been generated from a 4-cell multinomial distribution based on sample size n and probabilities $\mathbf{p}$ (we will say more about this assumption later). The natural estimate of p_i is the observed proportion of observations in the ith row, $\hat{p}_{i\cdot} = n_{i\cdot}/n$, while that of $p_{\cdot j}$ is the observed proportion of observations in the jth column, $\hat{p}_{\cdot j} = n_{\cdot j}/n$. Then, under independence, the estimated expected count for the (i,j)th cell is

$$\hat{e}_{ij} = n\hat{p}_{i\cdot}\hat{p}_{\cdot j} = \frac{n_{i\cdot}n_{\cdot j}}{n}. \tag{6.3}$$

The degrees of freedom for the tests equal $4-2-1 = 1$, since two additional parameters are estimated from the data (one marginal row probability and one marginal column probability, since the marginal probabilities must sum to 1).

Predicting the Relapse of Leukemia (continued)

The data given previously concerning leukemia relapse can be represented as the following table:

Followup status

PCR status	Relapse	No relapse	Total
Traces of cancer	30	45	75
Cancer free	8	95	103
Total	38	140	178

The estimated expected counts under independence are thus

Followup status

PCR status	Relapse	No relapse	Total
Traces of cancer	16.01	58.99	75
Cancer free	21.99	81.01	103
Total	38	140	178

Goodness-of-fit tests strongly reject independence: $X^2 = 26.85$ and $G^2 = 27.40$, both with tail probabilities less than 2.3×10^{-7}. We can use other χ^2 goodness-of-fit test statistics as well, such as the power-divergence statistic (which is $2nI^{2/3} = 26.79$ here), but we will focus on X^2 and G^2. Thus, PCR status provides information about a child's leukemia status three years later. Examination of the Pearson residuals shows where the departures from independence occur:

	Followup status	
PCR status	Relapse	No relapse
Traces of cancer	3.50	−1.82
Cancer free	−2.98	1.55

The largest contribution to the Pearson X^2 statistic comes from the (Traces of cancer, Relapse) cell, where the observed number of children (30) is much larger than expected (16). Similarly, the number of children in the (Cancer free, Relapse) cell is smaller than expected. That is (as we already knew), children with traces of cancer had a higher relapse rate than would have been expected, while cancer-free children had a lower relapse rate (there are more observations than expected on the main diagonal, and fewer than expected on the off-diagonal).

A reasonable question to ask at this point is how these tests take into account the way the data were actually gathered. The fitted values for the leukemia relapse contingency table are based on a multinomial distribution and unconditional probabilities for the entire population of children in the study (for example, the (unconditional) probability of being cancer free and having a relapse). A seemingly more direct (and correct) approach is the one that was used in Section 6.1.1, where the conditional probabilities of relapse given PCR status are compared. Does this lead to a different test?

Fortunately, no. Consider, for example, the expected count of children who suffered a relapse when they were initially cancer free. A conditional approach would result in a value of

$$e_{\text{Relapse, Cancer free}} = n_{\text{Cancer free}} \times p_{\text{Relapse} \mid \text{Cancer free}},$$

where $n_{\text{Cancer free}}$ is the number of children who were initially cancer free and $p_{\text{Relapse} \mid \text{Cancer free}}$ is the probability of a suffering a relapse given they were initially cancer free. That is, the expected count of cancer-free children who suffered a relapse is the size of the cancer-free group multiplied by the chances that a member of that group would suffer a relapse.

Under the hypothesis of equal binomial probabilities,

$$p_{\text{Relapse} \mid \text{Cancer free}} = p_{\text{Relapse} \mid \text{Traces of cancer}},$$

so this probability would be estimated by a value that ignores screening status; in other words, the pooled probability estimate, n_{Relapse}/n. Therefore, the estimated expected count of children suffering a relapse who were initially cancer free is

$$\hat{e}_{\text{Relapse, Cancer free}} = n_{\text{Cancer free}} \times \hat{p}_{\text{Relapse} \mid \text{Cancer free}}$$
$$= \frac{n_{\text{Cancer free}} \, n_{\text{Relapse}}}{n}.$$

This is of course identical to the earlier version based on (6.3).

An obvious followup question is to wonder how these tests relate to the two-sample binomial z-test constructed earlier. In fact, there is a close connection. Some algebraic manipulation yields that $X^2 = z^2$; that is, the Pearson test of independence is exactly the square of the two-sample binomial z-test. Since the χ^2 distribution on one degree of freedom is the square of a standard normal distribution, tail probabilities from the two tests will be identical, so there is no difference in using either test for inference.

6.1.3   The Odds Ratio and the Relative Risk

An alternative representation of association in a 2×2 table is through the use of odds ratios. The odds of an event occurring is the ratio of the probability of the event occurring to the probability of the event not occurring. So, for example, if an event has probability $1/2$ of occurring, the odds of it occurring are $.5/(1 - .5) = .5/.5 = 1$ (sometimes phrased 1:1), while if it has probability $3/4$ of occurring, the odds of it occurring are $.75/(1 - .75) = .75/.25 = 3$ (or 3:1).

The odds ratio (as the name implies) is a ratio of odds. Consider again the two characteristics X and Y used earlier to construct tables of counts and probabilities. The odds of $Y = 1$ relative to $Y \neq 1$ (or, equivalently, $Y = 2$) if $X = 1$ are

$$\frac{P(Y = 1 \mid X = 1)}{P(Y = 2 \mid X = 1)} = \frac{P(Y = 1 \text{ and } X = 1)/P(X = 1)}{P(Y = 2 \text{ and } X = 1)/P(X = 1)}$$
$$= \frac{p_{11}/p_{1.}}{p_{12}/p_{1.}}$$
$$= \frac{p_{11}}{p_{12}}.$$

Similarly, the odds of $Y = 1$ relative to $Y \neq 1$ (or, equivalently, $Y = 2$) if $X = 2$ are

$$\frac{P(Y = 1 \mid X = 2)}{P(Y = 2 \mid X = 2)} = \frac{P(Y = 1 \text{ and } X = 2)/P(X = 2)}{P(Y = 2 \text{ and } X = 2)/P(X = 2)}$$
$$= \frac{p_{21}/p_{2.}}{p_{22}/p_{2.}}$$
$$= \frac{p_{21}}{p_{22}}.$$

The odds ratio is then

$$\theta = \frac{p_{11}/p_{12}}{p_{21}/p_{22}} = \frac{p_{11}p_{22}}{p_{12}p_{21}}. \tag{6.4}$$

An odds ratio of 2.0, for example, means that the odds of $Y = 1$ relative to $Y = 2$ are twice as high if $X = 1$ than if $X = 2$. The odds ratio is sometimes called the cross-product ratio, since it is the ratio of the product of the probabilities on the main diagonal of a 2×2 probability table to the product of the probabilities on the off-diagonal.

The odds ratio is useful because of its connection to independence. If the characteristics X and Y are independent, then the odds ratio satisfies

$$\theta = \frac{p_{11}p_{22}}{p_{12}p_{21}} = \frac{p_{1.}p_{.1}p_{2.}p_{.2}}{p_{1.}p_{.2}p_{2.}p_{.1}} = 1.$$

That is, the odds of $Y = 1$ relative to $Y = 2$ if $X = 1$ is the same as the odds of $Y = 1$ relative to $Y = 2$ if $X = 2$. A value of θ greater than one refers to a positive (direct) association between X and Y, in the sense that falling in the first row is associated with falling in the first column and falling in the second row is associated with falling in the second column, given the overall marginal probabilities. A value of θ between 0 and 1 refers to a negative (inverse) association, in the sense that falling in the first row is associated with falling in the second column and falling in the second row is associated with falling in the first column. Since the odds ratio has an inherent asymmetry between inverse association $(0 < \theta < 1)$ and direct association $(1 < \theta < \infty)$, the logarithm of θ is often used as an index of association instead $(\log \theta = 0$ corresponding to independence). The odds ratio typically is estimated using the empirical odds ratio,

$$\hat{\theta} = \frac{n_{11}n_{22}}{n_{12}n_{21}}. \tag{6.5}$$

In order to avoid problems with cells with zero counts, a small positive value, such as $1/2$, is sometimes added to each count before $\hat{\theta}$ is calculated. The odds ratio is crucial in the construction of logistic regression models, which we will discuss in Chapter 9.

The odds ratio is sometimes confused with a different measure, the relative risk. A weakness of the focus in Section 6.1.1 on the difference between two probabilities is that a particular difference is likely to have more importance when both probabilities are near zero or one than when they are near .5. For example, if the incidence rates of a disease are .01 and .001 in two groups, respectively, this represents a tenfold difference in risk, which seems of more clinical importance than the difference between rates of .46 and .451 (which is of course the same absolute difference in incidence), which only represents a 1.8% difference. The relative risk is just this ratio of success probabilities, p_1/p_2. Note that since the designation of "success" and "failure" to the two outcomes in a binary trial is arbitrary, a relative risk of failure probabilities, $(1 - p_1)/(1 - p_2)$, could also be a meaningful measure; generally, the relative risk is defined with respect to the smaller of the two probabilities.

Referring back to (6.4), it is clear that the odds ratio satisfies

$$\theta = \frac{p_1/(1 - p_1)}{p_2/(1 - p_2)} = (\text{Relative risk}) \times \left(\frac{1 - p_2}{1 - p_1}\right).$$

Note that if the relative risk is greater than one, the odds ratio is greater than it, while if the relative risk is less than one, the odds ratio is less than it. If the success probabilities are close to zero for both groups, the second term on the right-hand side of the equation is close to one, and the odds ratio is close to the relative risk. This can be useful in situations where the relative risk is difficult (or impossible) to estimate, but the odds ratio is not.

Consider the following situation. Say we were interested in building a model for the probability that a business will go bankrupt as a function of the initial debt carried by the business. There are two ways that we might imagine constructing a sample (say of size 200) of businesses to analyze:

1. Randomly sample 200 businesses from the population of interest. Record the initial debt, and whether or not the business eventually went bankrupt. This is conceptually consistent with following the 200 businesses through time until they either go bankrupt or don't go bankrupt, and is called a *prospective* sampling scheme for this reason. In the biomedical literature this is often called a *cohort* study.

2. First consider the set of all businesses in the population that didn't go bankrupt; randomly sample 100 of them and record the initial debt. Then consider the set of all businesses in the population that did go bankrupt; randomly sample 100 of them and record the initial debt. This is conceptually consistent with seeing the final state of the businesses first (bankrupt or not bankrupt), and then going backwards in time to record the initial debt, and is called a *retrospective* sampling scheme for this reason. In the biomedical literature this is often called a *case-control* study.

Each of these sampling schemes has advantages and disadvantages. The prospective study is more consistent with the actual physical process that we are interested in; for example, the observed sample proportion of businesses that go bankrupt is an estimate of the actual probability that a randomly chosen business from this population will go bankrupt, a number that cannot be estimated using data from a retrospective study (recall that we arbitrarily decided that half the sample would be bankrupt businesses, and half would be nonbankrupt businesses). A prospective study also can simultaneously examine multiple outcomes (that is, the relationship between debt level and outcomes other than bankruptcy can be easily studied). On the other hand, if bankruptcy rates are low, a sample of size 200 is only going to have a few bankrupt businesses in it, which makes it more difficult to model the probability of bankruptcy (this is a particularly important consideration in medical studies, where disease incidence is often very low).

Assume that initial debt is recorded only as Low or High. This implies that the data take the form of a 2×2 contingency table (whatever the sampling scheme):

Bankrupt

Debt level	Yes	No	Total
Low	n_{LY}	n_{LN}	n_L
High ·	n_{HY}	n_{HN}	n_H
Total	n_Y	n_N	n

Even though the data have the same form for each sampling scheme, the way these data are generated is very different. The following two tables give the expected counts in the four data cells, depending on the sampling scheme. The p values are conditional probabilities (so, for example, $p_{Y|L}$ is the probability of a business going bankrupt given that it has low initial debt):

PROSPECTIVE SAMPLE

Bankrupt

Debt level	Yes	No
Low	$n_L p_{Y|L}$	$n_L p_{N|L}$
High	$n_H p_{Y|H}$	$n_H p_{N|H}$

RETROSPECTIVE SAMPLE

Bankrupt

Debt level	Yes	No
Low	$n_Y p_{L|Y}$	$n_N p_{L|N}$
High	$n_Y p_{H|Y}$	$n_N p_{H|N}$

There is a fundamental difference between the probabilities that can be estimated using the two sampling schemes. For example, what is the probability that a business goes bankrupt given that it has low initial debt? As noted above, this is $p_{Y|L}$. It is easily estimated from a prospective sample ($\hat{p}_{Y|L} = n_{LY}/n_L$), as can be seen from the left table, but it is impossible to estimate it from a retrospective sample. (Note that it is estimable from a prospective sample even if n_L and n_H are chosen before sampling.) For the same reason, the relative risk of going bankrupt given low initial debt versus high initial debt is easily estimated in a prospective sample, but is impossible to estimate in a retrospective sample. On the other hand, given that a business went bankrupt, what is the probability that it had low initial debt? This is $p_{L|Y}$, which is estimable from a retrospective sample $(\hat{p}_{L|Y} = n_{LY}/n_Y)$, but not necessarily from a prospective sample (it is not if n_L and n_H are chosen before sampling).

The comparison between odds for the groupings that are relevant for that sampling scheme, however, is identical in either scheme, and is always estimable. Consider first a prospective study. The odds are $p/(1-p)$, where here p is the probability of a business going bankrupt. For the low debt businesses, $p = p_{Y|L}$, so the odds are

$$\frac{p_{Y|L}}{1 - p_{Y|L}} = \frac{p_{Y|L}}{p_{N|L}}.$$

For the high debt businesses, the odds are

$$\frac{p_{Y|H}}{1 - \pi_{Y|H}} = \frac{p_{Y|H}}{p_{N|H}}.$$

The odds ratio equals

$$\frac{p_{Y|L}/p_{N|L}}{p_{Y|H}/p_{N|H}},$$

which, by the definition of conditional probabilities, simplifies to

$$\frac{p_{LY}p_{HN}}{p_{HY}p_{LN}},$$

the usual cross-product ratio (which can be estimated using the empirical cross-product ratio). Similarly, in a retrospective study, the odds are $p/(1-p)$, where here p is the probability of a business having low debt (remember that the proportion that are bankrupt is set by the design). For the bankrupt businesses, $p = p_{L|Y}$, so the odds are

$$\frac{p_{L|Y}}{1 - p_{L|Y}} = \frac{p_{L|Y}}{p_{H|Y}}.$$

For the nonbankrupt businesses, the odds are

$$\frac{p_{L|N}}{1 - p_{L|N}} = \frac{p_{L|N}}{p_{H|N}}.$$

The odds ratio equals

$$\frac{p_{L|Y}/p_{H|Y}}{p_{L|N}/p_{H|N}},$$

which, by the definition of conditional probabilities, simplifies to

$$\frac{p_{LY}p_{HN}}{p_{HY}p_{LN}},$$

once again the usual cross-product ratio. Say the probability of bankruptcy is small. Then, in a prospective study, the odds ratio would be close to the risk ratio of bankruptcy. In a retrospective study this risk ratio is not directly estimable, but the odds ratio is, and is identical to the prospective odds ratio (and thus approximates the prospective risk ratio).

Predicting the Relapse of Leukemia (continued)

The empirical odds ratio for the leukemia relapse data is

$$\hat{\theta} = \frac{(30)(95)}{(8)(45)} = 7.92,$$

with $\log \hat{\theta} = 2.07$. This is consistent with a strong association between being cancer free and not having a relapse, and having traces of cancer and having a relapse, in that the odds of having a relapse versus not having a relapse are almost eight times higher among children with traces of cancer than children who were cancer free. The estimated relative risk of relapse is

$$\frac{\overline{p_1}}{\overline{p_2}} = \frac{30/75}{8/103} = 5.15.$$

That is, the relapse rate was more than five times higher among children with traces of cancer than children who were cancer free. Note that since in this case the relapse rates are not very small, the relative risk and the odds ratio are considerably different from each other, although (of course) both lead to the same general impression.

6.2 Loglinear Models for Two-Way Tables

6.2.1 2×2 Tables

Recall the definition of independence: X and Y are independent if

$$P(X = i \text{ and } Y = j) = P(X = i) \times P(Y = j)$$

for all i and j, or, using notation for a contingency table,

$$p_{ij} = p_{i \cdot} p_{\cdot j}.$$

Another way to represent this is in terms of logged expected values; since $e_{ij} = np_{ij}$, under independence

$$\log e_{ij} = \log n + \log(p_{i\cdot}p_{\cdot j})$$
$$= \log n + \log p_{i\cdot} + \log p_{\cdot j}. \tag{6.6}$$

Independence is consistent with an additive model for the logged expected cell count, which can be written as

$$\log e_{ij} = \lambda + \lambda_i^X + \lambda_j^Y, \tag{6.7}$$

where X and Y are labels to represent the two variables X (rows) and Y (columns). This is just a loglinear model based on two categorical predictors, and to make the model identifiable a constraint must be put on the λ_i^X and λ_j^Y parameters, such as $\lambda_1^X + \lambda_2^X = 0$ and $\lambda_1^Y + \lambda_2^Y = 0$. Thus, the loglinear model is simply a count regression with four observations, based on two effect coding predictors representing the two marginal effects.

The nonuniqueness of the parameter constraints implies that the λ parameters are also not unique, so their interpretation depends on the constraint chosen. On the other hand, differences between them are meaningful (and unique), providing information about odds. For example, consider the odds of falling in the first column versus falling in the second column. Under independence the odds are the same for both rows, since the odds ratio is one, so these odds equal

$$\frac{P(Y = 1 | X = 1)}{P(Y = 2 | X = 1)} = \frac{p_{11}/p_{1\cdot}}{p_{12}/p_{1\cdot}} = \frac{p_{11}}{p_{12}}.$$

Since the ratio of probabilities equals the ratio of expected counts, the log of the odds equals

$$\log \left[\frac{P(Y = 1 | X = 1)}{P(Y = 2 | X = 1)} \right] = \log \left[\frac{e_{11}}{e_{12}} \right]$$
$$= \log(e_{11}) - \log(e_{12})$$
$$= (\lambda + \lambda_1^X + \lambda_1^Y) - (\lambda + \lambda_1^X + \lambda_2^Y)$$
$$= \lambda_1^Y - \lambda_2^Y.$$

Thus, under independence, in each row the odds of being in column 1 relative to being in column 2 are $\exp(\lambda_1^Y - \lambda_2^Y)$, no matter what constraint is put on the parameters. This difference represents the column effect, which is the same for both rows under independence. Of course, the corresponding result also holds for columns (that is, in each column the odds of being in row 1 relative to being in row 2 are $\exp(\lambda_1^X - \lambda_2^X)$, the row effect, which is the same for both columns under independence). If the difference between the two row parameters (or two column parameters) is zero, the odds of being in one row (or column) versus the other are one; that is, the marginal distribution for rows (or columns) is uniform (the probability of falling in each category is the same). If uniformity of rows, for example, was of interest, its presence could thus be tested via a test of the hypotheses

$$H_0 : \lambda_1^X = \lambda_2^X = 0$$

versus

$$H_a : \text{at least one } \lambda_i^X \neq 0.$$

This is identical to a test for a binomial proportion equaling .5 when there are two rows.

Tests of independence also can be represented using the loglinear model. If $p_{ij} \neq p_{i\cdot} p_{\cdot j}$, the logged expected cell count can be written

$$\log e_{ij} = \lambda + \lambda_i^X + \lambda_j^Y + \lambda_{ij}^{XY}, \tag{6.8}$$

where

$$\lambda_{ij}^{XY} = \log \left(\frac{p_{ij}}{p_{i\cdot} p_{\cdot j}} \right).$$

This term corresponds to an interaction effect; just as in the case of analysis of variance models (Section 3.2.3), the interaction reflects that the main effect for one factor is different depending on the level of the other factor (a lack of independence). Without any further restrictions on the λ_{ij}^{XY} terms, the model (6.8) is *saturated*, in the sense that it fits the table perfectly ($\hat{e}_{ij} = n_{ij}$ for all (i, j)). The test of the adequacy of the independence model is just a test of model (6.7) versus (6.8), or equivalently

$$H_0 : \boldsymbol{\lambda}^{XY} = \mathbf{0}$$

versus

$$H_a : \boldsymbol{\lambda}^{XY} \neq \mathbf{0}.$$

Unlike for the count regression models in Chapter 5, lack of fit is not necessarily viewed as problematic in this context; rather, deviations from independence are viewed as representing interesting interrelationships between the row variable and column variable.

6.2.2 $I \times J$ Tables

The great benefit of loglinear modeling comes when contingency tables are generalized to those with more than two rows and/or more than two columns, since models (6.7) and (6.8) generalize directly. Consider an $I \times J$ contingency table:

X	1	2	$\cdots$	J	Total
1	n_{11}	n_{12}	$\cdots$	n_{1J}	$n_{1\cdot}$
2	n_{21}	n_{22}	$\cdots$	n_{2J}	$n_{2\cdot}$
$\vdots$		$\vdots$		$\vdots$	
I	n_{I1}	n_{I2}	$\cdots$	n_{IJ}	$n_{I\cdot}$
Total	$n_{\cdot 1}$	$n_{\cdot 2}$	$\cdots$	$n_{\cdot J}$	n

Y

Independence is still defined as it was before, as in equations (6.6) and (6.7), since there is nothing special about the 2×2 situation. The model still can be viewed as a count regression model, now based on IJ observations and two sets of effect coding predictors ($I-1$ for rows, $J-1$ for columns). Under independence, the fitted values satisfy

$$\hat{e}_{ij} = \frac{n_{i\cdot}n_{\cdot j}}{n}, \tag{6.9}$$

just as in the 2×2 situation. Assume multinomial sampling; that is, $\mathbf{n} \sim \mathrm{Mult}(n, \mathbf{p})$. This implies that $\hat{e}_{ij} = n\hat{p}_{ij}$, which equals $n\hat{p}_{i\cdot}\hat{p}_{\cdot j}$ under independence. Omitting constants, the log-likelihood is

$$\begin{aligned} L &= \sum_{ij} n_{ij} \log p_{ij} \\ &= \sum_{ij} n_{ij} \log[p_{i\cdot}p_{\cdot j}] \\ &= \sum_{ij} n_{ij}[\log p_{i\cdot} + \log p_{\cdot j}] \\ &= \sum_{i} n_{i\cdot} \log p_{i\cdot} + \sum_{j} n_{\cdot j} \log p_{\cdot j} \end{aligned}$$

That is, the log-likelihood factors into two separate parts that refer to the two marginal distributions. Each can thus be treated as multinomials in their own right, with maximum likelihood estimates $\hat{p}_{i\cdot} = n_{i\cdot}/n$ and $\hat{p}_{\cdot j} = n_{\cdot j}/n$, respectively. Thus, the MLE fitted values under independence are as in (6.9). The model (6.7) is based on estimating $1+(I-1)+(J-1) = I+J-1$ parameters, so χ^2 tests of goodness-of-fit of the model (such as (4.13) and (4.14), only now with observed and fitted counts double-subscripted) are based on $IJ - (I + J - 1) = (I - 1)(J - 1)$ degrees of freedom.

When Do Business Students Choose Their Major?

During the Spring 1996 and Fall 1996 semesters, a survey was given to undergraduate students in the Leonard N. Stern School of Business of New York University as part of a program to assess and improve the undergraduate marketing program (LaBarbera and Simonoff, 1999). The following table summarizes the responses to two questions: the student's major (major or plan to major in marketing, don't plan to major in marketing, or undecided), and when the student chose or will choose their major (in high school, as a freshman, as a sophomore, as a junior, or as a senior):

	Major in marketing?			
Time when major chosen	Yes	No	Undecided	Total
High school	16	41	8	65
Freshman	21	29	8	58
Sophomore	38	63	12	113
Junior	34	30	8	72
Senior	9	12	10	31
Total	118	175	46	339

Is there a relationship between major and when the major is chosen? χ^2 tests of independence say yes, as $X^2 = 19.04$ on 8 degrees of freedom ($p = .015$) and $G^2 = 16.87$ ($p = .031$). What accounts for the rejection of independence? One way to see this is to look at the pattern of column percentages (the percentages of the total observations in a column that fall in a given row, for the different columns) or the pattern of row percentages (the percentages of the total observations in a row that fall in a given column, for the different rows) to see how they differ, since under independence these patterns would be roughly the same for all columns or rows (under independence the jth entry in the row percentage vector has expected value $p_{.j}$ for all rows, while the ith entry in the column percentage vector has expected value $p_{i.}$ for all columns). There's no inherent advantage to using row versus column percentages, so the choice can be made based on the data context.

Column percentages (that is, percentages of students of the three major categories that made their choices at each of the different times) are given on the next page. The percentages show that (as would be expected) people who are undecided about whether they will major in marketing make their decision later, with almost 40% deciding as juniors or seniors. In contrast, the nonmarketing majors decide relatively early, with more than 75% having decided by their sophomore years. Marketing majors fall somewhere in between, deciding later than nonmarketing majors. This reflects an ongoing concern of marketing educators, that marketing is viewed as a "fallback" major that is chosen if a student can't handle a more difficult major. The problem is compounded by Association to Advance Collegiate Schools of Business (AACSB) accreditation standards that call for the majority of business courses to be taken during the last two years of the undergraduate career.

Major in marketing?

Major chosen	Yes	No	Undecided	Total
High school	13.6	23.4	17.4	19.2
Freshman	17.8	16.6	17.4	17.1
Sophomore	32.2	36.0	26.1	33.3
Junior	28.8	17.1	17.4	21.2
Senior	7.6	6.9	21.7	9.1
Total	100.0	100.0	100.0	

A graphical representation of these percentages that can help to show what's going on is a *mosaic plot*. In this plot row or column percentages are represented by the heights of a box, where the width of the boxes in a given column is proportional to the percentage of observations that fall in the column (column percentages are used here, to correspond to the table given earlier). Under independence, all of the boxes in a given row would be the same height, so different patterns looking down the columns illustrate violations of independence. Figure 6.1 gives a mosaic plot for these data.

The display shows that there are more nonmarketing majors than any other group (the middle column is the widest of the three), and they tend to choose their majors earlier (reflected in the larger heights of the top boxes in that column). The relatively large height of the bottom box in the third column reflects that many undecideds choose late.

The marketing major data set is reasonably viewed as coming from an initial Poisson sampling scheme, since the total sample size was unknown until after the sampling was done. Analyzing it as a multinomial random variable after conditioning on the observed sample size is perfectly appropriate, since (as was noted in Section 4.3.3), Poisson sampling conditional

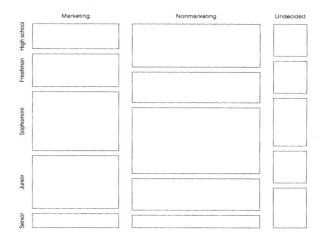

FIGURE 6.1. Mosaic plot for marketing major table.

on the total sample size is equivalent to multinomial sampling. The multinomial analysis is equivalent to a Poisson regression analysis, as long as an intercept term is included in the model.

A common sampling structure that leads to contingency tables is one where several populations are sampled separately, with the responses from the members of each population (defined by a variable X, say) forming a multinomial vector (defined by a variable Y, say). Both prospective studies and retrospective studies can have this form, for example, the distinction depending on whether a possible cause or possible effect are the populations being sampled. This sampling scheme also results from Poisson or multinomial sampling if inference is going to proceed conditioning on marginal totals (for example, the conditional probabilities for a characteristic for men are compared to those for women). Since the fitted values for the independence model condition on the marginal totals, there is no difference in fitting or testing for this product multinomial (sometimes called independent multinomial) sampling situation. In this context the hypothesis of interest is whether the probability vectors for Y given different values of X are the same; that is, the equality of conditional probabilities hypothesis, also called the homogeneity hypothesis. This is equivalent to independence, however, since conditional probability equality,

$$P(Y = j \mid X = i) = P(Y = j \mid X = i') = P(Y = j), \qquad i \neq i', \text{ for all } j,$$

holds if and only if $P(Y = j \text{ and } X = i) = P(Y = j)P(X = i)$ for all i and j.

Does Antibiotic Prescription Reduction Affect Diagnosis?

Medical evidence has accumulated in recent years that prescribing antibiotics to treat patients with acute bronchitis is of modest benefit. Since prescribing antibiotics needlessly can lead to drug-resistant strains of bacteria, overmedication, and increased cost, a group of doctors at the Medical University of South Carolina instituted a quality improvement project that provided educational programs to clinicians in the practice and recorded the use of antibiotic agents in patients with acute bronchitis over a 14-month period (Hueston and Slott, 2000). During this time there was a marked reduction in the percentage of patients diagnosed with acute bronchitis who were prescribed antibiotics. While this could reflect improved patient care, an alternative possibility is that physicians continued to prescribe antibiotics, but assigned a diagnosis other than acute bronchitis. In order to investigate this, the pattern of diagnosis of respiratory tract conditions was examined for 3-month periods, to see if there was a change in this pattern; movement away from the diagnosis of acute bronchitis towards other diagnoses would be consistent with physicians changing documentation, rather than actual practice. The results were as follows (URI refers to upper respiratory tract infection):

| | Time period | | | | |
Diagnosis	1–3/96	4–6/96	7–9/96	10–12/96	1–2/97
Acute bronchitis	113	58	40	108	100
Acute sinusitis	99	37	23	50	32
URI	410	228	125	366	304
Pneumonia	60	43	30	56	45
Total	682	366	218	580	481

As would be expected, diagnosis of respiratory tract conditions has a distinct seasonal component, peaking in the winter and dropping during the summer (note that the last time period corresponds to only 2 months, which accounts for the decrease in diagnoses). The mosaic plot for this table (top plot of Figure 6.2) doesn't look that different from that for the table of expected counts from an independence model (bottom plot of Figure 6.2), but there are differences, which are reflected in significant lack of fit of independence ($X^2 = 30.17$, $G^2 = 29.59$, $df = 12$, $p \approx .002$).

A way to see why independence is rejected is to look at the standardized Pearson residuals (often called the *adjusted residuals* in the contingency table context) from the independence fit. For the independence model they

Observed counts

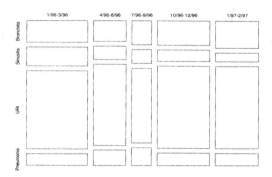

Independence model

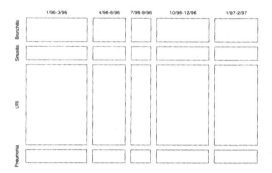

FIGURE 6.2. Mosaic plots of respiratory tract condition table. Observed counts (top plot) and fitted counts based on independence (bottom plot).

have the form

$$r_{ij} = \frac{n_{ij} - \hat{e}_{ij}}{\sqrt{\hat{e}_{ij}(1 - n_{i.}/n)(1 - n_{.j}/n)}}.$$

The standardized residuals are given on the next page. They (and their signs in particular) show that the lack of fit comes primarily from the acute sinusitis diagnosis, which occurred less often over time (dropping from 15% of the diagnoses at the beginning of the study to 7% at the end). There is a corresponding increase in the acute bronchitis diagnosis (both of these patterns can be seen from the change in the heights of the boxes in the first two rows of the mosaic plot). Note that this would not reflect physicians attempting to "cheat" the quality improvement project, since that would result in a decrease in acute bronchitis diagnoses and an increase in acute sinusitis diagnoses, rather than the observed (opposite) pattern. The authors of the study speculate that the intervention to discourage the use of antibiotics to treat acute bronchitis might have had the unintended effect of diffusing to acute sinusitis, another disorder that is often treated with antibiotics, resulting in the physicians being more careful of how often they diagnosis sinusitis.

| | Time period | | | | |
Diagnosis	1–3/96	4–6/96	7–9/96	10–12/96	1–2/97
Acute bronchitis	−1.16	−1.17	0.14	0.44	1.78
Acute sinusitis	4.24	−0.17	0.10	−1.58	−2.99
URI	−0.93	0.31	−1.35	0.87	0.82
Pneumonia	−1.30	1.17	1.91	−0.37	−0.57

Retrospective studies suffer from the inability to directly estimate the probabilities of interest from the table, but this does not mean that the association between a possible cause and effect cannot be investigated using χ^2 tests of independence. Independence implies constant conditional probabilities, whether those are row probabilities given column categories, or column probabilities given row categories, and the lack of independence implies a relationship between putative cause and effect, even if the sampling design conditions on the effect rather than the cause.

Is There a Relationship Between Smoking and Skin Cancer?

It is well-known that smoking has a deleterious effect on the health of internal organs, but the relationship with different forms of skin cancer is less well-understood. A hospital-based case-control (retrospective) study was performed in the Netherlands to investigate this relationship (De Hertog et al., 2001). The study included 161 patients with squamous cell carcinoma and 386 controls. The smoking status of each person was ascertained, yielding the table on the next page.

There is a strong association between smoking status and disease status ($X^2 = 19.8$, $G^2 = 20.0$, $p \approx .00005$). The empirical odds ratio for the 2×2 subtable of people who never smoked and former smokers is $(136)(78)/(180)(31) = 1.9$. While skin cancer is the most common form of cancer in the United States, it is still relatively rare, so this odds ratio provides a reasonable estimate of the relative risk of suffering from squamous cell carcinoma if a person is a former smoker relative to never having smoked (that is, the risk is estimated to be 90% higher for former smokers). Similarly, the empirical odds ratio for the 2×2 subtable of people who never smoked and current smokers is $(136)(52)/(70)(31) = 3.3$, indicating more than three times the risk for current smokers compared to people who have never smoked.

| | Disease status | |
Smoking status	Controls	Squamous cell carcinoma
Never smoked	136	31
Former smoker	180	78
Current smoker	70	52

6.3   Conditional Analyses

In the last section we discussed analysis of contingency tables arising from Poisson, multinomial, and product multinomial sampling schemes. A final possible sampling structure is one where both marginal totals are fixed by sampling. This can no longer be written as a multinomial (or product of multinomials) distribution. We will consider first the 2×2 situation.

6.3.1   Two-by-Two Tables and Fisher's Exact Test

Consider a 2×2 table, where the row totals $\{n_{1.}, n_{2.}\}$ and column totals $\{n_{.1}, n_{.2}\}$ are fixed and known. In this case any one entry in the table completely determines all of the others, so there is no need for a test statistic to test independence; rather, all that is needed is knowledge of the distribution of any one cell under independence. To fix ideas, we will focus on n_{11}, although any cell can be used. The distribution for n_{11} under independence, given the row and column margins, is *hypergeometric*, which has probability function

$$P(n_{11} = x) = \frac{\binom{n_{1.}}{x}\binom{n_{2.}}{n_{.1}-x}}{\binom{n}{n_{.1}}}, \quad \max(0, n_{1.} + n_{.1} - n) \le x \le \min(n_{1.}, n_{.1})$$

$$= \frac{n_{1.}!n_{2.}!n_{.1}!n_{.2}!}{n!n_{11}!n_{12}!n_{21}!n_{22}!}. \tag{6.10}$$

Note that this distribution does not involve the unknown parameters of the relative risk for rows or columns, because the marginal totals are sufficient statistics for those values.

Fisher (1935) discussed the use of this distribution to test independence in a 2×2 contingency table using what has come to be known as Fisher's exact test, based on an incident that took place in the late 1920s. A colleague of Fisher's claimed that she could tell the difference between cups of tea where milk was added before tea and cups where tea was added before milk. To test this claim, Fisher randomly presented the woman with eight cups of tea, four of which had milk poured first and four of which had tea poured first (the woman knew that there were four cups of each type). Say the woman correctly identified three of the four cups of each type. This would lead to the following observed table:

Woman's guess

True property	Tea poured first	Milk poured First	Total
Tea poured first	3	1	4
Milk poured first	1	3	4
Total	4	4	8

Does this result indicate that Fisher's colleague can distinguish between the two types of cups of tea? If, in fact, she could not, the rows and columns of the table would be independent, and n_{11} would follow a hypergeometric distribution. The probability that she would correctly identify as many as three of the cups that had tea poured first correctly is

$$P(n_{11} \geq 3) = P(n_{11} = 3) + P(n_{11} = 4)$$

$$= \frac{\binom{4}{3}\binom{4}{1}}{\binom{8}{4}} + \frac{\binom{4}{4}\binom{4}{0}}{\binom{8}{4}}$$

$$= .229 + .014 = .243.$$

Note that this is a one-sided alternative to independence, since we are only examining possible tables that show agreement at least as good as the observed table. The observed probability is not very small, of course, which would lead us to say that the observed evidence is not sufficient to reject the hypothesis that the woman cannot tell the difference between the two types of cups. A two-sided tail probability would be defined by summing all of the hypergeometric probabilities that are not larger than the probability for the observed value of n_{11}.

Fisher's exact test has the great advantage of being exact (since the test distribution does not involve any nuisance parameters), and therefore applicable to tables with very small cell counts where asymptotic arguments fail. Technically, it is only valid if both the row and column margins are fixed by the sampling scheme, but it is widely applied to tables with small observed counts even when that is not the case, based on the argument that significance is being assessed conditional on the observed margins, which contain little or no information about the association in the table. Determination of p-values requires enumeration of the appropriate hypergeometric distribution, which can be computationally intensive (the existence of high-speed computing makes this less of an issue). The test suffers from the same problem as the exact tests described in Chapter 4, however, in that it can be very conservative. For that reason, using a mid-p-value is advisable, since conservativeness is reduced (at the expense of exactness, of course). For the tea-tasting table the mid-p-value is $(.229)(.5) + .014 = .129$, still not enough to reject the null hypothesis of independence.

Interestingly, when the original tea-tasting experiment is described in the literature, the situation where $n_{11} = 3$ seems to be the one typically discussed (see, for example Agresti, 1996, page 40). Fisher (1935) never gave the results of the actual experiment, but according to Salsburg (2001, page 8), H.F. Smith (who was there) claimed that the woman correctly identified all eight cups. The mid-p-value for this outcome ($n_{11} = 4$) is .007, providing reasonably strong evidence that Fisher's colleague could, in fact, correctly distinguish the two types of cups of tea.

Can You Hide Your Lyin' Eyes (or Face)?

Technology that can be used to quickly screen many individuals for evidence of deceit is important in preventing terrorism. Pavlidis et al. (2002) investigated the use of thermal imaging for this purpose, based on the observation that people who are startled exhibit instantaneous warming around the eyes, presumably as a result of an unconscious "fright or flight" response. Twenty volunteers were split into two groups: one where the volunteer committed a mock crime (stabbing a mannequin and robbing it of $20) and then denied committing the "crime," and the other a control group. A high-definition thermal imaging technique was used to try to identify the people who were lying about committing the crime, with the following results:

Classified group

True group	Guilty	Innocent	Total
Guilty	6	2	8
Innocent	1	11	12
Total	7	13	20

χ^2 tests strongly reject the hypothesis of independence ($X^2 = 9.38$, $G^2 = 10.01$, $p \approx .002$), but the small expected cell counts cast doubt on the validity of these results. Fisher's exact test has mid-p equal to .002, so the asymptotic tests are not problematic. Thus, the thermal imaging technique is apparently effective at distinguishing the guilty individuals from the innocent ones.

The subjects were also evaluated using a polygraph, with the following results:

	Classified group		
True group	Guilty	Innocent	Total
Guilty	6	2	8
Innocent	4	8	12
Total	10	10	20

χ^2 tests marginally reject the hypothesis of independence ($X^2 = 3.33$, $p = .068$; $G^2 = 3.45$, $p = .063$), but the exact test suggests less evidence against independence (mid-$p = .095$). Thus, for these data, the standard polygraph diagnostic is, if anything, less effective than the newer, less time-consuming, method.

6.3.2 $I \times J$ Tables

Conditional analysis can be generalized to $I \times J$ tables, although in this case computational issues become more important. Once again, conditioning on the row and column marginal totals results in a null distribution free of any nuisance parameters, allowing exact inference. The hypergeometric distribution (6.10) generalizes to a *multiple hypergeometric distribution* for the observed table $\{n_{ij}\}$ under independence,

$$P(\mathbf{n}) = \frac{(n_1.! \cdots n_I.!)(n_{.1}! \cdots n_{.J}!)}{n! \prod_{ij} n_{ij}!}. \tag{6.11}$$

Since this distribution depends on $(I-1)(J-1)$ free values, a test statistic is needed to define an exact test; any of the usual χ^2 tests can be used. The exact p-value is determined by ordering all of the possible tables with observed marginal totals based on the chosen statistic, and then summing the multiple hypergeometric probabilities for all tables with test statistics at least as extreme as that of the observed table. This only is feasible for relatively small tables with relatively small observed counts, as the number of possible tables gets large very quickly (although clever computational algorithms can cut computing time considerably). An alternative is to calculate a Monte Carlo–based p-value. In this situation thousands of tables are generated at random based on the distribution (6.11). The chosen statistic is calculated for each generated table, with the estimated p-value being the proportion of generated tables with values of the statistic at least as large as that for the actual table.

The next example illustrates a situation where a conditional analysis is clearly appropriate based on the actual sampling scheme, although exact analysis is ultimately infeasible without appealing to a Monte Carlo approximation.

The 1970 Draft Lottery

In an attempt to equalize the risks of being drafted, a lottery was held on December 1, 1969, to allocate birthdates at random for the 1970 draft (Starr, 1997). The process was as follows: 366 capsules, each corresponding to a unique (1952) birthdate, were successively drawn from a container. The first date drawn, September 14, was assigned rank 1, and those eligible for the draft born on that date were called first for physicals. This process was then repeated until all of the capsules were chosen. The capsules were put in the container month by month (January through December), leading to the possibility of an ordering in the capsules being chosen related to birth month. The following table formalizes this by converting the results to a contingency table, separating the dates by month. In this table the first 31 capsules chosen are considered "January" dates in the lottery, the next 29 chosen are considered "February," and so on.

Actual month	Month in draft lottery												Total
	Jan	Feb	Mar	Apr	May	Jun	Jul	Aug	Sep	Oct	Nov	Dec	
Jan	1	2	2	2	1	3	4	2	4	3	2	5	31
Feb	3	1	1	1	4	1	2	5	1	1	4	5	29
Mar	1	4	0	5	2	4	2	1	3	3	3	3	31
Apr	4	0	1	0	2	3	4	5	2	2	3	4	30
May	1	3	3	3	2	2	0	3	2	2	6	4	31
Jun	2	2	3	0	2	1	2	3	5	2	3	5	30
Jul	4	4	2	2	2	2	4	1	4	4	2	0	31
Aug	5	3	3	3	1	2	1	0	4	7	1	1	31
Sep	1	0	6	7	1	4	2	2	4	2	1	0	30
Oct	2	6	3	0	5	1	6	2	0	3	2	1	31
Nov	4	0	4	2	6	1	3	4	1	0	2	3	30
Dec	3	4	3	5	3	6	1	3	0	2	1	0	31
Total	31	29	31	30	31	30	31	31	30	31	30	31	366

Immediately after the lottery was held controversy erupted, as it appeared that the pattern of birthdate choices was not fair. Interestingly, a scatter plot (Figure 6.3) of the lottery order by the birthdate order shows

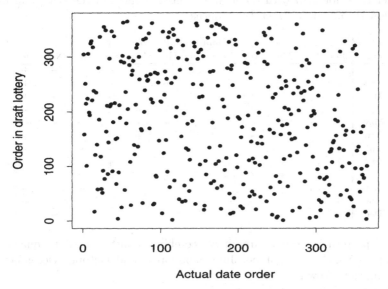

FIGURE 6.3. Scatter plot of order in which date was chosen in 1970 draft lottery versus actual date order.

little apparent evidence of a problem. What then was the source of accusations of unfairness? The capsules were put in month-by-month, which suggests a connection to month. The sampling structure fixes both marginal totals to be the numbers of dates in each month, so a conditional analysis is appropriate. Unfortunately, it is computationally infeasible to get an exact p-value, but Monte Carlo tail probabilities are available. The Pearson statistic X^2 equals 155.9, with Monte Carlo p-value (based on 100,000 replications) $p = .017$ (the asymptotic p-value is .018), while the likelihood ratio statistic $G^2 = 180.3$, with Monte Carlo p-value .009 (the asymptotic p-value based on $df = 121$ is noticeably smaller at $p = .0004$, reflecting that the χ^2 approximation for G^2 is more problematic than that for X^2).

The signs of the residuals from the independence fit (given on the next page) highlight what is going on here. Minus signs (reflecting fewer observations than expected) cluster in the upper left and lower right of the table, while plus signs (corresponding to more observations than expected) cluster in the upper right and lower left. That is, earlier birthdates, which were put in the container first, were drawn later than expected, while later birthdates, which were put in the container last, were drawn earlier than expected. A hypothesis that the capsules were not mixed sufficiently is clearly indicated. The potential human cost of this nonrandomness is clear, and apparently actually occurred: Sommers (2003) examined fatal casualties in Vietnam of men who were subject to the draft lottery, and found higher rates for men with birthdates in later months of the year than for those with birthdates in earlier months.

Actual	*Month in draft lottery*											
month	Jan	Feb	Mar	Apr	May	Jun	Jul	Aug	Sep	Oct	Nov	Dec
Jan	−	−	−	−	−	+	+	−	+	+	−	+
Feb	+	−	−	−	+	−	−	+	−	−	+	+
Mar	−	+	−	+	−	+	−	−	+	+	+	+
Apr	+	−	−	−	−	+	+	+	−	−	+	+
May	−	+	+	+	−	−	−	+	−	−	+	+
Jun	−	−	+	−	−	−	−	+	+	−	+	+
Jul	+	+	−	−	−	−	+	−	+	+	−	−
Aug	+	+	+	+	−	−	−	−	+	+	−	−
Sep	−	−	+	+	−	+	−	−	+	−	−	−
Oct	−	+	+	−	+	−	+	−	−	+	−	−
Nov	+	−	+	−	+	−	+	+	−	−	−	+
Dec	+	+	+	+	+	+	−	+	−	−	−	−

The pattern is even clearer if we combine months into four quarters (January–March, April–June, July–September, and October–December), yielding the following table:

Actual	*Quarter in draft lottery*				
quarter	1st quarter	2nd quarter	3rd quarter	4th quarter	Total
1st quarter	15	23	24	29	91
2nd quarter	19	15	26	31	91
3rd quarter	28	24	22	18	92
4th quarter	29	29	20	14	92
Total	91	91	92	92	366

The nonindependence ($X^2 = 20.4$, $p = .015$; $G^2 = 21.1$, $p = .012$, both with $df = 9$) is now much clearer, as a mosaic plot (Figure 6.4) shows. Note that in this plot the row percentages determine the boxes, as that reinforces the concept that the actual birth quarter is a predictor of the lottery quarter, rather than the other way around. Thus, the discretization of effectively continuous data into four categories, rather than obscuring the relationship (as might be expected), makes it much clearer.

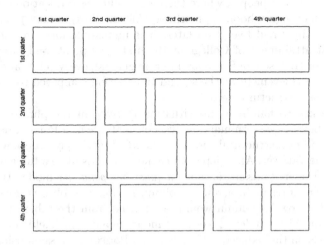

FIGURE 6.4. Mosaic plot of 1970 draft lottery table for data put in quarterly form.

6.4 Structural Zeroes and Quasi-Independence

It sometimes happens that a contingency table has cells with zero observations in them. There are two different kinds of mechanisms that can cause this to happen: a random occurrence, because of a small probability of the event occurring that corresponds to that cell (called a *random zero*); or a nonrandom occurrence, because it is impossible for the combination of levels of the factors to occur (called a *structural zero*).

Random zeroes are generally not a problem, unless they lead to problems in calculating the maximum likelihood estimates for the model. For the independence model, this only happens if all of the cells in a row or column have zero observations, in which case that row or column would be omitted from the table. Random zeroes typically correspond to small expected counts, however, so they can indicate that the usual asymptotic approximations for goodness-of-fit tests, tests of significance, etc., might not be valid.

Structural zeroes are a very different matter. These are cells that cannot possibly have any observations in them, so they are effectively "not there." How might a structural zero arise? Consider, for example, a cross-classification of people, where the person's highest educational attainment (Less than high school, High school, College, Postgraduate) is recorded at a given time, and five years later. It is impossible for someone to have a highest attainment of College at the first time point, and Less than high school or High school five years later; in general, every cell that corresponds to lower achievement at the second time period compared to the first time period is a structural zero.

The key to handling this situation is given in the phrase "not there" used above. If we formulate analysis of the table as a Poisson regression on effect codings, structural zeroes are handled by simply dropping those cells from the data set. An independence model that is fit to a table with certain cells considered missing is called a *quasi-independence* model. Unless structural zeroes result in a row or column having zero observations (in which case that row or column would be omitted from the table, just as in the situation with random zeroes), standard maximum likelihood estimation proceeds on the reduced table (that is, a Poisson regression using the usual variables that define independence, with the observations corresponding to the structural zeros omitted from the data set). Note that structural zeroes affect the degrees of freedom of goodness-of-fit tests; since there are only $IJ - s$ "observations" in the Poisson regression (where s is the number of structural zeroes), degrees of freedom for tests of fit are reduced by s.

An equivalent way to fit the model is to add s indicator variables as predictors, where each indicator identifies a particular outlier cell. Since this adds s predictors to the regression model, the degrees of freedom for the test are reduced by s, as should be the case. The fitted values for the structural zero cells are forced to be zero in this regression, effectively removing them from the model fitting.

New Cancer Incidence and Gender

The Division of Cancer Prevention and Control of the National Cancer Institute provides counts of new cases of cancer separated by various demographic and geographical factors (actually, these are estimated counts of new cases based on data from ten sites that are applied to other locations). The table on the next page gives counts of different cancers, separated by gender, for New York State in 1989. The cells corresponding to (Ovarian cancer, Male) and (Prostate cancer, Female) are given as "—," since these cells correspond to structural zeroes; these combinations of gender and cancer type cannot occur.

Analysis of this table with those cells considered missing shows that there is a strong association between gender and type cancer, as tests of quasi-independence reject it strongly ($G^2 = 40.0$, $X^2 = 40.4$; degrees of freedom for the tests are $(5 - 1)(2 - 1) - 2 = 4 - 2 = 2$, so both tests have $p < .00001$). Note that since this table has only two columns, once the structural zeroes are taken into account the (Ovarian cancer, Female) and (Prostate cancer, Male) cells are perfectly determined, and provide no information about independence. Thus, the table is effectively collapsed down to the 3×2 table that omits the rows corresponding to ovarian and prostate cancers.

Gender

Type of cancer	Male	Female	Total
Lung	7355	4831	12,186
Melanoma	1104	964	2068
Ovarian	—	1563	1563
Prostate	9986	—	9986
Stomach	1014	618	1632
Total	19,459	7976	27,435

A table of Pearson residuals demonstrates that the violation of quasi-independence comes from relatively lower percentages of lung cancer and stomach cancer incidence among women, and relatively higher percentage of melanoma incidence among women. This could easily reflect behavioral differences between the genders, since lung cancer is associated with smoking, stomach cancer with diet, and melanoma with overexposure to the sun.

Gender

Type of cancer	Male	Female
Lung	1.04	−1.26
Melanoma	−3.68	4.47
Ovarian		0.00
Prostate	0.00	
Stomach	1.31	−1.59

The following table gives similar data, this time for Alaska, a much smaller state in population. A key point when examining this table is that the zero entry for female incidence of stomach cancer is a random zero, not a structural zero, so it is included in the analysis.

Gender

Type of cancer	Male	Female	Total
Lung	90	38	128
Melanoma	15	15	30
Ovarian	—	18	18
Prostate	111	—	111
Stomach	5	0	5
Total	221	71	292

Evidence against quasi-independence is much weaker in this table (no doubt the smaller sample size has a lot to do with this): $X^2 = 7.1$ ($df = 2$, $p = .03$) and $G^2 = 8.3$ ($p = .02$). Pearson residuals show that the main deviation from independence comes from a higher incidence of stomach and lung cancer in men and melanoma in women than expected, just as was true for the New York data.

Gender

Type of cancer	Male	Female
Lung	0.39	−0.56
Melanoma	−1.17	1.68
Ovarian		0.00
Prostate	0.00	
Stomach	0.88	−1.27

6.5   Outlier Identification and Robust Estimation

In all of the examples that we've considered, rejection of the null hypothesis was viewed as evidence of what we might call a structural violation of the model; that is, rejection of independence in a table was evidence of a general interaction between the two variables defining it. What if, instead, independence fit the table reasonably, except for a few cells, where independence was violated? This is a different kind of situation, since we could then summarize the relationship in the table as being consistent with independence, except for a few unusual cells.

Such cells are *outlier cells*. The concept of an outlier in contingency tables is, not surprisingly, similar to that in Poisson regression. In the contingency table context, the cells are the observations, rather than the individual members of the population that make up the counts in the cells. As such, it is the count in a cell that might be outlying relative to a null model, just as in Poisson regression it is the count in an observation that might be outlying (recall, for example, the outlier months in the killer tornadoes data of Section 5.2).

What if a table does contain outlier cells? In the least squares and Poisson regression contexts outliers could be removed from the data, and the model could be refit to assess the effect of the outliers. How can we "remove" observations from a contingency table? The answer, of course, is to treat them as structural zeroes, since that effectively removes them from the table. From the regression point of view, this corresponds to literally removing the lines in the data matrix that correspond to the outlying cells, and is no different from omitting outliers in any other Poisson regression model.

Outlier cells can be identified using any type of residual. The potential benefit of omitting an outlier cell then can be informally tested using a likelihood ratio test based on (5.11), since the difference between G^2 values with and without the outlier(s) is a test of whether omitting the cell(s) improves the fit of the model significantly (the test would be based on k degrees of freedom, where k is the number of cells omitted). It should be noted that since the outliers are chosen based on observed residuals (rather than *a priori*), tail probabilities are not strictly valid (they will be smaller than they should be), but tests of this type are still useful as guidelines. An informal way to see the effect of omitting outlier cells is to compare the observed count in the outlier cell to the fitted value from the quasi-independence model that omits that cell, say $\hat{e}_{ij}^*$, since $\hat{e}_{ij}^*$ is unaffected by the outlier itself. The difference between n_{ij} and $\hat{e}_{ij}^*$ is called the *deleted residual* (see page 53), and a (Pearson) standardized version is

$$\frac{n_{ij} - \hat{e}_{ij}^*}{\sqrt{\hat{e}_{ij}^*}}.$$

The Location of Prehistoric Artifacts

As a result of archaeological excavations in Ruby Valley, Nevada, various prehistoric artifacts were discovered. Archaeologists were interested in the relationship between the type of artifacts found and the distance to permanent water, since the type of artifact discovered describes the type of site used by prehistoric hunters. It was presumed that some tools were more difficult to move from place to place, and would thus be more likely to be discovered near permanent water. The following table is based on a subset of the artifacts discovered in Nevada (Mosteller and Parunak, 1985).

Distance from permanent water

Artifact type	Immediate vicinity	Within .25 miles	.25–.5 miles	.5–1 mile	Total
Drills	2	10	4	2	18
Pots	3	8	4	6	21
Grinding stones	13	5	3	9	30
Point fragments	20	36	19	20	95
Total	38	59	30	37	164

Independence is marginally rejected for this table ($X^2 = 16.1$, $p = .065$; $G^2 = 16.3$, $p = .062$, both with $df = 9$). Adjusted residuals show that the cells corresponding to (Grinding stones, Immediate vicinity) and (Grinding stones, Within .25 miles) are most extreme. The former cell is particularly interesting to the archaeologists, since grinding stones were too large to move very easily.

Distance from permanent water

Artifact type	Immediate vicinity	Within .25 miles	.25–.5 miles	.5–1 mile
Drills	−1.28	1.83	0.46	−1.23
Pots	−1.03	0.22	0.10	0.71
Grinding stones	2.90	−2.44	−1.30	1.08
Point fragments	−0.75	0.60	0.66	−0.54

The quasi-independence model with these cells omitted fits the table well ($X^2 = 5.9$, $p = .55$; $G^2 = 6.3$, $p = .51$), with the drop in G^2 being $16.3 - 6.3 = 10.0$ on 2 degrees of freedom. The expected number of grinding stone artifacts in the immediate vicinity of water and within .25 miles of water, respectively, under the quasi-independence model are 5.45 and 11.78, respectively. This implies standardized deleted residuals of 3.23 and −1.98, respectively. While the first residual is noticeably large, the second is not particularly large (in absolute value), suggesting that it is not, in fact, an outlier; rather, its count (in the second column of the third row) appeared too low because of the unusually high count in the first column of the same row.

A quasi-independence model with only the (Grinding stones, Immediate vicinity) cell omitted fits well ($X^2 = 8.9$, $p = .35$; $G^2 = 8.7$, $p = .37$). The expected number of grinding stone artifacts in the immediate vicinity of water is 3.9, yielding a standardized deleted residual of 4.6. Thus, there appears to be little evidence of a relationship between the type of artifact and the distance from water, except that there were more grinding stones immediately next to water supplies than would have been expected.

Deleted residuals for outlying cells, which are based on the cell being omitted from the analysis, tend to be more extreme than Pearson residuals, which are based on including the cell. This is not surprising, as the unusual cell count affects the estimated expected cell count, thereby warping the residual from that expected cell count. The deleted residuals correct for this, but a problem is that they only can be calculated after the potentially outlying cell has been identified. This is sometimes difficult, especially if there are several outlying cells in the table. In that circumstance, residuals can suffer from *masking* and *swamping* effects. Masking occurs when two or more outliers effectively "hide" each other, so that they are not identified. Swamping occurs when one or more outliers draw the model fit towards themselves enough so that nonoutlying cells are mistakenly identified as outliers (this is what happened in the prehistoric artifacts data).

A way around this problem is to fit the independence model to the table in such a way that outliers have relatively little effect; that is, to construct a *robust* independence fit. There are several ways to do this, but a direct way is to work by analogy with analysis of variance (ANOVA) fitting. There are two key points that guide the robust fitting. First, recall that the independence model is a loglinear model; that is, the log of the expected count is a linear function, which is reminiscent of a two-way ANOVA model with no interaction term,

$$\log[E(n_{ij})] = \lambda + \lambda_i^X + \lambda_j^Y. \tag{6.12}$$

To first order $E[\log(n_{ij})] \approx \log[E(n_{ij})]$, so the first part of the robust fitting is to use $\log(n_{ij})$ as the target in fitting the model (6.12). An ANOVA model fit to $\log(n_{ij})$ would roughly correspond to the MLE fit (although this would ignore the nonconstant variance inherent in $\log(n_{ij})$), and would not be robust. The second key point is to fit the independence model to $\log(n_{ij})$ using a robust fitting method, *median polish*. Just as ANOVA corresponds to a fitting of row and column means, median polish corresponds to a fitting of row and column medians (based on iteratively "sweeping out" row and column medians) and is highly robust. Since the fitting of the model is done in the logged scale, the robust fitted values are also in that scale, and need to be exponentiated. Scaled (Pearson) residuals then can be determined, allowing identification of outlier cells (these residuals should not be compared to a Gaussian reference, but values above 3 or 4 in absolute value are indicative of unusual cells).

The Location of Prehistoric Artifacts (continued)

The median polish fit immediately identifies the outlying cell in the prehistoric artifacts table. The Pearson median polish residual for the (Grinding stones, Immediate vicinity) cell is much larger than that for any other cell, including the adjoining cell that was originally misidentified by the adjusted residuals.

Distance from permanent water

Artifact type	Immediate vicinity	Within .25 miles	.25–.5 miles	.5–1 mile
Drills	−0.41	1.72	0.58	−0.91
Pots	−0.32	0.04	−0.03	0.40
Grinding stones	4.38	−1.34	−0.76	1.32
Point fragments	0.82	−0.08	0.06	−0.73

The median polish fit is also resistant to masking effects, as the following example shows.

How Reliable are Different Classes of Automobiles?

The April 1997 issue of *Consumer Reports* magazine included predicted reliability ratings for 1997 automobiles that were not new models (*Consumer Reports*, 1997). The automobiles were classified into eight types. Is there an association between auto type and reliability rating? The table on the next page summarizes the data.

Predicted reliability

Auto type	Below average	Average	Above average	Total
Small	4	3	10	17
Medium	4	19	9	32
Large	1	8	4	13
Sporty	5	3	6	14
Coupe	4	3	2	9
Minivan	6	5	3	14
Sports utility vehicle	3	10	5	18
Total	27	51	39	117

Independence is marginally rejected for this table ($X^2 = 21.4$, $p = .045$; $G^2 = 21.5$, $p = .043$, both with $df = 12$). The Pearson residuals do not indicate any outlier cells, and the nonindependence seems to be spread over many cells in the table, with eight cells having absolute residuals greater than 1.15 but none having absolute residuals greater than 2.

Predicted reliability

Auto type	Below average	Average	Above average
Small	0.04	−1.62	1.82
Medium	−1.25	1.35	−0.51
Large	−1.15	0.98	−0.16
Sporty	0.98	−1.26	0.62
Coupe	1.33	−0.47	−0.58
Minivan	1.54	−0.45	−0.77
Sports utility vehicle	−0.57	0.77	−0.41

Pearson median polish residuals tell a very different story, as the residuals for (Small, Above average) and (Medium, Average) are both large and positive. That is, according to median polish, small cars are noticeably more likely to be predicted to have good reliability (this type of auto includes many from Japanese imprint manufacturers), while medium cars are noticeably more likely to be predicted to be average (perhaps in keeping with a stolid, family car, image). These unusual cells are not identified by the usual residuals because of masking.

Predicted reliability

Auto type	Below average	Average	Above average
Small	0.00	−0.50	3.00
Medium	−1.67	3.33	0.00
Large	−1.50	2.00	0.00
Sporty	0.00	−0.89	0.45
Coupe	0.58	0.00	−0.58
Minivan	0.45	0.00	−0.89
Sports utility vehicle	−0.89	2.24	0.00

The quasi-independence model that takes these two cells as outliers (missing) fits the table reasonably well ($X^2 = 12.48$, $p = .25$; $G^2 = 12.97$, $p = .23$, both with $df = 10$), with the drop in G^2 due to omission of the cells being 8.53 on 2 degrees of freedom. The fits to the two cells from the quasi-independence model are 3.33 and 8.98, respectively, implying Pearson deleted residuals of 3.66 and 3.34, respectively. Thus, except for small cars having higher predicted reliability than average and medium cars having average predicted reliability, there appears to be no relationship between the type of auto and *Consumer Reports*' predicted reliability.

What is not apparent from these residuals is that median polish has fit a simplified version of the independence model that sets the parameters for columns (predicted reliability) to zero. That is, the model is consistent with a uniform distribution for the conditional probability of predicted reliability given type of auto. This model also fits the data reasonably well (again removing the two outlier cells), with $G^2 = 15.38$ ($p = .22$), and the full quasi-independence model does not fit significantly better ($G^2 = 15.38 − 12.97 = 2.41$ on 2 degrees of freedom, with $p = .3$). The (Small, Above average) cell is lightly less outlying with respect to the latter model (Pearson deleted residual 3.47), while the (Medium, Average) cell is slightly more outlying (Pearson deleted residual 4.90).

6.6 Background Material

The early history of tests and estimators for association in contingency tables was marked by controversy, and great conflict between the leading lights of the day, such as R.A. Fisher, Karl Pearson, and G. Udny

Yule. Stigler (1999, Chapter 19) discusses the dispute over degrees of freedom between Fisher and Pearson, including the connection with quasi-independence models.

Hartigan and Kleiner (1981) proposed the mosaic plot as a way of graphically representing association in contingency tables. Friendly (2002) discusses the history of mosaic displays and their variants.

Constructing exact tests by conditioning on the observed margins when this is not consistent with the actual sampling scheme has been controversial since the introduction of Fisher's exact test. Barnard (1947) proposed an exact test that conditions on only one set of margins, although from 1949 on he withdrew support for this test; see Martin Andrés and Tapia García (1999) and Tapia García and Martin Andrés (2000) for further discussion.

Tukey (1977, Chapter 11) and Emerson and Hoaglin (1983) give details on the construction and use of median polish fits to two-way tables. Mosteller and Parunak (1985) proposed the use of median polish to identify outlier cells in a contingency table. Simonoff (1988), Yick and Lee (1998), and Shane and Simonoff (2001) discussed other, more general, approaches to identifying outlier cells.

6.7 Exercises

1. The prostate specific antigen (PSA) test is a blood test designed to detect if a man has prostate cancer. While use of the PSA test is widespread, there has been little investigation of whether it reduces deaths (part of the problem is that while some prostate cancers are lethal, many will remain confined to the prostate, causing no problem; doctors cannot tell which cancer is potentially lethal and which is not). In a presentation to the American Society of Clinical Oncology meeting on May 18, 1998, Dr. Fernand Labrie presented the results of an experiment designed to investigate this question (Ochs, 1998). A total of 46,193 men aged 45–80 from metropolitan Québec City were divided into two groups, those who would annually be given a PSA test and those who would not. After eight years, the proportion of screened men who died from prostate cancer was 0.061% (5 of 8137), while the proportion of unscreened men who died from prostate cancer was 0.36% (137 of 38,056).

   (a) Is there evidence of a significant gain in survival rate from annual administration of the PSA test?

   (b) Only 23% of the men asked to join the PSA screening group agreed to do so (those that refused were assigned to the unscreened group, which accounts for it being almost five times larger than the screened group). How might this affect an inference that PSA test administration causes a reduction in prostate cancer death rate?

2. A 1972 experiment in Canada examined the claim that using vitamin C can prevent the common cold (Anderson et al., 1972; see also Ramsey and Schafer, 1997, page 517). At the beginning of the winter, 818 volunteers were randomly divided into two groups. The members of one of the groups received a supply of vitamin C pills for the winter, while the others received a placebo. At the end of the winter, the volunteers were interviewed to determine if they had had a cold during the period. While 335 of the 411 people who received a placebo had had a cold during the winter, 302 of the 407 people who received vitamin C had had a cold.

   (a) Is there a significant difference in the rates of suffering a cold between the two groups?

   (b) Construct a 95% confidence interval for the difference in rates of suffering colds in the two groups.

3. It is often of interest to estimate the size of a population that is difficult to count. For example, the size of the populations of endangered species is a crucial piece of information in assessing the need for, and effectiveness of, environmental laws. Another example is the U.S. Census, where it is well-known that millions of people are missed during the actual count. One way to estimate a population size using a sample is the capture/recapture method. The method proceeds by first taking a random sample from the area, considering the objects in the sample as "captured." Then, another random sample is taken, and the number of captured objects that are picked for the second time ("recaptured") are counted.

   (a) Say the first sample consists of n_1 objects, while the second sample consists of n_2 objects, and a objects from the first sample are recaptured in the second sample. Assume that the population is closed, in the sense that its size is fixed at N for the duration of the sampling. Show that this process can be represented using the following 2×2 table:

Second sample

First sample	In sample	Not in sample	Total
In sample	a	$n_1 - a$	n_1
Not in sample	$n_2 - a$	$N - n_1 - n_2 + a$	$N - n_1$
Total	n_2	$N - n_2$	N

(b) Since the two samples are done separately, knowing that a person was in the first sample gives no information about being in the second sample; that is, the two samples are independent. Thus, the expected odds ratio for this table is one. The capture/recapture estimate of N sets the observed odds ratio equal to the expected odds ratio and solves for N. Show that this equals the *Peterson estimate*,

$$\hat{N} = \frac{n_1 n_2}{a}.$$

Sekar and Deming (1949) proposed estimating the standard error of $\hat{N}$ using

$$\widehat{s.e.}(\hat{N}) = \sqrt{\frac{n_1 n_2 (n_1 - a)(n_2 - a)}{a^3}}.$$

(c) Consider fitting a quasi-independence model to this table, taking the cell in the second row and second column as missing. Argue why the estimate $\hat{N} = \hat{e}_{22} + n_1 + n_2 - a$ is equivalent to the estimate given above for N, where $\hat{e}_{22}$ is the fitted value from the quasi-independence model for the missing cell. What is the deviance for this quasi-independence model?

4. The Salk polio vaccine was the subject of clinical trials of its effectiveness in 1954. In one trial, children of consenting adults were given the Salk vaccine or a placebo at random. Of the 200,000 children who received the placebo, 142 suffered infantile paralysis; of the 200,000 children who received the Salk vaccine, 56 suffered the disease (Ramsey and Schafer, 1997, page 534).

(a) Does the vaccine offer significantly more protection from infantile paralysis than the placebo?

(b) What is the relative risk of suffering paralysis for people who received the placebo compared to those people who received the vaccine?

(c) Put these data into the form of a 2×2 table. What is the odds ratio for that table? Is it close to the relative risk?

5. The logarithm of the estimated odds ratio (6.5) is asymptotically Gaussian with mean $\log \theta$ and asymptotic standard error

$$\widehat{s.e.}(\log \hat{\theta}) = \sqrt{\frac{1}{n_{11}} + \frac{1}{n_{12}} + \frac{1}{n_{21}} + \frac{1}{n_{22}}}.$$

Construct 95% confidence intervals for the logarithm of the odds ratio using (2.3) and this estimate of the standard error for each of the data sets in the three previous questions. Do these intervals result in the same implications regarding independence as the earlier analyses did?

6. A phase II cancer clinical trial is a pilot study designed to investigate whether a new treatment produces favorable results compared to a standard treatment; often, a randomized trial is performed, where patients are randomized to receive one of two therapies. Because it is a pilot study, sample sizes are usually small. Parzen et al. (2002) gave the results of a randomized phase II trial on patients with advanced large bowel cancer. The two treatments investigated were Homoharringtonine and Caracemide, and the outcome of the trial was success (shrinkage of the tumor of at least 50%) or failure. The following table gives the results.

	Outcome	
Treatment	Success	Failure
Homoharringtonine	62	40
Caracemide	0	14
Total	0	11

The observed odds ratio for this table is obviously undefined. Add 0.5 to each cell count and construct the estimated log odds ratio $\log \hat{\theta}$ and asymptotic 95% confidence interval for $\log \theta$. Convert the estimate and interval to ones for θ. What does this interval imply about the evidence regarding the effectiveness of the two drugs? (See Parzen et al., 2002, for an approach that does not require adding artificial data to the table, and results in a similar estimate.)

7. Show that the odds ratios for a 2×2 table based on prospective and retrospective studies do, in fact, simplify to the same cross-product ratio, as is stated on page 207.

8. Consider fitting independence to a 2×2 table.

   (a) Show that the Pearson residual for any cell is smaller in absolute value than the adjusted residual (and therefore has variance less than one).

   (b) Show that all four adjusted residuals have the same absolute value, which supports the notion that there is only one piece of independent information in the table, given the margins.

   (c) Show that the square of each adjusted residual equals the Pearson χ^2 statistic, X^2.

(Agresti, 1996, pages 51–52).

9. Trice, Holland, and Gagné (2000) (see also Holcomb, 2002, page 114) surveyed 120 sophomores and juniors enrolled in college psychology courses to explore the relationship between cutting classes ("I have missed a class during the past month for no valid reason") and other behavior. The following table summarizes responses to the statement "I visited family in the past month."

| | Missed class | |
Visited family	Yes	No
Yes	62	40
No	6	12
Total	68	52

(a) Construct a 90% confidence interval for the difference in probabilities of a respondent visiting family in the last month for people who missed class versus those who didn't. Is there evidence supporting a difference in those probabilities?

(b) Use χ^2 tests of independence to decide if these data provide evidence of a difference in behavior between the two groups. Would you use a one-sided or two-sided test here?

(c) Answer the question in part (b) using Fisher's exact test. Are the implications different from those in parts (a) and (b)?

10. In the case Sheehan vs. *Daily Racing Form*, the plaintiff, Jim Sheehan, alleged that his discharge (after a corporate acquisition) was discriminatory, as it was based solely on age (Good, 2001, page 143). He offered as evidence a list showing that while 9 of 11 of employees 48 years of age or more were discharged, none of the six employees under the age of 48 were discharged.

(a) Use χ^2 tests of independence to decide if these data provide evidence of age discrimination.

(b) Construct a 95% confidence interval for the difference in the probabilities that an older worker is discharged, and that a younger worker is discharged, respectively. Does this interval lead to the same conclusion as the tests in part (a) do?

(c) Use Fisher's exact test to assess the evidence. Does the test support the possibility of age discrimination?

(d) Two additional older employees affected by the acquisition who were not discharged were not on the original list. Does including these two employees change the results in parts (a)–(c)?

11. (a) Agresti and Caffo (2000) extended the Agresti–Coull one-sample
confidence interval for p (page 103) to the two-sample situation.
They suggested constructing the Wald interval (6.2), adding one
"fake" success and one "fake" failure to each group. Does the
use of this interval result in different inferences in the previous
two questions compared with the Wald interval?

(b) Yates (1934) suggested a continuity correction to X^2 that results
in χ^2-based tail probabilities that are closer to the hypergeometric tail probabilities from Fisher's exact test. The corrected test
has the form

$$X_Y^2 = \sum_{ij} \frac{(|n_{ij} - \hat{e}_{ij}| - .5)^2}{\hat{e}_{ij}}.$$

Does the use of this test result in different inferences in the
previous two questions compared with the standard χ^2 tests?
Does it yield p-values that are closer to those of Fisher's exact
test?

(c) Seneta and Phipps (2001) discussed what they called the Lieber-
meister p-value as an alternative to use of the mid-p-value for
Fisher's exact test. This p-value is the exact p-value for Fisher's
test based on the adjusted table with diagonal elements $n_{11} + 1$
and $n_{22} + 1$. How does this p-value compare with the mid-p-value
for the tests in the previous two examples?

12. Abzug et al. (2000) argued that the top ten employers in any city
represent a sizable proportion of employment in cities, and thus have
a strong effect on the culture, character, and economic health of the
city. In addition, research on nonprofit organizations in cities implies
that there are geographic patterns to the prominence of nonprofits
in cities, with more large nonprofit organizations in the northeast-
ern and midwestern parts of the United States. The following table
summarizes the relationship between the number of nonprofit health
organizations among the top ten employers in a city and region of the
country for 72 of the largest cities in the United States.

Number in top ten

Region	Zero	One	Two	Three	Four
Northeast	3	4	6	2	2
South	8	13	6	1	0
Midwest	2	3	5	2	1
West	1	5	4	3	1

(a) Does the evidence support the patterns predicted by nonprofit research?

(b) Do you get the same result if you examine this table using exact methods (using Monte Carlo tail probabilities if necessary)?

(c) The following table gives similar results for nonprofit educational institutions. Are your conclusions the same for these organizations as for the health organizations?

Number in top ten

Region	Zero	One	Two
Northeast	9	6	2
South	22	6	0
Midwest	8	4	1
West	14	0	0

13. The following table is a cross-classification of publicly traded companies in the database of Media General Financial Services of the region of the company (by its corporate headquarters) by its change in stock price during the third quarter of 1998, as given in *Newsday* (Associated Press and Media General Financial Services, 1998).

Change in stock price

Region	Up	Unchanged	Down
New England	100	7	561
Middle Atlantic	283	20	1314
South Atlantic	118	15	797
East North Central	190	15	798
East South Central	17	2	213
West North Central	88	5	375
West South Central	60	3	645
Mountain	49	4	337
Pacific	155	15	1110

Does the distribution of third-quarter stock price change differ based on the location of corporate headquarters? How might you explain the pattern you see?

14. The file **draft** gives information for the 1971 draft lottery (Starr, 1997). Is there evidence of any of the problems with the drawing for 1970 arising again in 1971? One of the changes made in 1971 was that both the date and the assigned lottery order number were randomized, resulting in two random processes instead of one. Are the results consistent with what you might expect from a system of this type?

15. Recall the data reported by Stelfox et al. (1998) regarding the possible relationship between the results of trials reported in journals and whether a researcher's research was supported by a drug company (page 105). Is there evidence of an association between drug company support and the result of the study?

16. On September 12, 2000, the U.S. Department of Justice released an initial report summarizing the results of its investigation into application of the federal death penalty since 1988 (when the Anti-Drug Abuse Act made the death penalty available for certain drug-related offenses). A supplemental report was released on June 6, 2001 (Department of Justice, 2001). According to the latter report, the Attorney General ultimately decided to seek the death penalty for 44 out of 166 white defendants, 71 out of 408 black defendants, and 32 out of 350 Hispanic defendants. Do these data provide evidence of different rates at which the death penalty is sought based on race? The U.S. Constitution prohibits the prosecutor from engaging in discrimination or favoritism based on invidious factors, such as race or ethnicity, in deciding whether to seek a capital sentence. Do you think these data suggest that this might be happening anyway?

17. Random number generation is an essential aspect of many statistical endeavors, including simulation and sampling. It is well established that people make poor random number generators, but why this is the case is not fully understood. Boland and Hutchinson (2000) reported the results of an experiment investigating this. Four groups of university students (first-year psychology students in their introductory statistics course, first-year science students in their mathematics course, first-year arts students taking an elective introductory statistics course, and upper-level students in an advanced statistics course, respectively) were asked to randomly generate 25 digits from $\{0, 1, 2, \ldots, 8, 9\}$. The resultant data are given in the file **randomgen**.

    (a) Do the different types of students exhibit different patterns in the way they choose random digits?

    (b) Test the hypothesis that the students generate random digits consistent with a uniform distribution. (*Hint:* Compare the fit of the model implying consistent patterns across groups to that implying uniform patterns across groups, taking into account the different sizes of the groups.)

18. The table cross-classifying smoking status and skin cancer given on page 218 came from a case-control study that examined cancers other than squamous cell carcinoma. As part of the study, the skin type of each person in the study was recorded. The results are summarized in the following table (De Hertog et al., 2001).

	Skin type				
Type of cancer	I	II	III	IV	Total
Controls	25	156	181	24	386
Squamous cell carcinoma	27	80	47	7	161
Nodular basal cell carcinoma	45	131	110	15	301
Superficial multifocal basal cell carcinoma	25	66	52	10	153
Malignant melanoma	12	76	34	3	125

Skin types are graded on the following scale: I, very light skin; II, light to medium skin; III, medium to olive skin; IV, dark olive to light brown skin. (There are two further levels corresponding to very dark skin, but these were not relevant for this study.)

(a) Combine all of the cancer types into one category. Is there evidence of a relationship between skin type and cancer incidence?

(b) Estimate the relative risk of having skin cancer if a person has skin type I compared to having skin type II. What about the relative risk if a person has skin type II compared to having skin type III? Skin type III versus skin type IV?

(c) Now, omit the control group. Is there evidence of a relationship between skin type and the incidence of the different types of cancer?

19. The 14 × 14 table given in the file kotzhawk is an artificial table created by T.J. v W. Kotze and D.M. Hawkins (Kotze and Hawkins, 1984). This table was constructed to demonstrate the effects of masking and swamping.

(a) Fit an independence model to this table. Does independence seem to fit?

(b) Construct the adjusted residuals from the independence fit. Do they identify any cells as outliers?

(c) Use median polish on the logged counts in the table to identify outlier cells. Do the median polish residuals identify different cells than do the adjusted residuals?

20. There has been a good deal of controversy in recent years over the use of Native American team names and mascots for sports teams in the United States. *Sports Illustrated* commissioned a survey of Indians living on reservations, Indians living off of reservations, and fans to determine their feelings on this issue (Price, 2002). One question asked what respondents thought of the "tomahawk chop" chant used at home games of the Atlanta Braves baseball team, and resulted in the following pattern of responses:

| | | *Group* | |
| | | Indians on | Indians off |
Response	Fans	reservation	reservation
Like it	208	24	44
Don't care	379	100	66
Find it objectionable	156	85	24
Total	743	209	134

(a) Is there evidence that the three groups feel differently about the tomahawk chop chant?

(b) Examine residuals from the independence fit to this table. Can you isolate the lack of fit of independence to a small number of cells? Can you characterize these cells in any way (for example, people who have strong feelings)?

(c) Fit a quasi-independence model to the table, omitting the outlier cells identified in part (b). Does this model fit the table? What does it say about how the opinions of fans and of Indians living off the reservation compare to each other? What about the opinions of Indians living on the reservation?

21. Yick and Lee (1998) gave a table cross-classifying student enrollment figures from seven community schools in the Northern Territory, Australia, for eight different time periods during the school year. The data are given in the data file **enrollment**.

(a) Is there evidence that the enrollment patterns over time are different for the seven schools?

(b) Examine adjusted residuals from the independence fit. Are any cells identified as outlying? What about using median polish residuals? During time periods 5 and 6, a group of transient fruit picker families moved into the area near school 1; is this consistent with what you find? Does a quasi-independence model omitting the outlier cells fit the data well? What about a model hypothesizing that the average enrollment within schools is constant over all of the time periods?

(c) Yick and Lee reported that during time periods 4 and 5, a traditional aboriginal funeral procession (which lasted about three months) brought people from a different region into the area around school 2, inflating the school enrollment there during that time. Is that reflected in the residuals constructed in part (b)? Does a quasi-independence model omitting the cells mentioned here and in part (b) fit the table well? (This is the set of cells identified as outlying by Yick and Lee.) What about a constant average enrollment within schools model?

7

Tables with More Structure

The hypothesis of independence is often just a straw man — we don't really think that two factors are independent of each other, so rejection of independence isn't very interesting. What would be more promising would be to be able to fit models that allow for some structure in the table. Depending on the form of the table, many such models are possible. These models allow for a general interaction term to be summarized using fewer than $(I-1)(J-1)$ degrees of freedom (often, far fewer).

In this chapter we examine models for tables with ordered categories, square tables, and data that arise in the form of matched pairs. Many of these models are nested within one another, allowing the construction of likelihood ratio tests of the gain of more complex models over simpler ones based on the change in deviance for the models. Of course, as always, different models can be compared using information measures such as AIC_C, even if the models are not nested.

7.1 Models for Tables with Ordered Categories

Many tables are formed by cross-classifying variables with ordered categories. These can be categorical but ordinal, such as Likert scales (for example, Strongly disagree — Disagree — Neutral — Agree — Strongly agree), or continuous variables that have been discretized, such as income formed into intervals of width $20,000.

Tables with ordered categories allow for models with different types of association built in, since concepts of direct and inverse relationships make sense. This permits parsimonious representation of a lack of independence. Since association is represented by a small number of parameters, more focused (and hence more powerful) tests for lack of independence can be constructed. This is done by fitting the usual loglinear model

$$\log e_{ij} = \lambda + \lambda_i^X + \lambda_j^Y + \lambda_{ij}^{XY}$$

with specific structure for λ_{ij}^{XY} that reflects ordering in the categories.

7.1.1 Linear-by-Linear (Uniform) Association

Consider a table with rows and columns with ordinal categories, and assume that there exist known scores $\{u_i\}$ (for rows) and $\{v_j\}$ (for columns) that represent that ordering. These scores could be the actual values of a discrete underlying variable (the number of children in a family, for example), a score linked to an underlying continuous variable (the midpoint of salary intervals, for example), or an equispaced representation of a non-numerical, but ordinal scale (such as a Likert scale). Most typically $u_i = i$ and $v_j = j$. The *linear-by-linear* (or *uniform*) *association model U* is

$$\log e_{ij} = \lambda + \lambda_i^X + \lambda_j^Y + \theta(u_i - \overline{u})(v_j - \overline{v}). \tag{7.1}$$

The uniform association model adds only one parameter (θ) to the independence model, focusing all possible lack of independence on that one parameter. If $\theta = 0$ independence holds. If $\theta > 0$, equation (7.1) implies that a higher expected cell count occurs when u_i and v_j either go up together or go down together, so there is a direct association relationship; if $\theta < 0$, the model implies that higher expected cell counts occur when u_i is high and v_j is low, or vice versa, so there is an inverse association relationship.

The parameter θ has a simple interpretation in terms of odds ratios. Say the scores are equispaced with a spacing of one, and sum to zero. Then the log of the odds ratio for any contiguous 2×2 subtable under model (7.1) is

$$
\begin{aligned}
\log\left(\frac{e_{ij}e_{i+1,j+1}}{e_{i,j+1}e_{i+1,j}}\right) &= \log e_{ij} + \log e_{i+1,j+1} - \log e_{i,j+1} - \log e_{i+1,j} \\
&= \theta[u_i v_j + u_{i+1}v_{j+1} - u_i v_{j+1} - u_{i+1}v_j] \\
&= \theta[u_i v_j + (u_i + 1)(v_j + 1) - u_i(v_j + 1) \\
&\quad - (u_i + 1)v_j] \\
&= \theta[u_i v_j + u_i v_j + v_j + u_i + 1 - u_i v_j - u_i \\
&\quad - u_i v_j - v_j] \\
&= \theta
\end{aligned}
$$

Thus, the uniform association model posits that the local association is constant throughout the table. This model is closely related to the most well-known continuous model with constant local association — the bivariate normal distribution. The density function for the bivariate normal distribution with means $\mu_x = \mu_y = 0$ is

$$f(x, y; \rho) = \frac{1}{2\pi\sqrt{1 - \rho^2}} \exp\left[-\frac{1}{2(1 - \rho^2)} \left(\frac{x^2}{\sigma_x^2} - \frac{2\rho xy}{\sigma_x \sigma_y} + \frac{y^2}{\sigma_y^2} \right) \right]$$

for a density with correlation ρ between X and Y. The log of the odds ratio for this density,

$$\frac{f(x, y; \rho)f(x + \sigma_x, y + \sigma_y; \rho)}{f(x + \sigma_x, y; \rho)f(x, y + \sigma_y; \rho)},$$

is $\rho/(1 - \rho^2)$; that is, the log odds for locations one standard deviation apart in both X and Y is $\rho/(1 - \rho^2)$, which is a constant that corresponds to θ if the scores $\mathbf{u}$ and $\mathbf{v}$ have unit standard deviations. Thus, a bivariate normal random variable that is discretized into a two-dimensional table should be consistent with a uniform association model.

More generally, under the uniform association model, for any pair of rows $r < s$ and columns $c < d$, the log of the odds ratio formed from the 2×2 table of those rows and columns is

$$\log\left(\frac{e_{rc}e_{sd}}{e_{rd}e_{sc}} \right) = \theta(u_s - u_r)(v_d - v_c).$$

Thus, the log odds ratio is directly proportional to the product of the distance between the rows and the distance between the columns.

A test of independence can be constructed based on the difference between G^2 values for the independence model and the uniform association model, since this corresponds to testing

$$H_0 : \theta = 0$$

versus

$$H_a : \theta \neq 0.$$

This is a one degree-of-freedom test of the adequacy of independence (I) given a uniform association (U) relationship,

$$G^2(I|U) = G^2(I) - G^2(U).$$

Since the test focuses nonindependence onto one parameter, it can be considerably more powerful than the omnibus goodness-of-fit tests. This test is strictly only valid if the uniform association model fits the table, although it is reasonably effective even when uniform association is not valid. It also tends to follow its χ_1^2 null distribution reasonably well, even when the $G^2(I)$ and $G^2(U)$ distributions are problematic because of a small sample size.

Does Fiscal Health Go With Nonprofit Employment?

Researchers into urban economics have long known that large organizations have a strong effect on a city's fiscal health. While attention has traditionally focused on the for-profit sector, the public and nonprofit sectors are also important, particularly because downsizing has reduced the size of many for-profit organizations in recent years. Large nonprofit organizations are also likely to become more intimately connected to the life of a city because, once institutionalized, they are more wedded to place than more agile market-driven entities.

The following table (based on one given in Simonoff and Tutz, 2000) gives a cross-classification of the fiscal health of 57 of the 75 largest U.S. cities, as measured by *Standard and Poor's* 1994 rating for their general obligation bonds, by the number of nonprofit organizations among a city's top ten employers. Bond ratings are classified by the broad ordinal ratings of AAA, AA, A, or BBB or below, resulting in a table with both ordered rows and ordered columns.

	Nonprofits among top ten employers					
Rating	0	1	2	3	4	5
BBB or below	1	0	1	2	0	2
A	1	1	3	5	2	0
AA	5	12	11	4	1	0
AAA	0	4	2	0	0	0

There is apparently a relationship between fiscal health and the prominence of the nonprofit sector, as independence is strongly rejected ($X^2 = 35.3$, $p = .002$; $G^2 = 29.9$, $p = .012$). The uniform association model based on equally spaced $\mathbf{u}$ and $\mathbf{v}$ scores is a better fit, with $X^2 = 22.2$ ($p = .07$) and $G^2 = 18.0$ ($p = .21$). The focused test of independence strongly rejects independence ($G^2(I|U) = 29.9 - 18.0 = 11.9$, $p < .001$). The estimated association parameter is $\hat{\theta} = -.534$, implying a constant local odds ratio of $\exp(-.534) = 0.59$. That is, the model implies that the odds of a city having $k + 1$ nonprofits among the top ten employers versus k nonprofits are 40% lower for cities with a rating class one level higher (AAA versus AA, AA versus A, and so on).

The connection to an underlying bivariate normal representation can be made, but first the scores have to be standardized to have unit standard deviation. This is done based on the marginal counts (for example, for rows there are 6 scores of 1, 12 scores of 2, 33 scores of 3, and 6 scores of 4; this yields a standard deviation of 0.805). The standard deviations for unit-spaced scores are 0.805 for rows and 1.231 for columns, respectively, so the standardized version of $\hat{\theta} = (-.534)(.805)(1.231) = -.529$. Since $\theta = \rho/(1-\rho^2)$, $\hat{\rho} = (\sqrt{1+4\hat{\theta}^2} - 1)/(2\hat{\theta})$, so the estimate of the underlying correlation for these data is $\hat{\rho} = -.431$, reinforcing the moderate inverse association between bond rating and number of nonprofits.

There are several possible reasons why a lower bond rating would be associated with more nonprofits among the top ten employers. Nonprofits do not contribute to a city's tax base, and thus constitute a potential drain on municipal resources. More nonprofits among the top ten means fewer public and for-profit institutions among the top ten, perhaps indicating a weaker municipal government, and weaker business sector. Most large nonprofit organizations are hospitals or health organizations, and the recent pressures on the health care industry might lead to a view in the bond market that dependence on such organizations is risky. If city planners believe in these causal links, the observed association would presumably imply active programs to lure large for-profit organizations into the city, rather than large nonprofits.

The fit of the uniform association model is not as good as we might like. Pearson residuals for the model show that there is one outlier cell, corresponding to the cell with no nonprofits among the top ten employers and a BBB or lower rating:

Nonprofits among top ten employers

Rating	0	1	2	3	4	5
BBB or below	3.45	−0.71	−0.30	0.01	−1.01	0.92
A	0.52	−0.96	−0.46	0.90	0.98	−0.78
AA	0.18	0.16	0.18	−0.55	0.07	−0.57
AAA	−1.31	0.95	0.58	−0.63	−0.20	−0.09

This cell corresponds to one city, Buffalo, New York. Buffalo has no nonprofits among the top ten, but it has an unusually high number of large government employers; given the argument given earlier, this could be viewed as evidence of weakness in the city's business sector.

This outlier cell can be addressed by omitting it from the model fitting. Once again, this can be accomplished by either omitting the cell from the underlying regression data set, or adding an indicator variable identifying it. The quasi-independence model does not fit the table ($X^2 = 36.0$, $p = .001$;

$G^2 = 29.8$, $p = .008$), but the *quasi-uniform association* model fits very well ($X^2 = 9.9$, $p = .70$; $G^2 = 13.2$, $p = .43$). The estimated association parameter is $\hat{\theta} = -.747$, implying an even stronger association between fiscal health and the number of large nonprofit employers ($\hat{\rho} = -.518$).

7.1.2   Row, Column, and Row + Column Effects Models

The uniform association model assumes prespecified row and column scores. Sometimes either the rows or columns (but not both) are not ordinal, so such scores don't exist for the nominal variable. Another possibility is that equispaced scores are not appropriate for a set of rows or columns, and it is sensible to estimate appropriate scores based on the observed data. So, for example, for Likert scaled rows or columns, it might be that "Strongly disagree" is closer to "Disagree" than "Disagree" is to "Neutral" for an observed table, in the sense that odds ratios are more similar for the former pair than for the latter pair. In that case the uniform association model is inappropriate. Models that can fit tables of this type are the *row effects* and *column effects* models. The row effects model R has the form

$$\log e_{ij} = \lambda + \lambda_i^X + \lambda_j^Y + \tau_i(v_j - \overline{v}), \qquad (7.2)$$

where the $\{\tau_i\}$ are parameters that sum to zero. The model adds $I - 1$ parameters to the independence model, so the degrees of freedom for the model are $(I - 1)(J - 1) - (I - 1) = (I - 1)(J - 2)$, and is unsaturated for $J > 2$. The row effects model treats the columns as ordinal with known scores, and rows as (potentially) nominal, since the τ parameters can take on any values that sum to zero.

Once again the parameters of the model can be understood through odds ratios. For any pair of rows $r < s$ and columns $c < d$, the log of the odds ratio formed from the 2×2 table of those rows and columns is

$$\log \left(\frac{e_{rc} e_{sd}}{e_{rd} e_{sc}} \right) = (\tau_s - \tau_r)(v_d - v_c).$$

The log odds ratio is proportional to the distance between the columns, with the constant of proportionality being $\tau_s - \tau_r$. Thus, $\tau_s - \tau_r$ is a measure of the closeness of the rows r and s with respect to the conditional distribution of the columns given the row. For a model with unit-spaced columns scores, the log odds for the 2×2 table for rows i and $i+1$ and columns j and $j+1$ is $\tau_{i+1} - \tau_i$. This reflects that the local log odds ratios are not constant (as in the uniform association model), but are determined by an effect for row i. Since a positive log odds ratio means that more observations fall on the main diagonal of the 2×2 subtable than expected (and fewer on the off-diagonal), a positive value of $\tau_{i+1} - \tau_i$ means that more of the conditional probability given membership in row $i + 1$ falls to the right, while more of the conditional probability given membership in row i falls to the left (that is, more of the probability in row $i + 1$ corresponds to the high end of the column scores, while more of the probability in row i corresponds to the low end of the column scores). Thus, if $\tau_{i+1} > \tau_i$, the conditional distribution for row $i + 1$ is *stochastically larger* than that for row i.

The column effects model C takes the form

$$\log e_{ij} = \lambda + \lambda_i^X + \lambda_j^Y + \rho_j(u_i - \overline{u}), \tag{7.3}$$

where the ρ_j parameters sum to zero (note that the $\{\rho_j\}$ have no connection to the bivariate normal correlation ρ). The model is based on $(I-2)(J-1)$ degrees of freedom. This model treats the rows as ordinal with known scores, and columns as (potentially) nominal. As would be expected from the earlier discussion, $\rho_d - \rho_c$ is a measure of the closeness of the columns c and d with respect to the conditional distribution of the rows given the column. The local log odds ratio for unit-spaced row scores is $\rho_{j+1} - \rho_j$, reflecting that they are determined by a column effect.

A generalization of the row and column effects models that allows for both row and column effects in the local log odds ratios is the *row + column effects* $(R + C)$ model,

$$\log e_{ij} = \lambda + \lambda_i^X + \lambda_j^Y + \tau_i(v_j - \overline{v}) + \rho_j(u_i - \overline{u}), \tag{7.4}$$

where the τ_i and ρ_j parameters are restricted to make the model identifiable (in addition to making them sum to zero, one set must have its scale specified). The model is based on $(I-2)(J-2)$ degrees of freedom. The local log odds ratio for unit-spaced row and column scores is $(\tau_{i+1} - \tau_i) + (\rho_{j+1} - \rho_j)$, incorporating row effects and column effects.

These association models form a natural progression of increasingly complex association patterns, since various models are special cases of one another. This can be represented graphically as follows:

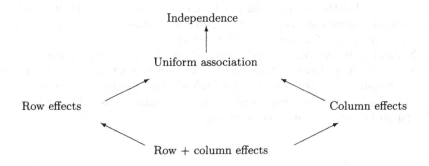

The diagram illustrates that models are special cases of the models below them. So, as was noted earlier, independence is a special case of the uniform association model (7.1) with $\theta = 0$; a test of independence given (7.1) holds is $G^2(I) - G^2(U)$ on 1 degree of freedom. Uniform association (7.1) is a special case of the row effects model (7.2) with $\tau_i = \theta(u_i - \bar{u})$; a test of uniform association given (7.2) holds is $G^2(U) - G^2(R)$ on $I - 2$ degrees of freedom. Uniform association (7.1) is also a special case of the column effects model (7.3) with $\rho_j = \theta(v_j - \bar{v})$; a test of uniform association given (7.3) holds is $G^2(U) - G^2(C)$ on $J - 2$ degrees of freedom. The row effects model (7.2) is a special case of the row + column effects model (7.4) with $\rho = 0$; a test of the row effects given (7.4) holds is $G^2(R) - G^2(R + C)$ on $J - 2$ degrees of freedom. The column effects model (7.3) is a special case of the row + column effects model (7.4) with $\tau = 0$; a test of the column effects given (7.4) holds is $G^2(C) - G^2(R + C)$ on $I - 2$ degrees of freedom. Thus, model selection can proceed based on subset tests, which are more powerful than omnibus goodness-of-fit tests. The information criterion AIC_C also can be used to choose among the models.

The following example is based on a nominal/ordinal table.

Student Perceptions of Statistics Class Assessment Methods

In recent years experts in statistical education have emphasized the need for students in introductory statistics classes to "get their hands dirty" by analyzing real data sets using statistical computer packages. There has been less investigation into the worth of such analyses in advanced statistics classes. O'Connell (2002) described an investigation into the perceptions of students in an advanced multivariate statistics class as to the worth of four different assessment strategies: structured data analyses (data analysis assignments where data are supplied to the student, and specific questions are provided for the student to answer), open-ended assignments (assignments where students choose a data set from several that are available, and analyze the data themselves), article analyses (where students are provided with an article to read and questions to answer regarding the article), and annotated output (where students run a particular analysis on the computer and label the resultant output based on an example given in class). It is clearly important to a teacher to gauge student responses to these four different methods of assessing understanding.

Student responses regarding the amount they learned using each method were tabulated as follows. Note that we will ignore that the same 14 students responded to the question for each method (other than for article analysis, where one student didn't answer), treating this table as being based on product multinomial sampling.

Type of assignment

Response	Structured computer assignments	Open-ended assignments	Article analysis	Annotating output
Didn't learn anything	0	0	1	0
Learned a little bit	3	1	6	4
Learned enough to be comfortable with topic	8	7	4	8
Learned a great deal	3	6	2	2

This table has naturally ordered rows, but the columns are nominal. Table 7.1 summarizes model fitting for this table. Independence is not rejected ($p = .24$), suggesting that there is no difference in student views of the four assessment methods. The table is sparse, but the lack of rejection of independence is not because of an inappropriate χ^2 approximation for G^2, as Fisher's exact test also doesn't reject the null, with $p = .23$. The column effects model fits well (note that the uniform association, row effects, and row + column effects models are not meaningful here, due to the nominal nature of the columns). The table labels the fitted model with smaller value of AIC_C as having zero value, with the other AIC_C entry representing the difference from this minimal value, as was done in earlier chapters, and the column effects model has the lower value. Perhaps more importantly, the focused (subset) test of independence given the column effects model $G^2(I|C)$ is statistically significant with $p = .03$, reinforcing that the focused test of independence can have considerably more power than the omnibus test.

The estimated ρ parameters are $(0.1069, 0.8866, -0.8137, -0.1798)$, implying a preference ordering (from least favorable to most favorable) of {Article analysis, Annotating output, Structured computer assignments,

Model	Subset test	G^2	df	p	AIC_C
Independence		11.6	9	.24	0.38
Column effects		2.6	6	.86	0.00
	$(I|C)$	9.0	3	.03	

TABLE 7.1. Model fitting for student perceptions of assessment methods data.

Open-ended assignments}. These parameter estimates suggest that there could be, in fact, an ordering to the columns, corresponding to assignments ranging from most passive (from a data-analytic point of view) to most active, with more active assignments being associated with more learning. Thus, it seems that analyzing real data is a valuable learning tool not only in introductory classes, but also in advanced ones.

The next example is based on a table with both ordered rows and columns.

Nearsightedness and the Use of Night Lights

Myopia (nearsightedness) is a condition in which a person's eye is slightly longer than it should be, resulting in images of distant objects focusing short of the retina, leading to blurry images. Its cause is not well understood, although it is believed to involve both genetic and environmental factors, and its prevalence is increasing. Hyperopia (farsightedness) is the corresponding condition when a person's eye is slightly shorter than it should be, resulting in blurry images of near objects. Emmetropia is the technical term for perfect vision.

Quinn et al. (1999) reported evidence that nighttime ambient light exposure during sleep before the age of two is associated with myopia, based on a questionnaire given to parents of children aged 2–16 years that were seen as outpatients in a university ophthalmology clinic. The results of the questionnaire relating to nighttime ambient light and the present refraction (eyesight status) of the child are given in the following table, which is ordered both in rows (direction of refraction) and columns (amount of light).

	Nighttime ambient lighting		
Present refraction	Darkness	Night light	Room light
High hyperopia	2	1	1
Hyperopia	21	16	15
Emmetropia	66	50	29
Myopia	9	31	48
High myopia	1	3	7

Table 7.2 summarizes model fitting for this table. Independence is clearly rejected, but simple integer scores for both rows and columns are not suf-

Model	Subset test	G^2	df	p	AIC_C
Independence		51.2	8	< .001	45.72
Uniform association		22.7	7	.002	19.32
Row effects		1.9	4	.756	4.96
	$(I\|R)$	49.3	4	< .001	
	$(U\|R)$	20.8	3	< .001	
Column effects		22.6	6	.001	21.33
Row + Column effects		0.3	3	.951	5.59
	$(R\|R+C)$	1.6	1	.206	
	$(C\|R+C)$	22.3	3	< .001	
Myopia effect		3.4	7	.851	0.00
Myopia/high myopia effect		3.1	6	.792	1.90

TABLE 7.2. Model fitting for myopia and nighttime ambient light data.

ficient to represent the association in the table. In particular, while integer column scores are reasonable, they are not for rows, as a row effects model is significantly a better fit than uniform association. The $R + C$ model is not a significantly better fit than row effects, and the row effects model has smaller AIC_C value.

The estimated τ parameters are $(-.553, -.331, -.562, .612, .834)$. There is a clear separation between the two myopic conditions and the other three refraction statuses, with increasing ambient light associated with the child more likely to be myopic. Table 7.2 summarizes the fits of two "myopia effect" models that combine the effects of the nonmyopic conditions, one merging the effects of the two myopic conditions ("Myopia effect") and one treating them separately ("Myopia/high myopia effect"). Merging the myopic condition effects results in smaller AIC_C value, with $\hat{\tau} = (-.567, .567)$. That is, the odds of being myopic versus not being myopic are estimated as being more than three times higher $(\exp(1.134) = 3.11)$ for children who slept with a night light compared to those who slept in darkness, and more than three times higher for children who slept in room light compared with sleeping with a night light. Figure 7.1 gives mosaic plots for the observed counts and fitted counts based on row effects model using the merged myopia conditions; it is apparent that the implied associations from the model are very close to the ones observed in the table.

As a result of this study, the authors stated that "it seems prudent that infants and young children sleep at night without artificial lighting in the

Observed counts

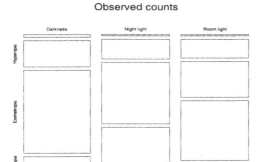

Myopia effect model

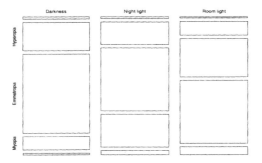

FIGURE 7.1. Mosaic plots of nighttime ambient lighting table. Observed counts (top plot) and fitted counts based on row effects model with merged myopia conditions (bottom plot).

bedroom," but that recommendation was challenged in Zadnik et al. (2000) and Gwiazda et al. (2000). Zadnik et al. (2000) reported their own results examining the relationship between nighttime ambient lighting and myopia:

	Nighttime ambient lighting		
Present refraction	Darkness	Night light	Room light
Not myopic	333	630	35
Myopic	84	128	10

There is no evidence of any relationship between nighttime ambient light and myopia in this table, as independence fits the table well. Zadnik et al. (2000) noted a strong association between the number of myopic parents and nursery lighting in their study, also noted in Gwiazda et al. (2000). Given the known association between parental and child myopia, these researchers speculated that the observed association in the original study could be a result of it not having corrected for parent myopia. Indeed, one of the authors of Zadnik et al. (2000) was quoted in an article in the March 9, 2000, issue of the *Boston Globe* as saying "Surprise, surprise: those of us who can't see like to leave the lights on" (Hsu, 2000).

The final example of this section illustrates a situation where the $R + C$ model is an improvement over the simpler models.

Book Deterioration in the New York Public Library (Part 1)

One of the most serious problems facing large research libraries is the preservation of the materials that comprise their collections. These materials are deteriorating because of their chemical composition, the mechanics of their construction, and the effects of uncontrolled environmental conditions. In large university and public libraries, including the Library of Congress, it is estimated that millions of volumes have suffered severe damage (Walker et al., 1985).

Before appropriate preservation strategies can be devised, it's necessary to understand the scope of the problem. In 1983 a study was conducted by the New York Public Library into book deterioration in the library. The following table relates the condition of books sampled to the strength of the paper in the book (Simonoff and Tsai, 1991). Strength of paper is typically measured by the number of folds in the paper that the book can withstand without breaking. The rows are ordered in increasing strength of the paper, while the columns are ordered in increasing level of deterioration.

Degree of book deterioration

Strength	Intact	Slight	Moderate	Extreme
1 fold	181	14	18	43
2–4 folds	140	6	1	15
5–15 folds	44	2	0	0
> 15 folds	369	7	0	0

Table 7.3 summarizes model fitting for the table. Independence clearly does not fit the table. The uniform association and row effects models are marginally rejected in terms of fit. Both the column effects and $R + C$

Model	Subset test	G^2	df	p	AIC_C
Independence		143.4	9	< .001	130.57
Uniform association		16.4	8	.037	5.63
Row effects		12.4	6	.054	5.70
	$(I|R)$	131.0	3	< .001	
	$(U|R)$	4.0	2	.135	
Column effects		8.2	6	.226	1.51
	$(I|C)$	135.2	3	< .001	
	$(U|C)$	8.2	2	.016	
Row + Column effects		2.6	4	.635	0.00
	$(I|R+C)$	140.8	5	< .001	
	$(U|R+C)$	13.8	4	.008	
	$(R|R+C)$	9.8	2	.008	
	$(C|R+C)$	5.6	2	.061	

TABLE 7.3. Model fitting for book strength and book deterioration data.

models fit adequately, and the test of the fit of column effects given that $R + C$ holds is inconclusive ($p = .061$). AIC_C, on the other hand, comes down on the side of the $R + C$ model.

The column parameters in the column effects model are $\hat{\rho}_{\text{Intact}} = 1.15$, $\hat{\rho}_{\text{Slight}} = .70$, $\hat{\rho}_{\text{Moderate}} = -1.60$, and $\hat{\rho}_{\text{Extreme}} = -.25$. Thus, there is a generally inverse relationship between rows and columns (row scores are set at $\{1, 2, 3, 4\}$, while the column scores are decreasing), which says that books in worse condition have weaker paper, which makes sense. Interestingly, however, extremely deteriorated books have stronger paper than moderately deteriorated ones. This can be seen in the table, since 95% of the moderately deteriorated books have paper that breaks after one fold, but only 74% of the extremely deteriorated books do. Deterioration can come from many sources other than paper strength, and that is obviously the case here.

A weakness of the $R+C$ model (7.4) is that it can be difficult to interpret the coefficients, because of the use of two sets of row scores ($\mathbf{u}$ and $\boldsymbol{\tau}$) and two sets of column scores ($\mathbf{v}$ and $\boldsymbol{\rho}$), respectively. The column parameters for the $R + C$ model are $\hat{\rho}_{\text{Intact}} = -4.36$, $\hat{\rho}_{\text{Slight}} = -.91$, $\hat{\rho}_{\text{Moderate}} = 0$, and $\hat{\rho}_{\text{Extreme}} = 5.27$, while the row parameters are $\hat{\tau}_{1 \text{ fold}} = 5.73$, $\hat{\tau}_{2\text{-4 folds}} = 2.23$, $\hat{\tau}_{5\text{-15 folds}} = -1.98$, and $\hat{\tau}_{\text{more than 15 folds}} = -5.98$. This would seem

to imply that the slight and moderate deterioration categories are similar
to each other, but as was noted above, that is not the case. In the next
section we examine an alternative model based on row and column effects
that is much easier to interpret.

7.1.3 A Log-Multiplicative Row and Column Effects Model

As was just noted, a problem with the $R + C$ model is that two sets of
scores for rows and columns makes interpretation of coefficients problematic. Another weakness in the model is that while it allows for both row
and column effects, it is still based on prespecified row and column scores.
If those scores are unavailable, or are inappropriately set, this can make
the results from an $R + C$ fit difficult to interpret.

A model that does not require any scores, and that results in easily
interpretable coefficient estimates, is the RC model, which has the form

$$\log e_{ij} = \lambda + \lambda_i^X + \lambda_j^Y + \theta \tau_i \rho_j, \tag{7.5}$$

where the τ_i and ρ_j parameters are restricted to make the model identifiable
(typically the conditions $\sum_i \tau_i = \sum_j \rho_j = 0$ and $\sum_i \tau_i^2 = \sum_j \rho_j^2 = 1$ are
imposed). This model is most obviously similar to the uniform association
model (7.1), only now scores are estimated from the table; in this sense,
the RC model finds the best-fitting version of a linear-by-linear association
model, by allowing the rows and columns to be reordered and moved farther
apart or closer together. This model implies local log odds ratios of $\theta(\tau_{i+1} -
\tau_i)(\rho_{j+1} - \rho_j)$, so it includes row and column effects multiplicatively, rather
than additively (as is implied by the $R + C$ model).

The RC model is different from the other models discussed here in that
it is log-multiplicative, rather than loglinear. This makes it more difficult
to fit the model (loglinear code only can be used in an iterative way, and
degrees of freedom of tests must be adjusted to reflect the appropriate parameter restrictions), and also complicates inference. For example, the subset
test of independence given the RC model does not follow a χ^2 distribution,
because τ and ρ are undetermined if $\theta = 0$ (independence). Other subset
tests are still valid, however, since they correspond to the τ_i and ρ_j para-
meters equaling the prespecified row and column scores $\mathbf{u} - \overline{u}\mathbf{1}$ and $\mathbf{v} - \overline{v}\mathbf{1}$,
respectively.

Book Deterioration in the New York Public Library (Part 2)

The following table comes from the same New York Public Library study discussed earlier, and relates book condition to the preservation strategy that would be appropriate for that book (Simonoff and Tsai, 1991). Preservations strategies are given ordered from most to least serious, in terms of both cost and technical difficulty.

	Book condition (degree of deterioration)			
Strategy	Intact	Slight	Moderate	Extreme
Repair	27	1	2	0
Microfilm	50	6	3	34
Restore	7	0	1	30
No preservation needed	676	22	13	0

Table 7.4 summarizes model fitting for the table. Subset tests are not given in the table, because it is obvious that only the RC model fits the table. What is going on here? The parameter estimates from the RC model are as follows: $\hat{\theta} = 8.10$, the row estimates $\hat{\tau} = (-.20, .36, .55, -.72)$, and the column estimates $\hat{\rho} = (-.38, -.26, -.23, .86)$. Since the row and column effects are generally increasing, this says that there is a generally direct relationship between decreasing difficulty of preservation and worsening book condition. This might seem surprising (you might think that more seriously deteriorated books would need more help), but it actually isn't. For the first three rows, which are the ones referring to an active preservation strategy, the more deteriorated a book is, the less expensive the strategy. The reason for this is purely economic — preservation librarians don't want to spend lots of time and money on books that are already in terrible shape. The "No preservation needed" row is very different, since the vast majority of books that need no preserving are intact, so its $\hat{\tau}$ value effectively places it at the top of the table rather than the bottom.

The other noteworthy result from the RC fit is that the estimated ρ value for the fourth column (extremely deteriorated) makes it very distinct from the other three, implying that the chances of such a book getting the most expensive preservation strategy are very small. In fact, of the 64 extremely deteriorated books, none were marked for repair, and the estimated probability based on the RC model of an extremely deteriorated book being repaired is less than .0001.

Model	G^2	df	p	AIC_C
Independence	308.4	9	< .001	294.86
Uniform association	203.7	8	< .001	192.17
Row effects	31.2	6	< .001	23.75
Column effects	202.7	6	< .001	195.23
Row + Column effects	22.1	4	< .001	18.75
Row × Column effects	3.3	4	.50	0.00

TABLE 7.4. Model fitting for book repair strategy and book deterioration data.

In the final example, the *RC* model is the only model of interest (even if it doesn't fit very well), since the goal of the analysis is to explicitly order rows in a table with nominal columns.

Estimating Relative Archaeological Chronologies

A primary goal of archaeologists is to construct timescales based on excavated materials. Ideally, relative chronologies are constructed based on well-excavated vertical stratigraphic sequences (that is, clear strata from different time periods on top of each other), but sometimes this is not possible, because the site did not form strictly sequentially, or the material was partially mixed after deposition.

A typical data set used in such circumstances is called an abundance matrix, which is simply a contingency table of excavated features versus artefact types. The goal is to use the observed patterns of artefact types to order the excavated features, which would imply a chronological ordering. The table on the next page summarizes the results of excavations of seven types of mesolithic flint tools (called microliths) from six different sites in southern England (Buck and Sahu, 2000).

The rows of the table are assumed to have a natural (chronological) ordering, although this ordering is unknown, while the columns are nominal. The *RC* model doesn't fit the table ($G^2 = 119.4$, $df = 20$), but the estimated row parameters still provide a suggested ordering of the sites. The estimates are $\hat{\tau} = \{.3137, -.7242, .2248, .3251, -.3947, .2552\}$, implying an ordering of $\{2, 5, 3, 6, 1, 4\}$. This is the same ordering as was chosen by Buck and Sahu (2000) using a Bayesian Langevin diffusion model. Buck and Sahu (2000) also discussed the possible ordering $\{3, 6, 5, 2, 1, 4\}$, ultimately deciding that the former ordering was preferred. Clearly the *RC* analysis agrees, as the latter ordering is not close to the estimated parameter ordering.

Type of microlith

Site	A	B	C	D	E	F	G
1	20	3	4	42	18	0	13
2	85	3	12	0	0	0	0
3	26	40	8	0	0	26	0
4	20	1	4	13	58	0	4
5	67	10	23	0	0	0	0
6	26	29	8	3	0	33	1

7.1.4   Bivariate Discrete Distributions

The strength of association models is their flexibility in modeling tables with ordered categories, but that flexibility comes with a price. Even the uniform association model uses $I + J$ parameters in an $I \times J$ table, because the marginal distribution parameters $\boldsymbol{\lambda}^X$ and $\boldsymbol{\lambda}^Y$ are free (other than summing to zero), and must be estimated. The discrete distributions described in Chapter 4 provide a parsimonious way of representing discrete data, but are restricted to univariate data. Fortunately, such distributions can be generalized to more than one dimension. We will consider here one simple approach to fitting such distributions; see the references in Section 7.4 for more complex approaches.

In general, there is no unique generalization to higher dimensions of most distributions, including discrete ones (the Gaussian distribution is a notable exception). One approach, described by Arnold and Strauss (1991), is as follows. Let X and Y be two discrete random variables that are members of k- and ℓ-parameter exponential families, respectively, so that

$$f_1(x; \boldsymbol{\theta}) = r_1(x)\beta_1(\boldsymbol{\theta}) \exp\left[\sum_{i=1}^{k} \theta_i p_i(x)\right]$$

and

$$f_2(y; \boldsymbol{\tau}) = r_2(y)\beta_2(\boldsymbol{\tau}) \exp\left[\sum_{j=1}^{\ell} \tau_j q_j(y)\right]$$

(these forms are equivalent to (5.1), although they do not distinguish between the systematic component and a scale factor). A bivariate distribution with these marginals can be defined using the conditional distributions of $X|Y$ and $Y|X$. Specifically, the bivariate distribution satisfies two conditions: for every y for which $f(x|y)$ is defined,

$$f(x|y) = f_1[x; \theta(y)],$$

and for every x for which $f(y|x)$ is defined,

$$f(y|x) = f_2[y; \tau(x)].$$

That is, the conditional distributions are from the same families as the marginal distributions, with parameters that can depend on the conditioned value. So, for example, in a contingency table with Poisson marginals, the observations falling in each row are also Poisson distributed, with mean depending on the row, as are the observations falling in each column (with mean depending on the column). These conditions imply that the joint distribution satisfies

$$f(x,y) = r_1(x)r_2(y) \exp[\mathbf{p}(x)'M\mathbf{q}(y) + \mathbf{a}'\mathbf{p}(x) + \mathbf{b}'\mathbf{q}(y) + c]$$

for appropriate choices of M, $\mathbf{a}$, $\mathbf{b}$, and c. Note that X and Y are independent if and only if $M = 0$.

Consider now marginal distributions that would be reasonable for a contingency table. If the row and column marginals are both Poisson distributed, then

$$r_1(t) = r_2(t) = (t!)^{-1}$$

and

$$p(t) = q(t) = t$$

for $t = 0, 1, \ldots$. This gives the joint distribution

$$f(x,y) = \frac{\exp(mxy + ax + by + c)}{x!y!};$$

equivalently,

$$P(X = i, Y = j) = k(\mu_X, \mu_Y, \nu)\frac{\mu_X^i}{i!} \frac{\mu_Y^j}{j!}\nu^{ij}, \qquad (7.6)$$

where μ_X is the mean for rows, μ_Y is the mean for columns, $k(\mu_X, \mu_Y, \nu)$ is the normalization constant, and ν measures the association between rows and columns ($\nu > 0$, with $\nu = 1$ corresponding to independence). Similarly, the bivariate distribution with binomial conditional distributions (with the same number of trials but possibly different probabilities of success) has the form

$$f(x,y) = k\binom{n}{x}p_X^x(1 - p_X)^{n-x}\binom{n}{y}p_Y^y(1 - p_Y)^{n-y}\nu^{xy}.$$

These distributions can be fit using loglinear (Poisson regression) models. Consider, for example, the bivariate Poisson distribution. Let $n_{ij}, i = 0, \ldots, I - 1, j = 0, \ldots, J - 1$ be the number of observations with $X = i$ and $Y = j$, and assume that I and J are large enough to cover all frequencies with appreciable positive probabilities. By (7.6), n_{ij} can be viewed as Poisson distributed, with mean equal to

$$E(n_{ij}) = nk \frac{\mu_X^i}{i!} \frac{\mu_Y^i}{j!} \nu^{ij}.$$

Equivalently,

$$\log[E(n_{ij})] = \log(nk) - \log(i!) - \log(j!) + i \log(\mu_X) + j \log(\mu_Y) + ij\rho,$$

where $\rho = \log(\nu)$. Thus, the parameters (μ_X, μ_Y, ρ) can be estimated using a Poisson regression of the counts in each cell of the table on the corresponding row index i, column index j, and product of the row and column indices, with $- \log(i!) - \log(j!)$ as an offset (this same trick was used in the univariate case in an exercise on page 185). Note that the normalization constant is estimated as part of the intercept in the regression. Independence corresponds to $\rho = 0$, and $\rho > 0$ implies direct association between the variables, with $\rho < 0$ implying an inverse association. The bivariate Poisson distribution provides a very parsimonious description of a table when appropriate, since the independence model requires estimating only three parameters, rather than the usual $I + J - 1$.

Violence in G-Rated Animated Films

Data on violence in the 74 G-rated animated features released in movie theaters between 1937 and 1999, from Yokota and Thompson (2000), were discussed on page 107 as part of an exercise. As was stated there, we will assume that these data can be viewed as a sample from a stable ongoing process of the creation of G-rated animated features. The table on the next page cross-classifies the number of injuries to good or neutral characters versus the number of injuries to bad characters for each film.

Table 7.5 summarizes model fitting for this table. Note that the sparseness of the table (more than three-fourths of the cells have fewer than five observations) should encourage the use of simpler models, as AIC_C is designed to do. The first analyses, summarized in part (a) of the table, refer to the entire data set. Two classes of independent and uniform association models are compared: unstructured models (with arbitrary marginals) and bivariate Poisson models. None of the models fit the table very well, although the correlated Poisson model is best, suggesting that movies with more injuries to good or neutral characters also tend to have more injuries to bad characters, presumably explainable by the general level of violence in the movie or its length.

Model	G^2	df	p	AIC_C
(a) Injuries, full data set				
Independence	41.0	20	.004	9.62
Uniform association	33.4	19	.022	4.75
Independent Poisson	63.2	27	< .001	14.61
Correlated Poisson	46.3	26	.008	0.00
(b) Injuries, *The Swan Princess* omitted				
Independence	38.4	16	.001	18.61
Uniform association	26.5	15	.033	9.46
Row effects	17.7	12	.125	9.26
Column effects	24.8	12	.016	16.31
Row + Column effects	17.2	9	.045	18.35
Row × Column effects	12.5	9	.189	13.56
Independent Poisson	48.1	22	.001	13.87
Correlated Poisson	32.0	21	.058	0.00
Equal-mean correlated Poisson	36.4	22	.028	2.11

TABLE 7.5. Model fitting for animated G-rated movies injuries data.

Good character injuries	Bad character injuries				
	0	1	2	3	4
0	28	6	1	0	0
1	7	5	6	0	0
2	3	6	0	2	1
3	3	1	3	0	0
4	0	1	0	0	0
8	0	1	0	0	0

The problem with each of the models for this table is in the last row, where one movie with eight injuries to good or neutral characters appears. This movie is the 1994 release *The Swan Princess*. The movie is not that violent at an overall level (Yokota and Thompson, 2000, report roughly

13 minutes of violence in the movie, which is not unusually high for the
movies in the sample), but eight injuries to good or neutral characters
is very surprising (the average number for the sample is around 1). The
following table of Pearson residuals from the bivariate correlated Poisson
fit highlights just how surprising it is, as the residual for the corresponding
cell is more than 46.

| Good character injuries | Bad character injuries | | | | |
	0	1	2	3	4
0	1.83	−0.52	−0.33	−0.42	−0.13
1	−1.84	−1.15	2.23	−0.70	−0.27
2	−0.84	0.63	−1.49	1.52	2.01
3	1.76	−0.56	1.44	−0.83	−0.51
4	−0.44	0.74	−0.77	−0.70	−0.55
8	−0.01	46.79	−0.06	−0.14	−0.28

The results of reanalyzing the 5 × 5 table when *The Swan Princess* is
omitted from the sample appear in Table 7.5 under part (b). According to
the p-values, the row effects and RC association models fit the table, but
the bivariate Poisson model with $\hat{\rho} = .50$ is far superior based on AIC_C.
The reason for this, of course, is that it uses 9 and 12 fewer parameters,
respectively, but still reflects the positive association between good and bad
character injuries within a movie.

Recognizing the association between good and bad character injuries is
important in making inferences about these data. For example, we might
be interested in testing whether the mean number of injuries to good or
neutral characters is equal to the mean number of injuries to bad characters
in the underlying process of the creation of G-rated animated movies. If
we treated the number of injuries of the two types as independent, the
appropriate test statistic is the two-sample Gaussian-based test statistic

$$z = \frac{\overline{Y_G} - \overline{Y_B}}{\sqrt{\overline{Y_G}/n + \overline{Y_B}/n}},$$

where $\overline{Y_G}$ $(\overline{Y_B})$ is the sample mean of injuries to good or neutral (bad)
characters and n is the number of movies in the sample. Note that this
test statistic also can be constructed as the Wald statistic for the predictor
in a Poisson regression model with all $2n$ injury values (to both good or
neutral and bad characters) as the target variable and an indicator variable
identifying the counts corresponding to the good or neutral character injury

counts as the predictor. For these data,

$$z = \frac{.671 - .918}{\sqrt{.671/73 + .918/73}} = -1.67,$$

indicating weak evidence that the means are different ($p = .095$).

This is not the correct test statistic, however, since it does not take into account the association between the two variables. The appropriate equal-mean model is (7.6) with $\mu_X = \mu_Y = \mu$, or

$$P(X = i, Y = j) = k(\mu, \nu)\frac{\mu^i}{i!}\frac{\mu^j}{j!}\nu^{ij}.$$

This model can be fit using Poisson regression, since it implies

$$\log[E(n_{ij})] = \log(nk) - \log(i!) - \log(j!) + (i + j)\log(\mu) + ij\rho.$$

The entry for "Equal-mean correlated Poisson" in Table 7.5 summarizes the results of this fit, which is based on the two predictors $i + j$ and ij with offset $-\log(i!) - \log(j!)$. The difference in G^2 values for this model and the (general) correlated Poisson model is the likelihood ratio test for $\mu_X = \mu_Y$, and it clearly rejects equal means ($G^2 = 4.4$, $df = 1$, $p = .036$). This is also supported by the fact that the AIC_C value is smaller for the two-mean model than for the equal-means model. Thus, it appears that good or neutral characters suffer significantly more injuries than bad characters in animated G-rated movies. This might seem surprising at first glance, but presumably it reflects the initial trials and tribulations suffered by the hero before ultimate victory.

Table 7.6 summarizes results for the following table, which refers to fatal injuries in these movies.

Good character fatalities	Bad character fatalities			
	0	1	2	3
0	37	12	8	2
1	5	6	1	0
2	1	2	0	0

There is little evidence of any association between the two types of fatalities here, and independent Poisson distributions provides an effective, parsimonious, representation of the data. The hypothesis of equal means is strongly rejected ($G^2 = 10.5$, $p = .001$) in favor of separate means (an average of .24 good or neutral character fatalities and .59 bad character fatalities, reinforcing the earlier speculation that good characters might get beaten up in these movies, but they rarely die).

Model	G^2	df	p	AIC_C
Independence	7.3	6	.297	4.28
Uniform association	7.1	5	.210	6.59
Independent Poisson	9.9	9	.359	0.00
Correlated Poisson	9.7	8	.284	2.07
Equal-mean independent Poisson	20.4	10	.026	8.33

TABLE 7.6. Model fitting for animated G-rated movies fatalities data.

7.2   Square Tables

Another class of tables where more complicated probability structures than simple independence can be modeled is that of $I \times I$ square tables, where rows and columns have the same levels. This might reflect the same characteristic at different times or places, or on different people, or different characteristics measured on the same scale. In this case, various possible probability structures suggest themselves.

7.2.1   Quasi-Independence

Consider a situation where the same respondents are sampled at two different times and asked the same question each time. It is easy to imagine that in that case independence would be strongly rejected, since most people would answer the same way both times. On the other hand, it might be that given someone changed their opinion, the response the first time is independent of the response the second time. This model is simply a quasi-independence model discussed in Section 6.4, with the diagonal cells being treated as if they are structural zeroes,

$$\log e_{ij} = \lambda + \lambda_i^X + \lambda_j^Y + \delta_i I(i = j), \qquad (7.7)$$

where $I(\cdot)$ is the indicator function.

7.2.2   Symmetry

The symmetry model S is a particularly simple one, which states that the expected count in the (i, j)th cell is the same as that in the (j, i)th cell.

Written as a loglinear model, the symmetry model is

$$\log e_{ij} = \lambda + \lambda_i^X + \lambda_j^Y + \lambda_{ij}^{XY}, \tag{7.8}$$

subject to conditions

$$\lambda_i^X = \lambda_i^Y \text{ and } \lambda_{ij}^{XY} = \lambda_{ji}^{XY}. \tag{7.9}$$

For this model fitted values have a closed form,

$$\hat{e}_{ij} = \hat{e}_{ji} = \frac{n_{ij} + n_{ji}}{2}, \tag{7.10}$$

the average of the two observed counts on opposite sides of the main diagonal. Standardized (adjusted) residuals also have a closed form, $(n_{ij} - n_{ji})/\sqrt{n_{ij} + n_{ji}}$. The model also can be fit as a generalized linear model by regressing on indicator or effect coding variables that identify each cell on the diagonal (since they are fitted perfectly in the symmetry model), and ones that tie together each of the symmetric pairs of cells (that is, the $(1, 2)$ and $(2, 1)$ cell, the $(1, 3)$ and $(3, 1)$ cell, and so on). The model is based on estimating $I-1$ parameters for the $\lambda_i^X = \lambda_i^Y$ terms, and $I(I-1)/2$ parameters for the $\lambda_{ij}^{XY} = \lambda_{ji}^{XY}$ terms, leaving $I^2 - [(I-1) + I(I-1)/2] - 1 = I(I-1)/2$ degrees of freedom for goodness-of-fit.

Corporate Bond Rating Transitions

Rating agencies like Standard and Poor's examine outstanding corporate debt at the beginning of each year, and assign each company a specific rating. These are termed static pools, which are then used during the course of the year. These ratings reflect the agency's view of the likelihood that the company will default on its debt. Movement between the different rating classes from year to year, which is summarized in "transition matrices," is important because it reflects the health of the economy in general, and the corporate debt market in particular. Investors who are interested in stable ratings (for example, who are restricted by law to holding only top-quality bonds) would be alarmed by movement downwards, while those who are speculative investors (investing in lower-rated bonds in the hope that their rating will move up, making them more valuable) would want to identify movement upwards. Probabilities from one-year transition matrices also can be used to estimate cumulative default rates over different time horizons, allowing measurement of credit risk.

The following table gives the transition matrix for the Standard and Poor's 1997 static pool, summarizing the change in ratings of corporate debt from the beginning of 1997 to the beginning of 1998 (Standard and Poor's, 1998). "Speculative grade" refers to a rating of BB or lower (these are often called "junk" bonds).

1998 rating

1997 rating	AAA	AA	A	BBB	Speculative grade
AAA	111	5	0	0	0
AA	6	396	19	2	2
A	0	18	938	41	8
BBB	0	2	27	671	29
Speculative grade	0	0	4	48	930

As would be expected, independence is strongly rejected for this table ($G^2 = 7563.4$, $df = 16$). Examination of the table immediately shows that most corporate debt did not change rating class, as more than 93% of the observations fall on the main diagonal (good news for those who desire rate stability). This suggests fitting a quasi-independence model, omitting the diagonal, which implies that given a bond changed rating class, its 1997 rating gave no information about its 1998 rating. This model is a considerably better fit than the independence model, but it still fits poorly ($G^2 = 142.9$, $df = 11$).

Neither of these models reflects that this is a square table, but the symmetry model does. This model, in fact, fits the table very well ($G^2 = 11.9$, $df = 10$, $p = .29$). That is, given that a bond changed rating class, the probability of going from class i to class i' is the same as the probability of going from class i' to i. This reflects a general pattern that the perception of the general credit-worthiness of corporate debt issued in 1997 had not changed at the beginning of 1998, in the sense that (for example), it was just as likely for a bond to go from AAA to AA as it was for a bond to go from AA to AAA. Note, however, that symmetry does not imply that the probability of going from AA to AAA is the same as that of going from AA to A; it is obvious from the table that bonds were generally more likely to move down than to move up (for example, the estimated probability of going from AA to AAA is .002, while that of going from AA to A is .006). By pooling data in cells across the main diagonal, the model allows (hopefully) more accurate estimation of transition probabilities than the observed frequencies do.

Model	Condition
Independence	$\lambda_{ij}^{XY} = 0$
Quasi-independence	$\lambda_{ij}^{XY} = 0$ for $i \neq j$
Symmetry	$\lambda_i^X = \lambda_i^Y$

TABLE 7.7. Models that are special cases of the quasi-symmetry model.

7.2.3   Quasi-Symmetry

The symmetry model is often too simple to fit a table, because of the imposition of identical marginals (*marginal homogeneity*). A generalization of symmetry is the quasi-symmetry (QS) model, which generalizes (7.8)–(7.9) to allow for different marginal distributions. The model is

$$\log e_{ij} = \lambda + \lambda_i^X + \lambda_j^Y + \lambda_{ij}^{XY},$$

subject to conditions

$$\lambda_{ij}^{XY} = \lambda_{ji}^{XY}. \tag{7.11}$$

The main diagonal is again fit exactly. A good way to think about the meaning of the quasi-symmetry model is through odds ratios. The quasi-symmetry model implies that odds ratios on one side of the main diagonal are identical to corresponding ones on the other side of the diagonal. That is, for $i \neq i'$ and $j \neq j'$,

$$\frac{\hat{e}_{ij}\hat{e}_{i'j'}}{\hat{e}_{ij'}\hat{e}_{i'j}} = \frac{\hat{e}_{ji}\hat{e}_{j'i'}}{\hat{e}_{j'i}\hat{e}_{ji'}}.$$

Thus, adjusting for differing marginal distributions, there is a symmetric association pattern in the table. One way to fit the quasi-symmetry model is as a loglinear model that includes the same indicators used in the symmetry model to tie together symmetric cells, along with the usual marginal row and column effect parameters.

The usefulness of the quasi-symmetry model comes from the fact that many other models are special cases of it, as noted in Table 7.7 (the quasi-independence and quasi-symmetry models are equivalent when $I = 3$). These relationships allow for, among other things, the construction of a test of marginal homogeneity, even though the general marginal homogeneity model is not a loglinear model, and is difficult to fit. If quasi-symmetry holds, marginal homogeneity corresponds to

$$\lambda_i^X = \lambda_i^Y,$$

which is equivalent to the symmetry model (7.9). That is, for the quasi-symmetry model, marginal homogeneity is equivalent to symmetry. Thus, a test of whether the symmetry model is adequate given the quasi-symmetry model is also a test of marginal homogeneity (given quasi-symmetry). The appropriate likelihood ratio test is just

$$G^2(S|QS) = G^2(S) - G^2(QS),$$

which is asymptotically χ^2 on $I - 1$ degrees of freedom.

Occupational Mobility in the 1848 German Parliament

The following table is an occupational mobility table that summarizes the jobs of parliamentarians in the Frankfurter Nationalversammlung (the first democratic Parliament in Germany). The rows refer to the jobs of members of Parliament when they first entered the labor force, while the columns refer to their jobs at the beginning of the 1848 session (Greenacre, 2000). Column headings are identified by the first letter of job type in this square table.

1848 job

Initial job	J	A	E	M	C	F	L	S	Total
Justice	117	83	19	0	1	17	76	6	319
Administration	11	37	6	0	0	7	7	6	74
Education	4	3	67	1	4	3	5	13	100
Military	0	5	1	16	0	8	1	0	31
Church	0	1	11	0	26	0	1	0	39
Farmer	0	4	0	0	0	19	1	0	24
Lawyer	8	2	1	0	0	0	22	1	34
Self-employed	0	7	20	0	4	1	0	37	69
Total	140	142	125	17	35	55	113	63	690

Table 7.8 summarizes model fitting for this table. As is typical for mobility tables, the table is characterized by larger counts than expected along the diagonal, so independence is strongly rejected. Quasi-independence (omitting the diagonal) is a considerable improvement, but is still strongly rejected, reflecting that for members of parliament who changed their job category, knowing the initial job provides information about the job in 1848. A simple symmetric job mobility pattern is not adequate for this table, but quasi-symmetry fits moderately well; this implies strong rejection

Model	G^2	df	p	AIC_C
Independence	850.0	49	< .001	761.02
Quasi-independence	225.3	41	< .001	153.29
Symmetry	218.7	28	< .001	175.06
Quasi-symmetry	27.8	21	.14	0.00
Marginal homogeneity	190.8	7	< .001	

TABLE 7.8. Model fitting for German parliamentarian occupational mobility data.

of marginal homogeneity. Examination of the marginal totals shows migration away from justice and military jobs towards jobs in administration, the law, and farming.

The quasi-symmetry fit implies that, accounting for the overall change in job type distribution, the association pattern of movement from one job type to a second type was the same as that from the second to the first. The following table helps identify those associations. The entries are scaled differences between the fitted values of the quasi-symmetry model and those of the quasi-independence model (scaled by dividing by the square root of the fitted values of the quasi-independence model). Since quasi-independence is the special case of quasi-symmetry with $\lambda_{ij}^{XY} = 0$, these differences highlight the associations implied by those nonzero association parameters.

Initial job	1848 job							
	J	A	E	M	C	F	L	S
J	0.00	1.59	−2.74	−0.77	−1.95	−1.00	3.30	−2.80
A	2.00	0.00	−1.31	0.96	−0.75	1.68	−1.78	0.82
E	−0.95	−2.07	0.00	0.30	5.08	−0.66	−1.69	6.42
M	−1.40	0.03	−0.36	0.00	−0.59	5.51	−1.36	−1.02
C	−1.17	−1.67	4.99	−0.18	0.00	−1.09	−1.32	1.22
F	0.06	0.82	−0.15	2.14	−0.35	0.00	−0.78	−0.33
L	2.65	−0.74	−0.29	0.06	−0.21	−0.67	0.00	−0.80
S	−1.61	−0.75	6.13	−0.30	1.26	−1.28	−2.56	0.00

The positive entries correspond to movements that are expected under quasi-symmetry to occur more often than would be expected under quasi-independence, while negative entries correspond to fewer-than-expected movements. The job types split into three broad groups within which movement is more likely (although this is not a perfect pattern), and between which it is less likely. There is strong movement between the military and farmer job categories, and among the three types, education, church, and self-employed. The justice, administration, and lawyer connection is a little more complex, in that there is a good deal of movement between justice and administration and justice and lawyer, but noticeably little movement between administration and lawyer. Greenacre (2000) came to very similar conclusions when analyzing this table using correspondence analysis.

7.2.4 Tables with Ordered Categories

Many square tables also have ordered categories. In this situation, all of the models of Section 7.1 are applicable, in addition to symmetry-based models. A useful connection between the two classes of models is that the quasi-uniform association model (uniform association with the main diagonal omitted) is a special case of quasi-symmetry, since the former model implies a constant odds ratio (with an exact fit on the diagonal), which is obviously consistent with symmetric odds ratios on each side of the diagonal. The quasi-uniform association model is more parsimonious than the quasi-symmetry model, and allows the analyst to describe the symmetric pattern very simply.

Other parsimonious versions of quasi-symmetry that incorporate ordering are also possible. The *ordinal quasi-symmetry model* has the form

$$\log e_{ij} = \lambda + \lambda_i + \lambda_j + \beta u_j + \lambda_{ij}$$

with $\lambda_{ij} = \lambda_{ji}$, where $\mathbf{u}$ is the set of column (and row) scores. This model is a special case of quasi-symmetry with

$$\lambda_j^Y - \lambda_j^X = \beta u_j.$$

For this model the difference in the main effects for rows and columns for level j is a constant multiple of the score for level j. If $\beta > 0$, $P(X \leq j) > P(Y \leq j)$ for $j < J$, where J is the number of rows and columns in the table. That is, lower values are more likely for rows than columns, and the column marginal probability vector is stochastically larger than the row vector. If $\beta < 0$, the opposite pattern occurs. Note that since $\beta = 0$ corresponds to symmetry, a test of this constraint constitutes a focused test of marginal homogeneity (given ordinal quasi-symmetry). This model can be fit using a loglinear model for symmetry, adding the column score as an additional numerical predictor.

Occupational Mobility Between Fathers and Sons in Britain

Goodman (1979) analyzed the following British occupational mobility table, which appeared originally in Glass (1954). It represents the movement in occupational status from father to son, where the status variable is ordered, resulting in a square table with (identical) ordered categories.

Son's status

Father's status	1	2	3	4	5	6	7	Total
1	50	19	26	8	18	6	2	129
2	16	40	34	18	31	8	3	150
3	12	35	65	66	123	23	21	345
4	11	20	58	110	223	64	32	518
5	14	36	114	185	714	258	189	1510
6	0	6	19	40	179	143	71	458
7	0	3	14	32	141	91	106	387
Total	103	159	330	459	1429	593	424	3497

Table 7.9 summarizes model fitting for this table. Again, the main diagonal has more counts than expected, so independence is strongly rejected. Quasi-independence is also strongly rejected, implying information about the son's status in the father's status for father/son pairs with different status levels. The simple symmetric model fits surprisingly well, but quasi-symmetry fits very well, implying reasonably strong rejection of marginal homogeneity. Examination of the marginal totals suggests that the status distribution has shifted upwards (there are 172 more sons in categories 6 and 7 than there were fathers). The ordinal quasi-symmetry model is noticeably worse than the general version, implying strong rejection of main effects that follow a linear trend with row or column index.

While quasi-symmetry certainly fits the table adequately, we can do better. The quasi-uniform association model has the smallest AIC_C value, and the hypothesis that this model is adequate given the quasi-symmetry model is not rejected. Thus, other than the higher likelihood of father and son's status being the same (reflected in rejection of the uniform association model), there is a positive association between a father's status and son's status. Since $\hat{\theta} = .233$, the estimated odds of a son's status being in level $i + 1$ versus level i are 26% higher if the father's status was in level $i + 1$ versus level i ($\exp(.233) = 1.26$).

Model	G^2	df	p	AIC_C
Independence	897.5	36	$< .001$	850.96
Quasi-independence	408.4	29	$< .001$	375.95
Symmetry	54.0	21	$< .001$	37.82
Quasi-symmetry	15.6	15	.41	11.61
Marginal homogeneity	38.4	6	$< .001$	
Ordinal quasi-symmetry	35.5	20	.02	21.34
Uniform association	98.2	35	$< .001$	53.65
Quasi-uniform association	30.4	28	.34	0.00
QU | QS	14.8	13	.32	

TABLE 7.9. Model fitting for British father/son occupational mobility data.

7.2.5 Matched Pairs Data

Consider a situation where voters are being polled before an election regarding their support for a particular candidate. Two surveys are taken (one in June and one in August, say), with observed proportions $\overline{p_J}$ and $\overline{p_A}$ supporting the candidate. Clearly any change in opinion from June to August is of great interest to the pollster (and, presumably, the candidate). If two independent samples were taken on the two occasions, this is a standard two-sample binomial problem, and the methods of Section 6.1.1 are appropriate.

On the other hand, what if the same set of n people were sampled in the two surveys? We would expect that knowing someone's feelings in June would be informative about that person's opinions in August; that is, the responses at the two times are not independent (note that the responses of two different people at either time period are independent, however). That is, each response in June is matched with a response in August, resulting in a sample of *matched pairs*. In this case methods for independent samples are inappropriate, since the dependence must be taken into account.

This is easy to do using models for square tables. Consider Table 7.10, a table of expected counts that illustrates the correct representation of data of this type. The true proportion of people in the population who supported the candidate in June is $p_J \equiv p_1. = e_1./n$, while that in August is $p_A \equiv p_{.1} = e_{.1}/n$. The difference in proportions supporting the candidate

August opinion

June opinion	Support	Do not support	Total
Support	np_{11}	np_{12}	$e_1.$
Do not support	np_{21}	np_{22}	$e_2.$
Total	$e._1$	$e._2$	n

TABLE 7.10. Table of expected counts for results of June and August matched pairs surveys.

is thus

$$p_1. - p._1 = e_1./n - e._1/n$$
$$= (np_{11} + np_{12})/n - (np_{11} + np_{21})/n$$
$$= p_{12} - p_{21}$$

Thus, equal support for the candidate is equivalent to the off-diagonal elements being equal to each other, or symmetry. A test of the fit of the symmetry model (7.9), based on fitted values (7.10), is thus a test of the equality of the dependent proportions. The Pearson χ^2 test of this hypothesis is known as *McNemar's test*, and has the form

$$X^2 = \frac{(n_{12} - n_{21})^2}{n_{12} + n_{21}},$$

which is compared to a χ_1^2 critical value.

An alternative inferential tool is a confidence interval for the difference in proportions. It can be shown that the variance of the difference in sample proportions satisfies

$$V(\overline{p_1.} - \overline{p._1}) = [p_1.(1 - p_1.) + p._1(1 - p._1) - 2(p_{11}p_{22} - p_{12}p_{21})]/n, \quad (7.12)$$

so an approximate $100 \times (1 - \alpha)\%$ confidence interval has the form

$$\overline{p_1.} - \overline{p._1} \pm z_{\alpha/2}[\hat{V}(\overline{p_1.} - \overline{p._1})]^{1/2},$$

where $\hat{V}(\overline{p_1.} - \overline{p._1})$ substitutes the observed proportions for the probabilities in the right-hand side of (7.12). Note that while the test for symmetry is a function of only the off-diagonal cell counts, all of the cell counts are used when estimating and constructing a confidence interval for the difference in population proportions.

If the two proportions are positively correlated, as is usually the case in matched pairs data, the odds ratio $p_{11}p_{22} - p_{12}p_{21}$ is positive, implying that the variance of the difference in proportions is less than that from independent samples. This means that confidence intervals for the difference will be narrower, and tests for unequal probabilities will be more powerful than those for independent samples, with those benefits increasing as the samples become more highly correlated. This advantage of paired samples over independent samples is the same as was seen when the bivariate Poisson distribution was fit on page 268.

Can General Practitioners Effectively Screen for Diabetic Retinopathy?

Diabetes mellitus is a major public health problem. After 20 years of having the disease, most patients develop diabetic retinopathy (damage of the blood vessels in the retina), which, if left untreated, is likely to cause significant visual loss or blindness. If detected early, however, most patients can be saved from major vision loss.

Jackson et al. (2002) described the results of a small pilot study exploring whether Australian general practitioners can be effective as screeners for diabetic retinopathy after a short training session. While many strategies for increasing regular eye examinations for patients with diabetes can be expensive, particularly in remote regions of countries such as Australia, effective use of general practitioners has great potential benefit, given that they provide the bulk of adult diabetes diagnosis and care. In the study the eye skills of 17 general practitioners were evaluated before and after two "upskilling" workshops (which together took roughly three hours). They were asked to evaluate four patients who did, in fact, suffer from diabetic retinopathy, and four patients who did not, before and after the workshops. Note that the rate of successful identification of diabetic retinopathy in the first instance corresponds to the sensitivity of the screening, while successful identification of normal retinal conditions in the second instance corresponds to specificity (see page 66 for definitions of these terms). According to the Australian National Health and Medical Research Council, a sensitivity of 60% is high enough for a diagnostic tool to be useful as a successful screening mechanism, so the doctors were treated as having performed satisfactorily if their sensitivity or specificity values were over 60%.

The following table summarizes the sensitivity transitions pre- and post-workshop:

| | *Post-workshop* | | |
Pre-workshop	Not satisfactory	Satisfactory	Total
Not satisfactory	1	12	13
Satisfactory	0	4	4
Total	1	16	17

Change in sensitivity from pre- to post-workshop corresponds to a change in the marginal distribution, which is tested using McNemar's test; here $X^2 = 12.0$, which is highly statistically significant ($p < .001$). That is, the sensitivity of the doctors improved significantly after taking the workshops. A 95% confidence interval for the change in sensitivity is

$$.941 - .235 \pm$$
$$(1.96)\sqrt{[(.941)(.059) + (.235)(.765) - 2(.059)(.235)(.706)(0)]/17},$$

or $(.475, .937)$.

Specificity is explored using the transitions of evaluations of patients without diabetic retinopathy:

Post-workshop

Pre-workshop	Not satisfactory	Satisfactory	Total
Not satisfactory	3	10	13
Satisfactory	1	3	4
Total	4	13	17

McNemar's test for these data is $X^2 = 7.4$, with $p = .007$, now indicating a significant improvement in specificity after the workshops. A 95% confidence interval for this improvement is

$$.765 - .235 \pm$$
$$(1.96)\sqrt{[(.765)(.235) + (.235)(.765) - 2(.176)(.176)(.588)(.059)]/17},$$

or $(.246, .814)$. Thus, there is considerable evidence that with appropriate training general practitioners can make an important contribution to screening of diabetic retinopathy. In fact, the Royal Australian Colleges of General Practice and Ophthalmology implemented the workshop program across Queensland in 2000.

Matched pairs data are not restricted to the dependent proportions situation. Indeed, we have already examined three tables that correspond to matched pairs: the 5×5 corporate bond transition table in Section 7.2.2 (paired on the company issuing the debt), the 8×8 German Parliament job transition table in Section 7.2.3 (paired on the member of Parliament), and the 7×7 British occupational mobility table in Section 7.2.4 (paired on the father/son pair). Thus, any of the models for square tables are appropriate for the analysis of matched pairs data. Just as was true in the 2×2 case, a particularly important question is whether the two marginal distributions are equal to each other (that is, marginal homogeneity). The next example discusses an important application of matched pairs data, that of the matched case-control study.

Is Wearing a Head Covering Related to Cataract Occurrence?

We introduced the idea of a case-control (or retrospective) study in Section 6.1.3. In many such studies, people chosen to be in the control group (people without the disease of interest, for example) are not chosen completely randomly. Rather, each case (person with the disease of interest) is matched with a control on the basis of a set of control variables (variables that are not of direct interest, but might be related to incidence of the disease) — gender, age, income, and so on. This results in the cases and controls forming two dependent samples, rather than two independent ones.

The following table, which is based on data originally from Dolezal et al. (1989), comes from a study investigating a possible relationship between cataracts (clouding of the lens or capsule of the eye, which can lead to vision loss or blindness) and the use of head coverings during the summer. The cataract cases were formed from clinical patients receiving treatment for cataracts, who were matched with controls of the same gender and similar age who did not have cataracts. The subjects were asked to describe the amount of time that they used head coverings during the summer.

Cataract case

Control	Never	Occasionally	Frequently	(Almost) always	Total
Never	0	0	1	4	5
Occasionally	1	2	1	3	7
Frequently	3	0	0	3	6
(Almost) always	7	9	5	29	50
Total	11	11	7	39	68

First, we can recognize what is **not** the appropriate way to view these data — as a 2 × 4 table of case and control responses:

Use of head coverings

Group	Never	Occasionally	Frequently	(Almost) always
Controls	5	7	6	50
Cases	11	11	7	39

Model	G^2	df	p	AIC_C
Independence	11.1	9	.27	0.00
Symmetry	8.3	6	.22	5.15
Quasi-symmetry	3.8	3	.28	9.59
Ordinal quasi-symmetry	4.0	5	.56	3.67

TABLE 7.11. Model fitting for case-control cataract and head covering data.

A test of independence of rows and columns of this table corresponds to a test of equal probability distributions for cases and controls, but this is not appropriate, because of the pairing of cases and controls.

Table 7.11 summarizes the appropriate model fitting for the 4×4 matched pairs table. As can be seen, the relationship between head covering usage for controls and that for cases is not particularly strong, as the independence model is not rejected, and has the smallest value of AIC_C of the models examined. Note, however, that given the good fit of the ordinal quasi-symmetry model, a one degree-of-freedom likelihood ratio test of marginal homogeneity (given ordinal quasi-symmetry) is

$$G^2(\text{Symmetry}) - G^2(\text{Ordinal quasi-symmetry}) = 8.3 - 4.0 = 4.3,$$

with $p = .037$. Thus, the hypothesis of symmetry (and hence marginal homogeneity) is rejected. Since $\hat{\beta} = -.33$, the distribution of head covering usage for controls is stochastically larger than that for cases; that is, having cataracts is associated with less usage of head coverings. The estimated probabilities of using head coverings for controls is $\{.070, .104, .097, .729\}$, while that for cataract cases is $\{.165, .161, .095, .579\}$. This does not prove that wearing head coverings during the summer prevents the formation of cataracts, of course, but it is suggestive, since it is known that ultraviolet light causes damage to the lens of the eye. Agresti and Coull (1998b) also found evidence of this pattern using a test based on assuming monotone trends of odds ratios.

7.2.6   Rater Agreement Tables

A particular form of matched pairs data occurs when the same object is rated by (two) different evaluators. For example, in a clinical context, a patient might be evaluated for disease status by two different physicians. In this context, independence is not of any great interest — the presumption would be that the ratings of the two physicians are related to each other.

Rather, it is agreement between the two raters that is important, which requires association, but is a stronger condition (for example, the ratings of two physicians who always differ by one rating scale in the same direction are strongly associated, but agree poorly with each other). In this context models that put restrictions on the diagonals are particularly useful, since these cells correspond to agreement between the two raters. A compromise between fitting the diagonals exactly and not addressing them separately at all is to use a single parameter for all of the diagonal cells. So, for example, the general quasi-uniform association model, which has the form

$$\log e_{ij} = \lambda + \lambda_i^X + \lambda_j^Y + \theta(u_i - \overline{u})(v_j - \overline{v}) + \delta_i I(i = j), \qquad (7.13)$$

can be simplified by setting $\delta_i = \delta$ for all i. This model implies an inflation of the counts along the diagonal over those of a uniform association model by a constant multiple $\exp(\delta)$.

A popular summary of the amount of agreement in a table is *Cohen's kappa*. This measure is based on comparing the observed agreement in the table to what would be expected if the ratings were independent. Since agreement corresponds to being on the main diagonal, the excess probability of agreement over that implied by independence is

$$\sum_{i=1}^{I} p_{ii} - \sum_{i=1}^{I} p_{i\cdot} p_{\cdot i}.$$

Kappa scales this value by its maximum possible value (perfect agreement), yielding

$$\kappa = \frac{\sum_i p_{ii} - \sum_i p_{i\cdot} p_{\cdot i}}{1 - \sum_i p_{i\cdot} p_{\cdot i}}. \qquad (7.14)$$

The sample version of this measure, $\hat{\kappa}$, substitutes the observed frequencies for the probabilities in (7.14), and estimates the difference between observed and expected agreement as a proportion of the maximum possible difference. Note that negative values of κ are possible if agreement is worse than would be expected by chance, but this does not typically happen in practice. When $\kappa = 0$, agreement is what would be expected by random chance, and increasing values of κ represent increasing agreement. An asymptotic confidence interval for κ can be constructed using the estimated standard error of $\hat{\kappa}$,

$$\widehat{\text{s.e.}}(\hat{\kappa}) = \left[\frac{A + B - C}{n \left(1 - \sum_i \hat{p}_{i\cdot} \hat{p}_{\cdot i} \right)^2} \right]^{1/2}, \qquad (7.15)$$

where

$$A = \sum_i \hat{p}_{ii}[1 - (p_{i\cdot} + p_{\cdot i})(1 - \hat{\kappa})]^2,$$

$$B = (1 - \hat{\kappa})^2 \sum_{i \neq j} \hat{p}_{ij}(\hat{p}_{j\cdot} + \hat{p}_{\cdot i})^2,$$

and

$$C = \left[\hat{\kappa} - \left(\sum_i \hat{p}_{i\cdot}\hat{p}_{\cdot i}\right)(1 - \hat{\kappa})\right]^2.$$

Accuracy of the Pacifier Thermometer in Infants and Small Children

Rectal thermometers are generally accepted as the standard tool for measuring temperature in infants, although there is some controversy about this (see, for example, Chamberlain, 1998). Unfortunately, there are several disadvantages to rectal thermometry, including discomfort for the patient, upset for the patient and parent, the risk of traumatic injury, and the risk of transmission of pathogens. Because of this, alternative thermometric methods have been explored, but with limited success. Axillary (under the arm) thermometers are generally too inaccurate for routine clinical use, and tympanic (ear) thermometers, while fairly reliable in adults, have not proven adequate for infants and young children. The result has been a continuing exploration of alternative, less invasive, thermometers that closely agree with rectal temperature.

Press and Quinn (1997) reported the results of an investigation into the use of a supralingual (above the tongue), or pacifier, thermometer in determining the temperature of infants. The thermometer appears to be an ordinary pacifier, but has a thermistor in the nipple that, when the child has had the pacifier in his or her mouth long enough to result in a steady state temperature reading, results in the reading being reported in a digital display located in the front of the pacifier. A convenience sample of 100 infants from a Miami pediatric emergency department and inpatient ward were enrolled in the study, 50 of whom exhibited fever according to rectal temperature (temperature higher than 38 °C, or 100.4 °F). The table on the next page summarizes the results of the study.

If the pacifier thermometer is working as expected, the marginal distributions of the two variables should be the same, but this is apparently not the case, as McNemar's test strongly rejects marginal homogeneity ($X^2 = 11.3$, $p < .001$). Cohen's kappa (with 95% confidence interval) is $.7 \pm .135$, implying that the estimated difference between the observed agreement between the two types of thermometers and that expected by chance is 70% of the maximum possible difference.

Pacifier temperature

Rectal temperature	No fever	Fever	Total
No fever	49	1	50
Fever	14	36	50
Total	63	37	100

Treating rectal thermometer as being the correct measurement, the sensitivity of the pacifier thermometer is $36/50 = .72$, while the specificity is $49/50 = .98$, implying that the pacifier thermometer is not identifying fever cases as often as would be preferred. This is consistent with a generally lower temperature reading for the pacifier thermometer, which is indeed what was observed (it is generally accepted that rectal temperature is higher than oral temperature, although the precise relationship has not been established). Press and Quinn (1997) suggested adjusting the pacifier temperature by adding .5 °F, as the observed mean difference in temperatures was .49 °F. This results in the following agreement table:

Adjusted pacifier temperature

Rectal temperature	No fever	Fever	Total
No fever	38	12	50
Fever	4	46	50
Total	42	58	100

The two marginal distributions are now more similar, although there is still evidence against marginal homogeneity ($X^2 = 4.0$, $p = .046$). Cohen's kappa has *decreased* slightly to $.68 \pm .142$, since disagreements from lack of sensitivity for the unadjusted pacifier temperature readings (resulting in false negative readings) have been replaced with disagreements from lack of specificity for the adjusted settings (resulting in false positive readings). Based on these results, Press and Quinn suggested that while rectal thermometry was still preferable for very young children (90 days and younger), when accurate temperature measurement is crucial, for older children the pacifier thermometer (suitably adjusted) constitutes a reasonable noninvasive alternative, particularly at home and in low-volume ambulatory care settings, where the relatively lengthy waiting time for a steady-state temperature reading (roughly $3\frac{1}{2}$ minutes) is not a major problem.

Cohen's kappa treats all disagreements as being of equal importance, which might not be appropriate, especially for ordinal data (where presumably disagreements in neighboring categories would be considered less serious than ones separated by multiple categories). A weighted version of κ can address this by penalizing some disagreements more than others. Let $\{w_{ij}\}$ be a set of weights such that $0 \leq w_{ij} \leq 1$, $w_{ij} = w_{ji}$, and $w_{ii} = 1$. The weighted kappa is then defined as

$$\kappa_w = \frac{\sum_i \sum_j w_{ij} p_{ij} - \sum_i \sum_j w_{ij} p_{i\cdot} p_{\cdot j}}{1 - \sum_i \sum_j w_{ij} p_{i\cdot} p_{\cdot j}}.$$

Weights that decrease with increasing degree of disagreement are appropriate for an ordinal table; a common choice of weights is

$$w_{ij} = 1 - (i - j)^2 / (I - 1)^2. \tag{7.16}$$

The estimated asymptotic standard error of $\hat{\kappa}_w$ is

$$\widehat{s.e.}(\hat{\kappa}_w) = \left(\left\{ \sum_{ij} \hat{p}_{ij} \left[w_{ij} - (\overline{w}_{i\cdot} + \overline{w}_{\cdot j})(1 - \hat{\kappa}_w) \right]^2 \right.\right.$$
$$\left. - \left[\hat{\kappa}_w - \left(\sum_{ij} w_{ij} p_{i\cdot} p_{\cdot j} \right)(1 - \hat{\kappa}_w) \right]^2 \right\}$$
$$\left. \bigg/ \left\{ n \left[1 - \left(\sum_{ij} w_{ij} p_{i\cdot} p_{\cdot j} \right) \right]^2 \right\} \right)^{\frac{1}{2}}, \tag{7.17}$$

where $\overline{w}_{i\cdot} = \sum_j \hat{p}_{\cdot j} w_{ij}$ and $\overline{w}_{\cdot j} = \sum_i \hat{p}_{i\cdot} w_{ij}$.

The Sock Test for Evaluating Activity Limitation

Musculoskeletal pain, including back and lower body pain related to arthritis or traumatic injury, is common in both adults and children. The standard treatments for this pain are rest, medication, and physical therapy. Physical therapists and physicians working with patients require methods of measuring impairment, pain, and ability to perform daily activities for diagnosis, classification and prediction of ultimate outcome. While self-evaluation by patients is an important component of rehabilitation, clinician-derived assessments of physical function are also desirable. Thus, there is a need for a simple assessment tool that can be used by physical therapists to determine performance of a daily living task that is probably important to most patients with musculoskeletal pain.

Strand and Wie (1999) proposed and investigated such a tool, the Sock Test, which simulates the activity of putting on a sock. As would be true for any new clinical test, besides an ability to reflect perceived activity limitation in patients, it is also important to evaluate intertester reliability when the test is applied. That is, test results should agree when applied to the same person by different testers.

The Sock Test is performed as follows. The patient sits on a high bench, with his or her feet not touching the floor. He or she then lifts up one leg at a time and simultaneously reaches down toward the lifted foot with both hands, grabbing the toes with the fingertips of both hands. Scores on the Sock Test are given on an ordinal scale from 0 (can grab the toes with fingertips and perform the action with ease) to 3 (can hardly, if at all, reach as far as the malleoli (the rounded protuberances on each side of the ankle)). The leg with more limited performance is assigned the score. The following table summarizes the results of a study on 21 patients with musculoskeletal pain in Bergen, Norway, using evaluations from two therapists.

| | *Score for* | | | | |
| *Score for* | *therapist 2* | | | | |
therapist 1	0	1	2	3	Total
0	5	0	0	0	5
1	3	5	0	0	8
2	0	2	3	0	5
3	0	0	0	3	3
Total	8	7	3	3	21

A reliable test should result in similar marginal distributions for the two therapists. While formal testing for this table is somewhat problematic, given the small sample, the simple symmetry model fits adequately $(G^2 = 6.9, df = 6, p = .33)$, which is of course consistent with marginal homogeneity. The weighted kappa, using weights (7.16), is $\hat{\kappa}_w = .89$, with 95% confidence interval (.78, .99). Thus, the study implies that there is very strong intertester reliability for the Sock Test.

The use of kappa to measure rater agreement is controversial, for several reasons. Since κ is not a model parameter, it is difficult to interpret. Most recommendations agree that $\kappa > .75$ constitutes strong agreement and $\kappa < .4$ corresponds to fair-to-poor agreement (with values in between moderate-to-good), but this is obviously a very imprecise guideline. The value of kappa is also strongly dependent on the marginal distributions. For this reason, applications of the same rating system to populations with different characteristics can lead to very different values of κ.

The weighted kappa also has an arbitrariness to it from the choice of weights. For example, if weights $w_{ij} = 1 - |i - j|/(I - 1)$ are used in the Sock Test example, $\hat{\kappa}_w = .78$, with asymptotic 95% confidence interval $(.61, .97)$, with noticeably different implications for the strength of agreement. Model-based analyses of rater agreement tables do not suffer from these drawbacks, which is a distinct advantage over measures such as kappa.

7.3 Conditional Analyses

The test statistics and confidence intervals discussed in this chapter are asymptotic in nature, and are not necessarily valid for tables with small expected cell counts. Conditional analysis (that is, the construction of exact tests) is thus useful for these models. Just as was true in Section 6.3, the key is to condition on appropriate sufficient statistics, resulting in a null distribution free of any nuisance parameters, allowing exact inference. This is a harder problem here, since the row and column marginal totals are no longer necessarily the appropriate sufficient statistics.

In an important sense, conditional analysis is less important for the tables discussed in this chapter than the unstructured ones described in Chapter 6. The reason for this is that much of the testing here has been based on focused subset tests. For example, for ordered categorical data, the focused test of independence given the fit of the uniform association model is more powerful than the omnibus independence test. Similarly, in square tables, tests of marginal homogeneity are subset tests: symmetry given quasi-symmetry, or symmetry given ordinal quasi-symmetry. These subset tests, being based on (for large tables, far) fewer degrees of freedom, are much more likely to follow their asymptotic null distributions, so p-values are likely to be valid even in tables with small expected cell counts.

This does not mean, of course, that all asymptotic analyses of square or ordered tables are correct. A full discussion of how to adapt exact methods to such tables is beyond the scope of this book, but the flavor of the problem can be seen from one situation that is readily amenable to exact analysis: marginal homogeneity for matched binary data. Recall from Section 7.2.5 that McNemar's test of marginal homogeneity in a 2×2 table is actually just a test of equality of the two off-diagonal probabilities, $p_{12} = p_{21}$, and that it does not involve the diagonal cell counts at all. Let $K = n_{12} + n21$ be the total number of observations in the off-diagonal cells. Under the null hypothesis of marginal homogeneity, these K observations should be split evenly between the two cells. That is, under the null hypothesis, $n_{12} \sim \text{Bin}(K, .5)$ (obviously, n_{21} also has this null distribution). The exact

McNemar's test simply assigns exact binomial probabilities to each of the possible $\{n_{12}, n_{21}\}$ pairs, and hence to each of the possible values of X^2, and then derives an exact p-value based on that exact distribution.

Using Weight–Height Relationships to Assess Children's Nutritional Status

Anthropometry is the study of human body measurement for use in classification and comparison. Anthropometric indices, such as the weight–height relationship, are used by government and child welfare officials to monitor the nutritional status of children, with children that fall outside accepted parameters often being eligible for special government antimalnutrition intervention. The World Health Organization (WHO) adopted specific recommendations made by the U.S. National Center for Health Statistics (NCHS) for the reference population for anthropometric indices in 1983.

Forcheh (2002) investigated the usefulness of a different standard, the so-called Ehrenberg law–like relationship. This relationship states that the weight and height of children aged 5–13 years follows the relationship

$$\log_{10}(\text{Weight}) = .8 \times \text{Height} + .4 \; \pm .04,$$

regardless of gender, socioeconomic status, and race of the child. Using tabulated WHO–NCHS values, Forcheh showed that Ehrenberg's relationship fit the expected weight–height relationship closely, although it consistently predicted a slightly larger weight for height (the relationship

$$\log_{10}(\text{Weight}) = .8 \times \text{Height} + .39$$

seems to fit the data very well). These indices then can be used to classify children as being malnourished, adequately nourished, or overweight, with adequately nourished children having weight within $\mu \pm 2\sigma$, where μ and σ are the expected weight of an adequately nourished child of a given height, and σ is its standard deviation, respectively. The Ehrenberg-based index would have the advantage of being very easy to use in the field, requiring only a hand-held calculator, but it is important to verify that it is reliable, especially for very young children.

The table on the next page gives the results of classification of boys under 5 years old in Ngamiland, Botswana, on the basis of the WHO–NCHS standard and the Ehrenberg relationship (note that none of the children were rated overweight on either measure). McNemar's test is $X^2 = 4.0$, with $p = .046$, indicating that the Ehrenberg and WHO–NCHS indices lead to (barely) significantly different marginal distributions. The exact p-value, based on a Bin(4, .5) distribution, is $p = .125$, apparently implying less evidence against marginal homogeneity. Note, however, that the conservativeness of exact p-values noted in Chapters 4 and 6 still applies here, so the mid-p value of $p = .063$ is probably more reasonable (and closer to the asymptotic value).

Ehrenberg relationship

WHO–NCHS standard	Malnourished	Normal	Total
Malnourished	20	0	20
Normal	4	237	241
Total	24	237	261

Since the WHO–NCHS index is the standard, it is tempting to report that the Ehrenberg index is (marginally) significantly incorrect in its classification, but Forcheh reports that all four of the boys classified as malnourished using the Ehrenberg index but normal using the WHO–NCHS index were underweight (had a low weight for age), and three were stunted (had a low height for age), so it could be argued that the Ehrenberg index is correct.

The following table gives results for girls.

Ehrenberg relationship

WHO–NCHS standard	Malnourished	Normal	Total
Malnourished	9	0	9
Normal	14	229	243
Total	23	229	252

The evidence here is much clearer, as both the asymptotic and mid-p exact p-values are less than .001. Once again it is tempting to call the 14 off-diagonal observation misclassifications, but 8 of these girls were either underweight or stunted. Thus, the Ehrenberg relationship provides an easily applied and effective tool for health workers who need to make quick decisions about a child's nutritional status, particularly in rural areas.

7.4 Background Material

Goodman (1979) proposed and discussed the various association models. Agresti (1984) provides extensive discussion of these and other models for ordered category tables. Lindsey (1995, Section 6.4) and Lindsey (1997, Section 3.3.4) discuss using loglinear models to fit bivariate exponential family distributions; see Cameron and Trevidi (1998, Chapter 8) and Kocherlakota and Kocherlakota (1993) for other possibilities.

Aerts et al. (1997), Simonoff (1983, 1987, 1995, 1996, Chapter 6, 1998b) and Simonoff and Tutz (2000) described different approaches to extending the smoothing methods of Section 5.5.2 to two-dimensional tables with ordinal rows and columns. Just as was the case in that section, these methods have good properties under sparse asymptotics, and are especially appropriate (and useful) in application to sparse tables.

Correspondence analysis is a graphical technique designed to display the association pattern between rows and columns of a table, based on the estimation of canonical correlations. Greenacre (1984) provides a general discussion, while Beh (1997) and Greenacre (2000) discuss applications to the types of tables discussed in this chapter (tables with ordered categories and square tables, respectively).

The analysis of matched pairs tables that involve repeated measurement of the same person naturally focus on the transition of status of an individual from one time period to the next. Models that treat the diagonal cells as distinct from the off-diagonal cells, such as quasi-independence and quasi-symmetry, are examples of *mover–stayer* models in this context. Such models hypothesize that there are two types of individuals: those that are susceptible to change (movers) and those that are not (stayers). In a quasi-independence fit omitting the diagonal cells (7.7), for example, the number of movers from level i to level i is estimated using the estimated parameters $\hat{\lambda}$, $\hat{\lambda}_i^X$, and $\hat{\lambda}_i^Y$, with the number of stayers estimated to be the observed excess over this estimated value. The special case of this model where the probabilities of staying within each level are the same is called the loyalty model (see Lindsey, 1997, pages 32–33, for further discussion).

Schuster and von Eye (2001) discussed models appropriate for ordinal agreement tables, including (7.13); they refer to the λ terms as the association component of the model, and δ terms as the agreement component. Based on arguments from Darroch and McCloud (1986), they advocate only using models that satisfy quasi-symmetry. Cohen's kappa was introduced in Cohen (1960); the weighted version was introduced in Cohen (1968). Fleiss et al. (1981) gave the asymptotic standard errors for κ and κ_w. Altman (1991), Fleiss (1981), and Landis and Koch (1977) all gave (slightly different) benchmarks for the interpretation of different magnitudes of kappa.

The manual for the StatXact computer package (Cytel Software Corporation, 2001b) discusses conditional approaches to the analysis of tables with ordered categories (Chapters 20 and 21) and square tables (chapter 8). McDonald et al. (1998) described algorithms for the construction of exact test for models for square tables. Park (2002) suggested using the asymptotic form of McNemar's test over the exact version even for tables based on small sample sizes, arguing that the asymptotic version held its size well and was more powerful than the exact one. Haberman (1977) and Agresti and Yang (1987) showed that subset (focused) tests are much more likely to hold their asymptotic validity than overall tests, even for sparse tables.

7.5 Exercises

1. (a) The tables from Abzug et al. (2000) relating top ten employment with geographic location (page 240) are obviously singly ordered. Reanalyze these table using models that reflect that ordering. Are the results similar to what you see if you do not take this ordering into account?

   (b) The Kruskal–Wallis test is a popular nonparametric (rank) test that is used to compare populations. It also can be applied to singly ordered contingency tables. Consider a table with I nominal row categories and J ordinal column categories. Let

   $$w_j = \sum_{\ell < j} n_{.\ell} + \frac{n_{.j} + 1}{2}$$

   be the so-called midrank score for column j (it is the middle rank of observations falling in the jth column), and let

   $$r_i = \sum_{j=1}^{J} w_j n_{ij}$$

   be the weighted sum of observations in row i. Then the Kruskal–Wallis statistic is

   $$KW = \frac{12}{n(n+1)C} \sum_{i=1}^{I} \frac{[r_i - n_{i.}(n+1)/2]^2}{n_{i.}},$$

   where

   $$C = 1 - \frac{\sum_{j=1}^{J}(n_{.j}^3 - n_{.j})}{n^3 - n}$$

   is a correction for ties (this is the Wilcoxon–Mann–Whitney statistic when $I = 2$). The statistic KW can be compared to a χ^2_{I-1} critical value to test equality of the row probability vectors (that is, independence), and is directly comparable to the focused test of independence given the row effects model. Apply the Kruskal–Wallis test to the employment/geographical location tables. Are the resultant inferences similar to those you made in part (a)?

2. As was noted in Chapter 5, the home run data file **homerun** contains game-by-game information concerning the record-breaking seasons of Mark McGwire and Sammy Sosa (in 1998) and Barry Bonds (in 2001).

(a) It is natural to think that team success would be related to the ability of these sluggers to hit home runs. Is that true? Construct a cross-classification of whether the Cardinals won a game by the number of home runs hit by McGwire in that game. Is there evidence of an association between these two variables if you ignore the ordering of the number of home runs? What if you take that into account?

(b) Repeat part (a) for Sosa and the Cubs, and for Bonds and the Giants. Are the results similar?

(c) Now consider the relationship between the number of runs scored by the Cardinals in each game and the number of McGwire home runs. Fit a bivariate Poisson distribution to these data. Does this distribution provide a good fit?

(d) Repeat (c) for Sosa and the Cubs, and Bonds and the Giants. Are the results similar?

3. Consider again the 1970 draft lottery data discussed in Section 6.3.2. Fit association models to this table. Do row and/or column scores shed further light on the nonrandomness of the 1970 lottery? Construct a mosaic plot of the fitted counts from the model you choose. How does it compare to the mosaic plot of observed counts in Figure 6.4?

4. (a) Analyze the table on page 243 that relates skin type to skin cancer, taking advantage of the natural ordering of skin color. Is there evidence of a relationship between skin type and the incidence of the different types of cancer?

(b) Consider a partially ordered table, with nominal rows and ordinal columns. The row effects model hypothesizes a linear relationship between the logged probability and the score vector **v**. This pattern is tuned to a location effect across rows (conditional row probabilities that are shifted to the left or right from row to row), but is not well designed to identify a scale effect (conditional row probabilities that are more or less concentrated from row to row). Such an effect can be identified by partitioning the overall lack of fit of independence to represent location and scale effects (see, for example, McCullagh, 1980, and Nair, 1986). One way to do this is to add a quadratic term to the model,

$$\log e_{ij} = \lambda + \lambda_i^X + \lambda_j^Y + \tau_i(v_j - \overline{v}) + \psi_i(v_j - \overline{v})^2.$$

Is there evidence of a scale effect in the skin type/skin cancer table?

(c) The following table comes from the same survey of MBA students discussed on page 28. The table cross-classifies the score the student received on the GMAT (Graduate Management Admission Test), the standardized examination used by most U.S. business schools in the admission process, with whether or not they were an undergraduate business major.

Business major	GMAT score				
	< 660	[660, 670)	[670, 680)	[680, 690)	[690, 700)
No	1	2	2	5	3
Yes	0	0	0	0	3

	[700, 710)	[710, 720)	[720, 730)	≥ 730
No	3	3	2	1
Yes	4	5	3	0

i. Is there a relationship between performance on the GMAT and undergraduate major?

ii. Does the row effects model provide an adequate fit to the data? What does $\hat{\tau}$ imply about the effect of undergraduate major on GMAT performance? Is the pattern what you would have expected?

iii. Now include a quadratic term in the model. Does this improve the fit significantly? What does $\hat{\psi}$ imply about the relative variability of GMAT performance between business and nonbusiness majors? Is the pattern what you would have expected?

5. Between February 1998 and June 1999, the Serbian province of Kosovo was the site of a campaign of "ethnic cleansing" against ethnic Albanians that led to the deaths of more than 10,000 people and the forced displacement of more than 800,000 people. After roughly three months of air strikes, forces of the North Atlantic Treaty Organization (NATO) took control of the province in mid-June 1999. This was followed by reprisal attacks particularly targeting Serbs, resulting in more than half of the prewar Serbian population in Kosovo leaving the province. The health status of those remaining was of concern, and a survey was taken in Serbian residential areas in Kosovo to investigate this (Salama et al., 2000). One of the measures taken in

the survey was the respondent's body mass index (BMI) in kilograms of weight per meter^2 of height, with the following results:

	Age	
BMI	18–59 years	At least 60 years
≤ 17	0	2
$(17, 18.5)$	1	9
$[18.5, 20)$	2	8
$[20, 22)$	7	19
≥ 22	41	60

(a) Undernutrition is defined as BMI < 18.5. Convert this table into a 2×2 table, collapsing BMI into the groups BMI < 18.5 and BMI ≥ 18.5 (this is the analysis performed in the paper). Is there evidence of a difference in nutrition status based on age? Does Fisher's exact test give a different result from an asymptotic analysis?

(b) Consider now the full 5×2 table, ignoring the ordering in BMI value. Is there evidence of a difference in nutrition status based on age? Does Fisher's exact test give a different result from an asymptotic analysis?

(c) Now construct a focused test of independence given a column effects model. Is there evidence of a difference in nutrition status based on age? What pattern do you see? Does the column effects model provide a good summary of the data?

6. A kidney stone (nephrolithiasis or renal calculi) is a solid lump of crystals that separate from urine and build up on the inner surfaces of the kidney. Kidney stones can cause extreme pain, and if untreated can quickly lead to kidney failure. Although it is assumed that both kidneys secrete similar urine constituents, the vast majority of patients with recurrent stones exhibit them on only one side (unilaterally). Shekarriz et al. (2001) discussed an investigation into whether a patient's (self-reported) sleep posture is associated with unilateral stone formation, based on the hypothesis that sleep posture affects blood circulation to the kidneys, which affects stone formation. The results of the study are given on the next page.

(a) Is there evidence that sleep posture is associated with location of kidney stones?

(b) The authors of Shekarriz et al. (2001) ignored the patients without a sleep posture preference (rotisserie-like) when analyzing
these data, effectively arguing that these patients provided no
information about the association between sleep posture and
stone formation. Do you agree with this decision?

Stone location

Sleep posture	Right side	Left side
Right side down	31	13
Rotisserie-like	8	9
Left side down	9	40

7. The saturated model, with a deviance of zero and IJ exactly fitted
counts, has $AIC = 2IJ$ (using the definition in (5.12)). Thus, according to AIC, any model for which $G^2 > 2d$, where d is the degrees of
freedom of the model, is inferior to the saturated model. On the other
hand, Raftery (1986a, 1986b) showed that the Bayes factor $BF(M)$
for a model M relative to the saturated model (the ratio of the posterior probability that M is true to the posterior probability that the
saturated model is true) when using the Jeffreys prior satisfies

$$-2 \log BF(M) \approx BIC$$

for large samples. This implies that for large samples, any model for
which $G^2 > d \log(n)$ is less likely (in terms of posterior probability)
than the saturated model.

(a) Consider the table relating microlith type and site on page 264.
Are any association models better than the saturated model
based on AIC or BIC? Does your answer cast doubt on the use
of the RC model parameter estimates to order the sites?

(b) The table on the next page relates the number of previous pregnancies to the quality of prenatal care for a set of 36,354 black
live births and infant deaths in Washington, DC, from 1980 to
1985 (Ahmed et al., 1990). Fit whatever models seem appropriate for these data. How would you describe the relationship
between previous prenancies and the quality of prenatal care
obtained by the mother? Do any models fit the table, based on
p-values? Is this a surprise? Are any models considered better
than the saturated model based on BIC? What about based on
AIC?

Previous pregnancies	Prenatal care		
	Inadequate	Intermediate	Adequate
0	1654	3550	9983
1	1167	2191	6891
2	695	1236	4005
3	442	699	1886
4	455	726	1771

8. (a) Recall again the data on the relationship between the results of studies on calcium-channel blockers and whether or not the research was supported by a drug company (pages 105 and 242). Reanalyze these data, taking into account the ordering of the research result variable. Does your assessment of the relationship between drug company support and research results change?

   (b) Friedberg et al. (1999) conducted a similar study of research into the cost-effectiveness of cancer drugs, with the following results:

	Drug company sponsored	
Conclusion	Yes	No
Favorable	60	42
Neutral	35	21
Unfavorable	5	38

   Is there a relationship between the conclusion of the research and the type of sponsorship? Does taking the ordering of the conclusion variable into account change your opinion?

9. A change in the Chief Executive Officer (CEO) of a public firm is often viewed by the market as an important signal of the future direction of the firm. In particular, the hiring of an outsider sends an apparent message implying a change in policies or direction, which is often viewed positively. Borokhovich et al. (1996) examined 969 CEO successions at 588 large public firms in the United States between 1970 and 1988, and noted the possible relationship between the composition of the board of directors (the percentage of the members of the board from outside the firm) and the number of inside or outside CEO appointments:

Percentage of outside directors	Nature of appointment	
	Insider	Outsider
< 40	33	2
$[40, 50)$	47	7
$[50, 60)$	93	16
$[60, 70)$	35	163
$[70, 80)$	49	204
$[80, 90)$	61	200
≥ 90	17	42

(a) Is there an association between board composition and whether the new CEO is an insider or an outsider? Answer this question ignoring the ordering in the rows.

(b) In order to reflect the ordering in the rows, Borokhovich et al. (1996) constructed a test comparing the insider/outsider pattern for the first row to that for the last row, effectively omitting all of the rows between. Construct a more appropriate test of association that reflects the ordering in the rows. Do any of the association models provide a good fit to the data? What does the fitted model imply about outsider board composition and its relationship to outsider CEO appointment?

10. Cigarette smoking is known to be an important risk factor for the development of age-related cataracts, but does quitting smoking reduce the chances of cataract development? (This would suggest that damage to the lens of the eye is reversible.) Christen et al. (2000) examined this question based on U.S. male physicians participating in the Physicians' Health Study (a cohort study designed to investigate the use of aspirin and beta-carotene in the prevention of cardiovascular disease and cancer). The table on the next page refers to those physicians who had no diagnosis of cataract at the beginning of the study and provided information about cigarette smoking at that time.

(a) Fit the RC association model to this table. Does it provide an adequate fit to the data? What do the estimated row and column scores say about the apparent relationship between smoking status and cataract diagnosis? Do the results seem sensible to you?

(b) It is easy to imagine that smoking status, including the time since someone quit smoking, would be related to the respondent's age. This was not taken into account in this table. How might that affect the results? (Age, as well as many other potential predictive factors for cataract diagnosis were, in fact, taken into account by the researchers when they noted that an increased cumulative dose of smoking was associated with an increased risk of cataract diagnosis.)

Smoking status

		Years since quit smoking			
Cataract	Never				Current
status	smoked	> 20	10–20	< 10	smoker
No cataract	9582	2091	2534	1491	2022
Diagnosis, no extraction	361	207	109	64	95
Extraction	507	243	152	85	155

11. The file **devils** contains the game-by-game goal totals for the New Jersey Devils during the 1999–2000 season, which were examined in Section 4.3.1 from a univariate point of view. Fit a bivariate Poisson distribution to these data. Is there evidence that goals scored and goals given up are correlated? Estimate the probability that the Devils will win a randomly chosen game, based on this model. What about the probability of a game ending in a tie? How do these values compare to the team's actual win-loss-tie record of 45–29–8?

12. Show that the likelihood equations for the symmetry model are

$$n_{ij} + n_{ji} = 2np_{ij} \text{ for } i \neq j$$
$$n_{ii} = np_{ii}.$$

Show that this implies the estimated fitted values given in (7.10).

13. When a square table has ordered categories, it is often the case that either the entries above the diagonal or the entries below the diagonal are favored, in the sense that $p_{ij} > p_{ji}$ or $p_{ij} < p_{ji}$ for $i < j$. A generalization of symmetry that incorporates this property is the *conditional symmetry* model (McCullagh, 1978),

$$\log e_{ij} = \lambda + \lambda_i + \lambda_j + \lambda_{ij} + \tau I(i < j),$$

where $\lambda_{ij} = \lambda_{ji}$.

(a) Show that the expected cell counts for this model satisfy

$$e_{ij} = e_{ji} \times \exp(\tau)$$

for $i < j$.

(b) This model can be fit as a generalized linear model, adding the variable identifying cells with $i < j$ to the predictors for symmetry. Write down the likelihood equations, and show that the fitted values can be found in closed form, as follows:

$$\hat{\tau} = \log\left(\frac{\sum_{i<j} n_{ij}}{\sum_{i>j} n_{ij}}\right)$$

$$\hat{e}_{ii} = n_{ii} \text{ for } i = 1, \dots I,$$

$$\hat{e}_{ij} = \frac{n_{ij} + n_{ji}}{\exp(\hat{\tau}) + 1} \text{ for } i > j$$

$$\hat{e}_{ij} = \hat{e}_{ji} \times \exp(\hat{\tau}) \text{ for } i < j.$$

(c) Fit the conditional symmetry model to the case-control cataract data (page 282). Does this model provide significantly better fit than the symmetry model? (Note that this provides yet another test of marginal homogeneity.) Is this model a better choice than the ordinal quasi-symmetry model?

14. The following table is a reconstruction of the table formed by Sir Francis Galton in his study of the heritability of fingerprints (Stigler, 1995), and gives the characteristics of sibling pairs with regards to the broad type of pattern of fingerprints (arches, loops, or whorls).

Type for first child	Type for second child		
	Arches	Loops	Worls
Arches	5	4	1
Loops	12	42	14
Whorls	2	14	10

(a) Is there an association between siblings' fingerprint types?

(b) Is there evidence of too many observations (relative to independence) corresponding to siblings having the same type of fingerprints?

(c) Is the observed pattern consistent with a symmetric pattern? If not, why not?

15. The following table gives the social class mobility from 1971 to 1981 for a sample of men aged 45–64 years who were employed at the 1981 United Kingdom census and for whom social class was known at each time (Blane et al., 1999).

| | 1981 social class | | | | | |
1971 social class	I	II	IIIN	IIIM	IV	V
I	1759	553	141	130	22	2
II	541	6901	861	824	367	60
IIIN	248	1238	2562	346	308	56
IIIM	293	1409	527	12,054	1678	586
IV	132	419	461	1779	3565	461
V	37	53	88	582	569	813

Social class I are professionals (for example, doctors and lawyers); social class II are semi-professionals (for example, teachers and nurses) and managers; social class IIIN are (nonmanual) white collar workers (for example, retail and clerical workers); social class IIIM are (manual) skilled occupations (for example, electricians and plumbers); social class IV are semi-skilled manual workers (for example, assembly-line and agricultural workers); and social class V are unskilled manual workers (for example, general laborers).

(a) Attempt to model this table using the symmetry-based models of Section 7.2. Do any provide an adequate fit to the table? Are any better than the saturated model based on AIC or BIC (see page 297)?

(b) Attempt to model this table using the association models of Section 7.1 (presumably allowing for extra observations along the diagonal). Do any provide an adequate fit to the table? Are any better than the saturated model based on AIC or BIC?

16. Lapp et al. (1998) provided data from a study designed to examine the efficacy of fluvoxamine treatment of psychiatric symptoms possibly resulting from a dysregulation of serotonine in the brain. Patients were graded on a scale of one (worst) to four (best) with regard to the therapeutic effect and the side effects of the drug at three visits.

(a) The following table cross-classifies the therapeutic effect of the drug at the second and third visit.

Second-visit effect	Third-visit effect			
	1	2	3	4
1	13	2	0	0
2	37	40	8	4
3	13	58	18	4
4	1	13	36	21

What can you say about the change from the second time period to the third time period? Do symmetry-type models or association-type models provide a better representation of the pattern?

(b) The following table cross-classifies the side effects of the drug at the second and third visit.

Second-visit effects	Third-visit effects			
	1	2	3	4
1	2	0	1	3
2	2	10	7	2
3	1	7	80	34
4	0	0	14	105

What can you say about the side effects transitions from the second visit to the third visit?

17. (a) Show that the argument that McNemar's test is really just a test about a binomial proportion (made in Section 7.3 when describing an exact test) was implicitly made in the asymptotic analysis as well, in the sense that McNemar's test is just the square of the usual Gaussian-based test for whether a binomial probability equals .5.

(b) Show that McNemar's test is, indeed, the Pearson statistic X^2 testing the fit of symmetry in a 2×2 table.

18. Digital flexor tendinitis is very debilitating in racehorses because of the great stress put on tendons during racing. One method of treating this injury is to implant carbon fibers to support the tendon. Reed et al. (1994) reported the results of a clinical trial where 28 horses that had failed standard treatment (rest and drugs) had carbon fibers surgically implanted. Of the 28 horses, 16 had a successful result, returning to racing.

(a) Reed et al. (1994) demonstrated the effectiveness of the treatment by constructing a McNemar's test for homogeneity of the failure/success distributions before and after treatment. Construct such a test here. Write down the appropriate null hypothesis for the test. Do you reject the null hypothesis here? Construct a 95% confidence interval for the change in success probabilities. What does this interval imply?

(b) Note that if the initial status of the horse defines the rows of the 2×2 table, one row is necessarily a pair of zeroes. Show that for a table of this form, no matter what the total sample size is, McNemar's test will reject the null at the .05 level if four observations fall in the nonzero off-diagonal cell (that is, four of the horses return to racing).

(c) McNemar's test treats the observed proportions of successes for both rows and columns to be random. Given the argument in part (b), is that reasonable here? Do you agree with Reed et al.'s decision to use McNemar's test here? Would any hypothesis test have made more sense? Is this even a hypothesis testing problem?

19. Aickin (1990) proposed an alternative to Cohen's kappa based on the quasi-independence model (7.7) with $\delta_i = \delta$ for all i. Specifically,

$$\hat{\kappa} = \left(\sum_{i=1}^{I} \frac{n_{ii}}{n} \right) \left[1 - \frac{1}{\exp(\hat{\delta})} \right] .$$

(a) How does this value of κ compare to Cohen's kappa for the agreement tables given in this chapter?

(b) This measure can be generalized to the unrestricted quasi-independence model as

$$\hat{\kappa} = \sum_{i=1}^{I} \left\{ \left(\frac{n_{ii}}{n} \right) \left[1 - \frac{1}{\exp(\hat{\delta}_i)} \right] \right\}$$

(Guggenmoos-Holzmann and Vonk, 1998). How does this version of κ compare to the other versions for the agreement tables given in this chapter?

20. Coins (and other metallic foreign objects) that are ingested by children usually pass uneventfully into the stomach, but occasionally they can become lodged in the esophagus, leading to potentially serious complications, or even death. Clinical signs and symptoms are not reliable indicators of this, so radiographic (x-ray) examination is the standard diagnostic technique for localizing ingested objects. Seikel

et al. (1999) examined the usefulness of handheld metal detectors (HHMDs) for this purpose in a study from November 1995 through November 1996 in medical centers in Texas and Virginia.

(a) The following table compares the identification results of radiograph and HHMD, separated by experience level (with respect to HHMD use) of the investigator.

Experienced investigators			Inexperienced investigators		
	Radiograph			Radiograph	
HHMD	Positive	Negative	*HHMD*	Positive	Negative
Positive	60	6	Positive	67	20
Negative	0	73	Negative	3	85

Is the probability of a positive result for HHMD significantly different from that of radiography for experienced investigators? What about inexperienced investigators?

(b) Construct a 95% confidence interval for the difference in probability of positive results for each of the two subtables.

(c) Lehr (2000) showed that McNemar's test can be inverted to form an exact confidence interval for the difference in probabilities of success for matched pairs data in the form of a 2×2 table. An exact $100 \times (1 - \alpha)\%$ confidence interval has the form (LL, UL), where

$$LL = \left(\frac{n_{12} + n_{21}}{n}\right)\left[\frac{2n_{12}}{n_{12} + (n_{21} + 1)F_{\alpha/2}^{(2[n_{21}+1], 2n_{12})}} - 1\right],$$

$$UL = \left(\frac{n_{12} + n_{21}}{n}\right)\left[\frac{(2n_{12} + 2)F_{\alpha/2}^{(2[n_{12}+1], 2n_{21})}}{n_{21} + (n_{12} + 1)F_{\alpha/2}^{(2[n_{12}+1], 2n_{21})}} - 1\right],$$

and $F_{\alpha}^{(df_1, df_2)}$ is the α level critical value for an F distribution on (df_1, df_2) degrees of freedom. Use this formula to construct an exact 95% confidence interval for the difference in probability of positive results for each of the two subtables. Are these intervals very different from those you produced in part (b)?

(d) What are the values of Cohen's kappa for each table? Do these values indicate strong agreement between HHMDs and radiography?

(e) In a study with multiple strata (for example, the level of experience of the investigator in this example), the stratum-level values of $\hat{\kappa}$ can be combined to give an overall estimate of κ. Say there are q strata, and let $\hat{\kappa}_\ell$ be the value of $\hat{\kappa}$ for the ℓth stratum. Then

$$\hat{\kappa}_{\text{overall}} = \frac{\sum_{\ell=1}^q \hat{\kappa}_\ell / \widehat{\text{s.e.}}(\hat{\kappa}_\ell)^2}{\sum_{\ell=1}^q 1/\widehat{\text{s.e.}}(\hat{\kappa}_\ell)^2},$$

where $\widehat{\text{s.e.}}(\hat{\kappa}_\ell)$ is based on (7.15) for $\hat{\kappa}$ and (7.17) for $\hat{\kappa}_w$, respectively. The hypothesis of a common value of kappa for all strata can be tested using

$$\sum_{\ell=1}^q \frac{(\hat{\kappa}_\ell - \hat{\kappa}_{\text{overall}})^2}{\widehat{\text{s.e.}}(\hat{\kappa}_\ell)^2},$$

which is compared to a χ^2_{q-1} critical value (Fleiss, 1981). Estimate an overall kappa for the two subtables above. Is there evidence that rater agreement (as measured by kappa) is different for experienced investigators compared to inexperienced investigators?

(f) The following tables separate the data on the basis of whether or not the patient was actually found to have a metallic foreign object lodged in his or her esophagus by radiography (this table is based on only the 139 patients who were scanned by both experienced and inexperienced investigators).

With lodged foreign object

| Inexperienced | Experienced investigator | |
investigator	Positive	Negative
Positive	58	0
Negative	2	0

Without lodged foreign object

| Inexperienced | Experienced investigator | |
investigator	Positive	Negative
Positive	2	7
Negative	3	67

Repeat parts (a)–(e) for the tables above, now focusing on the difference in positive identification for experienced versus inexperienced investigators.

(g) Do you think that handheld metal detectors should be used routinely as an alternative to radiography for the detection of metallic objects lodged in the esophagus?

21. The file **reviewer** contains the ratings of several movie reviewers (Roger Ebert, Jeffrey Lyons, Michael Medved, Rex Reed, Gene Shalit, Joel Siegel, the late Gene Siskel, and Peter Travers) of movies released between January 1985 and March 1997, on the scale (Pro, Mixed, Con) (Agresti and Winner, 1997, with data for additional movies, kindly provided by Larry Winner).

    (a) Siskel and Ebert were very well known for their debates about movies on various television shows. Construct a cross-classification of their ratings. What can you say about the agreement between them? How does (weighted) kappa assess this agreement? Do their disagreements follow a (quasi-)symmetric pattern? What about an association-type pattern?

    (b) Michael Medved is somewhat iconoclastic in his opinions, having a very negative opinion of what he terms gratuitous sex and violence in movies and on television. Is this reflected in the data? That is, does Medved agree with his peers less often than they agree with one another? Is his overall opinion of the movies he reviewed (that is, his marginal distribution of ratings) different from that of other reviewers? How would you test this latter question?

22. The following data come from a study of the stability of responses to questions regarding atopic disease (allergic disorders caused by Type I hypersensitivity). In the study, mothers with newborn babies were asked their history with regards to atopic disease, with responses classified as (I) no atopy, (II) atopy, but no neurodermatitis (chronic skin illness accompanied by strong itching), or (III) neurodermatitis. The mothers were then asked the same question two years later (Guggenmoos-Holzmann and Vonk, 1998).

| | Second interview | | |
First interview	I	II	III
I	136	12	1
II	8	59	4
III	2	4	6

How would you characterize the agreement pattern here? Have the marginal distributions changed significantly over the two years? If they had, would that necessarily reflect a faulty memory on the part of the mothers? Can you think of any other reasons for such a change over the two-year time period?

8
Multidimensional Contingency Tables

Cross-classifications involving more than two variables are a natural generalization of the tables discussed in Chapters 6 and 7. From the generalized linear model (Poisson regression) point of view, this merely corresponds to incorporating more (nominal or ordinal) predictors into the model, and presents no particular difficulties. The structure in a multidimensional contingency table model, however, and the implied forms of association and independence that loglinear models imply, make it important to study in some detail the analysis of higher-dimensional categorical data. We start with what is in many ways the simplest generalization, the $2 \times 2 \times K$ table, since many of the general issues involved arise in this simpler context. We will then generalize results to $I \times J \times K$, and eventually four- and higher-dimensional tables.

8.1 $2 \times 2 \times K$ Tables

A three-dimensional table of counts $\{n_{ijk}\}$ often reflects a situation where the first two subscripts (which we might call row and column indices, respectively) refer to variables X and Y, respectively, whose association is of interest, while the third subscript (layers, or strata) indexes a control variable Z, the effect of which is controlled for in the analysis. A $2 \times 2 \times K$ table has indices that satisfy $i = 1, 2, j = 1, 2$, and $k = 1, \ldots, K$, and has the following form:

	$Z = 1$	
X	$Y = 1$	$Y = 2$
1	n_{111}	n_{121}
2	n_{211}	n_{221}

. . .

	$Z = k$	
X	$Y = 1$	$Y = 2$
1	n_{11k}	n_{12k}
2	n_{21k}	n_{22k}

8.1.1 Simpson's Paradox

The underlying 2×2 X/Y tables are termed *partial tables*, and the association patterns in each are called *partial*, or *conditional associations*. The 2×2 table that sums over all values of Z is called the *marginal table*, and the association pattern evident there is called the *marginal association*. A key point to remember in the analysis of any three-dimensional table is that the marginal association can be very different from the partial associations (which is precisely why it is important to account for the control variable Z).

Consider the following hypothetical example. The marketing director for a large corporation is considering whether to rehire its current advertising agency, or move to a new one. The head of the current agency sends the director the following e-mail message:

I'm happy to report the following results concerning the new national television advertising campaign. Over a two-week period ending ten days ago, we talked to consumers in both New York City and Los Angeles (2000 total consumers: 1000 consumers in each city, and 1000 consumers each week), to estimate the "recall rate" of the campaign. In New York City, the recall rates were 30% in the first week, and 40% in the second week; in Los Angeles, the recall rates were 80% in the first week and 90% in the second week. These increasing recall rates in both markets show that the campaign is catching on, just as we hoped.

Two weeks later, the CEO of the corporation comes into the marketing director's office. "I've got some bad news. I took a look at that data about the ad campaign, and I've found that, in fact, overall recall rates were **declining** in the data, not increasing." "Do you mean they lied to us? We'll sue!" the director shouts. "Well, no," she replies. "The numbers they gave you were correct, but what I'm saying is correct, too."

How can this be possible? Here are the detailed results of the market research study:

	New York	Los Angeles	Total
Week 1	60/200 (30%)	640/800 (80%)	700/1000 (70%)
Week 2	320/800 (40%)	180/200 (90%)	500/1000 (50%)

The recall rate is increasing within each city, but is decreasing overall. This association reversal is called *Simpson's paradox*, and is an example of how partial associations are not always consistent with the marginal association. The following 2 × 2 × 2 table is an equivalent representation of the data:

New York

	Recall	Don't recall
Week 1	60	140
Week 2	320	480

Los Angeles

	Recall	Don't recall
Week 1	640	160
Week 2	180	20

The partial association between recall of the campaign and time period (first or second week) in New York is summarized by the odds ratio in the left table, $\hat{\theta} = 0.64$, indicating an association between recalling the campaign and being in the second week (that is, an increasing recall rate). Similarly, the odds ratio in the Los Angeles table is $\hat{\theta} = 0.44$, again demonstrating an increasing recall rate. Simpson's paradox occurs when there is a strong association between the control variable (here the city) and each of the variables X and Y (here recall and time period, respectively). In this case, New York City had generally lower recall rates and was more heavily sampled the second week, while Los Angeles, with generally higher recall rates, was more heavily sampled in the first week. This results in the overall recall rate appearing higher in the first week (being closer to the Los Angeles rate) and lower in the second week (being closer to the New York rate).

Figure 8.1 represents the association reversal of Simpson's paradox in a graphical form. The two lines represent what the overall recall percentage in week 1 (solid line) and week 2 (dashed line) would have been, as a function of the percentage of the total sample that was taken in Los Angeles. So, for example, in week 1, if none of the survey had been taken in Los Angeles, the overall recall rate would have been the week 1 New York rate, or 30%. Similarly, if in week 1 all of the survey had been taken in Los Angeles, the overall recall rate would have been the week 1 Los Angeles rate, or 80%. In fact, 80% of the respondents in week 1 were from Los Angeles, so the overall recall rate was 70%, represented by the marked point on the week 1 line. The corresponding values for week 2 are also given in the figure. The consistently higher line for week 2 compared to week 1 reflects the conditional pattern that recall rates were generally higher in week 2, but the higher position of the marked point for week 1 compared to that for week 2 demonstrates the marginal pattern that the overall rate was higher in week 1 than in week 2.

This diagram shows that Simpson's paradox actually reflects a sampling problem, in the sense that the problem only can arise when the control variable Z is related to the variables of interest X and Y. This suggests

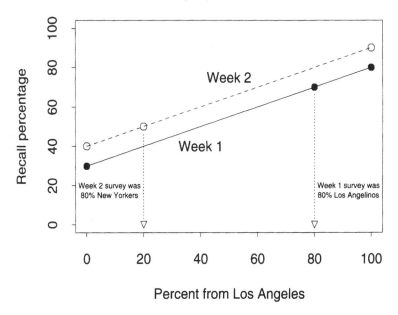

FIGURE 8.1. Graphical representation of Simpson's paradox for the commercial recall percentage example.

that when it is possible to control the level of the control variable for each observation, this should be done so as to minimize any chance of a relationship between it and the other variables. So, for example, if the marketing survey had been based on 500 respondents in each city for each week, city and time period would have been unrelated to each other, and Simpson's paradox could not have occurred (this corresponds to the week 2 line always being above the week 1 line in Figure 8.1 if the percentage of the survey from Los Angeles is the same in the two weeks). In a study of the relationship between dosage of a drug and survival, with gender a control variable, men and women should be assigned to dosage level randomly (and evenly), preventing the possibility of Simpson's paradox.

In situations where it is not possible to assign values of the control variable to respondents, the question then becomes which association to focus on — the marginal association or the partial associations. Almost invariably, it is the partial associations that are of interest. In the market research study, it seems clear that recall rates really *are* increasing, if they are doing so in both cities; the lack of balance of the survey is an annoyance, and should not be the driving force behind inference about the table. This is precisely the reason why in a multiple regression analysis the focus is on the coefficients from the appropriate multiple predictor model, rather than the coefficients from the simple regressions of the target variable on each predictor; the former coefficients represent the partial associations (given the other predictors), while the latter represent the marginal associations.

███

Smoking and Mortality in Whickham

In 1972–1974, a survey was taken in Whickham, a mixed urban and rural district near Newcastle upon Tyne, United Kingdom. Twenty years later a followup study was conducted. Among the information obtained originally was whether a person was a smoker or not. The following table summarizes the relationship between whether a person was a smoker at the time of the original survey, and whether they were still alive twenty years later (Appleton et al., 1996).

Mortality status

Smoking status	Alive	Dead	Total
Smoker	443	139	582
Nonsmoker	502	230	732
Total	945	369	1314

There is a significant association between smoking and mortality ($G^2 = 9.2$ on 1 degree of freedom, with $p = .002$), but it is not the one you might expect. The odds ratio for this table is $\hat{\theta} = 1.46$, implying that the odds of having lived were 46% higher for smokers than for nonsmokers. Does this mean that smoking is *good* for you?

Not very likely. There is a key "lurking variable" here — a variable that is related to both variables that is obscuring the actual relationship — namely, age. The table on the next page gives the mortality experience separated by initial age group. The odds ratios for each of these subtables (except the last one, where it is undefined) are as follows: 0.43, 1.33, 0.42, 0.69, 0.62, 0.87. Thus, for all age groups except 25–34 year olds, smoking is associated with higher mortality, as we would have expected. Why has Simpson's paradox occurred? Not surprisingly, there is a strong association between age and mortality, with mortality rates being very low for younger people (2.5% for 18–24 year olds) and increasing to 100% for 75+ year olds. There is also an association between age and smoking, with smoking rates peaking in the 45–54 year old range and then falling off rapidly. In particular, respondents who were over 65 at the time of the first survey had very low smoking rates (25.4%), but very high mortality rates (85.5%). Smoking was hardly the cause, however, since even among the 65–74 year olds mortality was higher among smokers (80.6%) than it was among nonsmokers (78.3%).

| | | Mortality status | |
Age	Smoking status	Alive	Dead
18–24	Smoker	53	2
	Nonsmoker	61	1
25–34	Smoker	121	3
	Nonsmoker	152	5
35–44	Smoker	95	14
	Nonsmoker	114	7
45–54	Smoker	103	27
	Nonsmoker	66	12
55–64	Smoker	64	51
	Nonsmoker	81	40
65–74	Smoker	7	29
	Nonsmoker	28	101
75+	Smoker	0	13
	Nonsmoker	0	64

8.1.2 Types of Independence

One implication of association reversal (Simpson's paradox) is that conditional independence, where two variables are independent given the level of a third variable, is very different from marginal independence, where two variables are independent and the third variable is ignored. Somewhat surprisingly, it can even happen that two variables can be conditionally independent given a third variable, yet when the third variable is ignored, the two variables are marginally dependent.

Freezer Ownership in Denmark

Respondents to the 1976 Danish Welfare Study were asked whether there was a freezer in the house or not. The respondents were classified by a social group classification devised by the Danish National Institute for Social Research into five social groups: I (academics and main executives), II (second-level executives), III (foremen and heads of small sections), IV (white collar workers and blue collar workers with special training), and V (blue collar workers). The respondents were also separated based on whether they owned or rented their dwelling (Andersen, 1997, page 63).

		Freezer	
Ownership	Social group	Yes	No
Owner	I–II	304	38
	III	666	85
	IV	894	93
	V	720	84
Renter	I–II	92	64
	III	174	113
	IV	379	321
	V	433	297

Freezer possession and social group are apparently independent for owners ($G^2 = 1.89$, $p = .60$), and are also apparently independent for renters ($G^2 = 5.49$, $p = .14$); that is, they are conditionally independent given ownership status. They are not, however, marginally independent, as the test for the table that ignores ownership status (by collapsing the table over that variable) is $G^2 = 16.36$ ($p < .001$). The lack of independence in the marginal table comes from greater possession rates of freezers among members of social groups I–III (roughly 80%) compared with groups IV–V (roughly 75%), but this is accounted for by the higher owner (versus renter) rates in the former group (roughly 70%) versus the latter group (roughly 55%).

8.1.3   Tests of Conditional Association

The conditional association of X and Y when $Z = k$ is described by the *conditional odds ratio*,

$$\theta_{XY(k)} = \frac{e_{11k}e_{22k}}{e_{12k}e_{21k}}, \tag{8.1}$$

where $e_{ijk} = E(n_{ijk})$. If the association pattern is the same for all levels of Z, *homogeneous association* is present, with

$$\theta_{XY(1)} = \cdots = \theta_{XY(K)}.$$

Conditional independence corresponds to the special case

$$\theta_{XY(1)} = \cdots = \theta_{XY(K)} = 1.$$

Homogeneous association is symmetric in the variables, in the sense that the conditional odds ratio between X and any two levels of Z is the same for both levels of Y, and the conditional odds ratio between Y and any two levels of Z is the same for both levels of X. Recall from the discussion in Section 3.2.3 that the absence of a (two-way) interaction effect between X and Y (for example) corresponds to the main effect of X being the same for each level of Y, and vice versa. Homogeneous association, where the association between two variables is the same for each level of a third, refers to the absence of a (three-way) interaction between X, Y, and Z.

The usual sampling schemes for $2 \times 2 \times K$ tables generalize those for two-dimensional tables: independent Poisson, multinomial with fixed overall sample size, multinomial with fixed sample size within each partial table and independence between tables, or independent binomials for each row (or column) within each partial table given the row (column) totals. For each of these sampling schemes, given the row and column totals in each partial table, any one cell count in each partial table determines all of the other counts, and under conditional independence follows a hypergeometric distribution.

When conditional independence holds, the mean of n_{11k} (based on the hypergeometric distribution) is

$$e_{11k} = \frac{n_{1 \cdot k} n_{\cdot 1 k}}{n_{\cdot \cdot k}},$$

while the variance is

$$V(n_{11k}) = n_{1 \cdot k} \left(\frac{n_{\cdot 1 k}}{n_{\cdot \cdot k}} \right) \left(1 - \frac{n_{\cdot 1 k}}{n_{\cdot \cdot k}} \right) \left(\frac{n_{\cdot \cdot k} - n_{1 \cdot k}}{n_{\cdot \cdot k} - 1} \right)$$

$$= \frac{n_{1 \cdot k} n_{2 \cdot k} n_{\cdot 1 k} n_{\cdot 2 k}}{n_{\cdot \cdot k}^2 (n_{\cdot \cdot k} - 1)}.$$

The *Cochran–Mantel–Haenszel statistic* compares the sum of all of the n_{11k} values to its expected value under conditional independence, standardizing by its variance,

$$CMH = \frac{[\sum_k (n_{11k} - e_{11k})]^2}{\sum_k V(n_{11k})},$$

and is compared to a χ_1^2 distribution. Note that it is the sum of the n_{11k} values that is centered and scaled, rather than the individual n_{11k} values. This means that if in some of the partial tables the association is direct (with larger than expected n_{11k}) while in others it is inverse (with smaller than expected n_{11k}), these will cancel out, implying a small value of CMH. That is, while the test is more powerful than separate tests on each partial table when the partial associations are similar across tables, it has little power to identify partial associations that differ greatly from one partial table to the next, particularly if they differ in direction.

If the partial associations seem reasonably stable (that is, homogeneous association is reasonable), an estimate of the common odds ratio θ is desirable. The *Mantel–Haenszel estimator* of θ is

$$\hat{\theta}_{MH} = \frac{\sum_k (n_{11k} n_{22k}/n_{..k})}{\sum_k (n_{12k} n_{21k}/n_{..k})}. \tag{8.2}$$

A large-sample confidence interval for θ is based on the estimated asymptotic variance of $\log \hat{\theta}$,

$$\hat{V}[\log(\hat{\theta}_{MH})] = \frac{\sum_k (n_{11k} + n_{22k})(n_{11k} n_{22k})/n_{..k}^2}{2 \left(\sum_k n_{11k} n_{22k}/n_{..k}\right)^2}$$
$$+ \frac{\sum_k [(n_{11k} + n_{22k})(n_{12k} n_{21k}) + (n_{12k} + n_{21k})(n_{11k} n_{22k})]/n_{..k}^2}{2 \left(\sum_k n_{11k} n_{22k}/n_{..k}\right)\left(\sum_k n_{12k} n_{21k}/n_{..k}\right)}$$
$$+ \frac{\sum_k (n_{12k} + n_{21k})(n_{12k} n_{21k})/n_{..k}^2}{2 \left(\sum_k n_{12k} n_{21k}/n_{..k}\right)^2}.$$

The presence of homogeneous association (a common odds ratio) can be tested using the *Breslow–Day statistic*, which is based on the Mantel–Haenszel estimator (8.2). Let $\hat{e}_k$ be the estimated expected frequency for n_{11k} given that the true odds ratio in the kth partial table is equal to $\hat{\theta}_{MH}$. This is obtained by equating $\hat{\theta}_{MH}$ to the empirical odds ratio in the table that has $\hat{e}_k$ as the $(1, 1)$th cell count and the same marginal totals as the actual kth partial table, and then solving the resultant quadratic equation

$$\frac{\hat{e}_k(n_{2 \cdot k} - n_{\cdot 1k} + \hat{e}_k)}{(n_{1 \cdot k} - \hat{e}_k)(n_{\cdot 1k} - \hat{e}_k)} = \hat{\theta}_{MH}$$

for $\hat{e}_k$. The estimated variance of n_{11k} given that the true odds ratio in the kth partial table is equal to $\hat{\theta}_{MH}$ is

$$\hat{V}(n_{11k}|\hat{\theta}_{MH}) = \left[\frac{1}{\hat{e}_k} + \frac{1}{n_{1 \cdot k} - \hat{e}_k} + \frac{1}{n_{\cdot 1k} - \hat{e}_k} + \frac{1}{n_{2 \cdot k} - n_{\cdot 1k} + \hat{e}_k}\right]^{-1}.$$

The Breslow–Day statistic is then

$$BD = \sum_{k=1}^{K} \frac{(n_{11k} - \hat{e}_k)^2}{\hat{V}(n_{11k}|\hat{\theta}_{MH})},$$

and is compared to a χ^2 distribution on $K - 1$ degrees of freedom. This approximation requires a reasonably large sample size in each partial table. Because $\hat{\theta}_{MH}$ is a consistent, but inefficient, estimator of the common θ, BD should actually be modified slightly to

$$BD^* = BD - \frac{[\sum_k (n_{11k} - \hat{e}_k)]^2}{\sum_k \hat{V}(n_{11k}|\hat{\theta}_{MH})},$$

although this adjustment is usually minor.

Smoking and Mortality in Whickham (continued)

As was noted earlier, the marginal table relating smoking to survival in Whickham does not provide a meaningful summary of the relationship, since it is the conditional association (given age) that is of interest. The Cochran–Mantel–Haenszel statistic for these data is $CMH = 5.84$, with $p = .016$, indicating strong rejection of conditional independence of smoking and survival given age. The estimated common conditional odds ratio is $\hat{\theta}_{MH} = 0.655$, implying that (given age) being a smoker is associated with a 35% lower odds of being alive 20 years later than being a nonsmoker. A 95% confidence interval for θ is $(.463, .925)$, reinforcing rejection of conditional independence $(\theta = 1)$. The Breslow–Day statistic $BD = 2.37$ on 5 degrees of freedom, which is not close to significance $(p = .80)$, implying that this common odds ratio is reasonable for all age groups $(BD^*$ also equals 2.37). Note that the degrees of freedom are 5, rather than 6, because the partial table corresponding to people aged 75 and over has a column total equal to zero and is not informative regarding the question of homogeneous association.

8.2   Loglinear Models for Three-Dimensional Tables

8.2.1   Hierarchical Loglinear Models

We will now generalize the loglinear models of the previous chapters to $I \times J \times K$ three-dimensional tables. Any three-dimensional contingency can be represented as a triply subscripted array of counts, $\{n_{ijk}\}$ ($i \in \{1, \ldots, I\}$, $j \in \{1, \ldots, J\}$, and $k \in \{1, \ldots, K\}$). In what follows, we will again call the first dimension X, the second dimension Y, and the third dimension Z. These counts can be generated based on a multinomial sampling scheme, a product multinomial scheme (fixing some margins), or a multivariate hypergeometric scheme (fixing all margins); this distinction doesn't matter for maximum likelihood estimation as long as the models being fit incorporate any margins that are set as fixed beforehand (that is, the models must yield estimated marginal distributions that equal the prespecified observed margins).

The saturated model for a three-dimensional table has the form

$$\log e_{ijk} = \lambda + \lambda_i^X + \lambda_j^Y + \lambda_k^Z + \lambda_{ij}^{XY} + \lambda_{ik}^{XZ} + \lambda_{jk}^{YZ} + \lambda_{ijk}^{XYZ}. \qquad (8.3)$$

Model (8.3) is made identifiable by imposing constraints on the parameters as follows:

$$\sum_i \lambda_i^X = \sum_i \lambda_{ij}^{XY} = \sum_i \lambda_{ik}^{XZ} = \sum_i \lambda_{ijk}^{XYZ} = 0$$

$$\sum_j \lambda_j^Y = \sum_j \lambda_{ij}^{XY} = \sum_j \lambda_{jk}^{YZ} = \sum_j \lambda_{ijk}^{XYZ} = 0$$

$$\sum_k \lambda_k^Z = \sum_k \lambda_{ik}^{XZ} = \sum_k \lambda_{jk}^{YZ} = \sum_k \lambda_{ijk}^{XYZ} = 0.$$

That is, the parameters sum to zero over all three indices. Using the "dot" notation, $\lambda_{jk}^{XYZ} = 0$ is identical to the statement $\sum_i \lambda_{ijk}^{XYZ} = 0$ (for example). This model can be fit using effect codings, using $I - 1$ variables for $\boldsymbol{\lambda}^X$, $J - 1$ variables for $\boldsymbol{\lambda}^Y$, and $K - 1$ variables for $\boldsymbol{\lambda}^Z$ (the main effects), $(I - 1)(J - 1)$ variables for $\boldsymbol{\lambda}^{XY}$, $(I - 1)(K - 1)$ variables for $\boldsymbol{\lambda}^{XZ}$, and $(J - 1)(K - 1)$ variables for $\boldsymbol{\lambda}^{YZ}$ (the two-way interactions), and $(I - 1)(J - 1)(K - 1)$ variables for $\boldsymbol{\lambda}^{XYZ}$ (the three-way interaction). Degrees of freedom for testing the fit of any model are just $IJK - 1 - p$, where p is the number of variables used to fit the model.

Models that include subsets of terms from (8.3) represent different forms of independence and dependence in the table. All of the models we will look at are hierarchical models. Recall that a hierarchical model is one where the presence of a higher-order term in the model implies that all lower-order terms based on the same variables are also in the model (see page 35). So, for example, if λ_{ij}^{XY} is in the model, then λ_i^X and λ_j^Y must also be in the model. An equivalent way to state this is that if a term is not in the model, all of its higher order "relatives" are also not in the model; so, for example, if λ_i^X is not in the model, then λ_{ij}^{XY}, λ_{ik}^{XZ}, and λ_{ijk}^{XYZ} are also not. Allowing only hierarchical models is analogous to requiring main effects in an ANOVA model that includes a higher-order interaction term.

Model	Symbol	Description
(1) λ		$X \perp Y \perp Z$,
		X, Y and Z uniform
(2) $\lambda + \lambda_i^X$	(X)	$X \perp Y \perp Z$,
		Y and Z uniform
(3) $\lambda + \lambda_i^X + \lambda_j^Y$	(X,Y)	$X \perp Y \perp Z$,
		Z uniform
(4) $\lambda + \lambda_i^X + \lambda_j^Y + \lambda_k^Z$	(X,Y,Z)	$X \perp Y \perp Z$
(5) $\lambda + \lambda_i^X + \lambda_j^Y + \lambda_{ij}^{XY}$	(XY)	$X \perp Z, Y \perp Z$,
		Z uniform
(6) $\lambda + \lambda_i^X + \lambda_j^Y + \lambda_k^Z + \lambda_{ij}^{XY}$	(XY,Z)	$X \perp Z, Y \perp Z$
(7) $\lambda + \lambda_i^X + \lambda_j^Y + \lambda_k^Z + \lambda_{ik}^{XZ}$ $+\lambda_{jk}^{YZ}$	(XZ,YZ)	$X \perp Y\|Z$
(8) $\lambda + \lambda_i^X + \lambda_j^Y + \lambda_k^Z + \lambda_{ij}^{XY}$ $+\lambda_{ik}^{XZ} + \lambda_{jk}^{YZ}$	(XY,XZ,YZ)	Homogeneous association

TABLE 8.1. Loglinear models for three-dimensional contingency tables.

Such a requirement is usually sensible from an intuitive point of view, and has advantages in estimation and testing.

Table 8.1 summarizes eight different types of models that are possible for a three-dimensional table (models of identical type based on different sets of variables are not given). For each model, two types of independence are referred to: marginal independence of X and Y (for example), denoted by $X \perp Y$, and conditional independence of X and Y given Z (for example), denoted by $X \perp Y|Z$. The symbol $X \perp Y \perp Z$ refers to complete independence of all three variables (sometimes called mutual independence). The "uniform" condition is stating that the marginal probability for those variables is uniform. So, for example, "X uniform" means that $p_{i..} = 1/I$, "X and Y uniform" means that $p_{ij.} = 1/(IJ)$ (which implies that X and Y are also uniform), and "X, Y and Z uniform" means that $p_{ijk} = 1/(IJK)$. Homogeneous association means that the conditional odds ratios of any two levels of any two variables are the same for all levels of the third variable.

Hierarchical loglinear models can be summarized by their highest-order interaction(s), so, for example, (XYZ) symbolizes the saturated model. A model's symbolic representation also defines the so-called *sufficient marginals* of the model, the sufficient statistics from which the MLEs can be computed. These correspond to the set of marginal totals of the table of fitted cell counts that equal those for the table of observed cell counts.

So, for example, the sufficient marginals for the model (X, YZ) are $n_{i..}$ and $n_{.jk}$, and fitted cell counts $\hat{e}_{ijk}$ based on this model satisfy $\hat{e}_{i..} = n_{i..}$ and $\hat{e}_{.jk} = n_{.jk}$, respectively. As was noted earlier, if the sampling scheme that generated the table was, for example, a product of independent multinomials for each level of X, the loglinear model should result in fitted counts that satisfy the marginal sampling counts $n_{i..}$; the model (X, YZ) does this, and would be used in favor of the model (YZ), even if uniformity of the X variable is plausible.

The sufficient marginals for a model identify the ways that the maximum likelihood estimates might not exist. Consider, for example, the model (XY, YZ). The sufficient marginals for this model are $\{n_{ij.}, n_{.jk}\}$. If there are random zeroes in the table (cells with a zero count that occurred by random chance) such that one of the $n_{ij.}$ margins equals zero, for example, the (XY, YZ) model cannot be fit exactly, because the maximum likelihood estimates do not exist (the exponential form of loglinear models implies that the fitted values for any model must be positive, so $\hat{e}_{ij.} > 0$ for all i and j, contradicting equality of the observed and fitted sufficient marginals). The model (X, YZ), on the other hand, would still be estimable if $n_{i..} > 0$ for all i. This problem becomes more serious as the dimension of the table increases (see Section 8.4). One way around this is to treat the random zeroes as if they were structural zeroes, but it is not universally agreed that this is appropriate. Another approach is to add a small positive value (such as .5) to any cells with zero counts (or all of the cells), although this is obviously somewhat arbitrary. If an iterative technique, such as iteratively reweighted least squares, is used to fit the model without any correction being made, certain parameter coefficients will be estimated as being large negative numbers, implying fitted values very close to zero for the cells corresponding to zero marginals.

Models (1)–(4) all correspond to complete independence, where the cell probabilities satisfy

$$p_{ijk} = p_{i..}p_{.j.}p_{..k}$$

(models (1)–(3) put additional constraints on the uniformity of the marginal probability distributions). Under models (5) and (6),

$$p_{ijk} = p_{ij.}p_{..k}.$$

For these models Z is jointly independent of X and Y, which means that Z is independent of a new variable defined by the IJ combinations of the levels of X and Y. Model (7) is conditional independence of X and Y given Z, where the conditional probability $p_{ij|k} \equiv P(X = i, Y = j \mid Z = k)$ satisfies

$$p_{ij|k} = p_{i.|k}p_{.j|k},$$

or

$$p_{ijk} = \frac{p_{i \cdot k} p_{\cdot jk}}{p_{\cdot \cdot k}}.$$

Loglinear models also can be defined using conditional odds ratios $\theta_{ij(k)}$, $\theta_{i(j)k}$, and $\theta_{(i)jk}$, where for example

$$\theta_{ij(k)} = \frac{e_{ijk} e_{i+1,j+1,k}}{e_{i,j+1,k} e_{i+1,j,k}}, \qquad i = 1, \ldots, I-1, \quad j = 1, \ldots, J-1.$$

These generalize the conditional odds ratios given in (8.1) for $2 \times 2 \times K$ tables to general $I \times J \times K$ tables. So, for example, model (4) (complete independence) is equivalent to

$$\log \theta_{ij(k)} = \log \theta_{i(j)k} = \log \theta_{(i)jk} = 0$$

for all i, j, and k, while model (6) (joint independence of Z with X and Y, sometimes called partial independence) is equivalent to

$$\log \theta_{i(j)k} = \log \theta_{(i)jk} = 0$$

and

$$\log \theta_{ij(k)} \equiv \log \theta_{ij} = \lambda_{ij}^{XY} + \lambda_{i+1,j+1}^{XY} - \lambda_{i+1,j}^{XY} - \lambda_{i,j+1}^{XY}$$

(the conditional odds ratio $\theta_{ij(k)}$ is the same for each k). Model (7) (conditional independence of X and Y given Z) is equivalent to

$$\log \theta_{ij(k)} = 0,$$
$$\log \theta_{i(j)k} \equiv \log \theta_{ik} = \lambda_{ik}^{XZ} + \lambda_{i+1,k+1}^{XZ} - \lambda_{i+1,k}^{XZ} - \lambda_{i,k+1}^{XZ},$$

and

$$\log \theta_{(i)jk} \equiv \log \theta_{jk} = \lambda_{jk}^{YZ} + \lambda_{j+1,k+1}^{YZ} - \lambda_{j+1,k}^{YZ} - \lambda_{j,k+1}^{YZ}.$$

Finally, model (8) (homogeneous association) is equivalent to

$$\log \theta_{ij(k)} \equiv \log \theta_{ij} = \lambda_{ij}^{XY} + \lambda_{i+1,j+1}^{XY} - \lambda_{i+1,j}^{XY} - \lambda_{i,j+1}^{XY},$$
$$\log \theta_{i(j)k} \equiv \log \theta_{ik} = \lambda_{ik}^{XZ} + \lambda_{i+1,k+1}^{XZ} - \lambda_{i+1,k}^{XZ} - \lambda_{i,k+1}^{XZ},$$

and

$$\log \theta_{(i)jk} \equiv \log \theta_{jk} = \lambda_{jk}^{YZ} + \lambda_{j+1,k+1}^{YZ} - \lambda_{j+1,k}^{YZ} - \lambda_{j,k+1}^{YZ}.$$

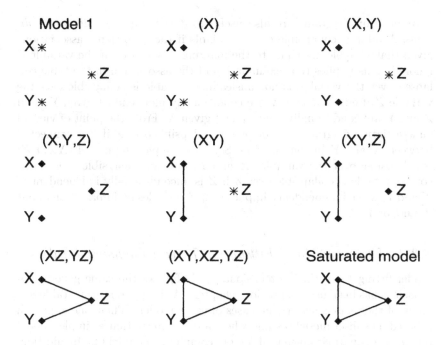

FIGURE 8.2. Association diagrams for three-dimensional loglinear models.

8.2.2 Association Diagrams, Conditional Independence, and Collapsibility

Hierarchical loglinear models can be represented graphically using *association diagrams*. In this diagram each variable is represented by a point (or a ∗ if the variable is uniform). Two points in the diagram are connected if there is an interaction between the two variables in the model that is not assumed to be zero. Figure 8.2 gives association diagrams for the hierarchical models for three-dimensional tables.

The association diagram of a model is useful to help the data analyst interpret the association patterns implied by the model. In the diagram for model (XZ, YZ), for example, conditional independence of X and Y given Z is represented by the fact that X and Y are not directly connected, but are only connected through the point for Z. By contrast, in the model (XY, Z), Z is completely disconnected from X and Y, implying that it is (unconditionally) independent of both variables (since X and Y are connected in the diagram, they are not independent of each other, so Z is jointly independent of X and Y, but there is not complete independence of X, Y, and Z). Note that since the homogeneous association model (XY, XZ, YZ) does not imply any independence conditions, its association diagram is identical to that of the saturated model.

Association diagrams are also useful to determine *collapsibility conditions*. A table is collapsible over a variable if the conditional associations given that variable are equal to the marginal association if the variable is ignored. This implies that examination of the association in the table collapsed over the variable is not misleading. A table is collapsible over the variable Z if either Z and X are conditionally independent given Y, or if Z and Y are conditionally independent given X. From the point of view of an association diagram, a model is not collapsible over Z if Z is connected between X and Y in the diagram. So, for example, the model (XZ, YZ) is collapsible over the variables X and Y, but not collapsible over Z. Of course, a table is collapsible over Z if Z is unconditionally independent of X and Y, since independence implies that Z provides no information about X and/or Y.

8.2.3 Loglinear Model Fitting for Three-Dimensional Tables

Model fitting for a three-dimensional table follows the same general approach as has been used in previous chapters. As is apparent from Table 8.1, many of the models are special cases of other models. Thus, one approach is based on subset model testing, where a test is made that a simpler model provides adequate fit given that a more complicated model fits. In addition, AIC_C can be used to choose the appropriate model. There are 18 different unsaturated hierarchical models possible for a three-dimensional table, but some might be able to be dismissed without examination. For example, if the model (X, Y, Z) does not fit the table, it isn't necessary to examine the six versions of models (1)–(3), since they cannot fit either. This suggests looking at more complicated models first, and then proceeding to simpler subsets if the more complicated model fits.

Certain tests are of particular interest. In a $2 \times 2 \times K$ table, for example, the test of conditional independence given the assumption of homogeneous association, $(XZ, YZ)|(XY, XZ, YZ)$, is directly comparable to the Cochran–Mantel–Haenszel test. If the sample sizes in each of the 2×2 subtables are reasonably large, the CMH statistic and the likelihood ratio statistic of conditional independence given homogeneous association will be similar, both being compared to a χ^2_{K-1} critical value. An advantage of CMH is that for that test, appeal to the χ^2 distribution only requires that the total number of off-diagonal elements in the underlying 2×2 partial tables be reasonably large, while the likelihood ratio test requires the set of marginal totals in each subtable to be reasonably large, which requires considerably more observations. Similarly, in this context the likelihood ratio test for the fit of the homogeneous association model is directly comparable to the Breslow–Day test.

Smoking and Mortality in Whickham (continued)

Table 8.2 summarizes model fitting for the smoking and mortality table. Here A represents Age, S represents Smoking status, and M represents Mortality status. The two models that are supported are (MS, MA, AS) (homogeneous association) and (MA, AS) (mortality and smoking are conditionally independent given age). The latter model seems dubious, given the well-known risks of smoking, and both the subset test for MS given homogenous association (that is, the test for conditional independence) and AIC_C agree that the homogeneous association model is the better choice (this is another example of how a focused test can have more power than an omnibus one).

The test statistic for conditional independence $G^2 = 5.95$ can be compared to the Cochran–Mantel–Haenszel statistic $CMH = 5.84$; as expected, the large sample sizes result in the two test statistics being very similar. The goodness-of-fit statistic for the homogeneous association model, $G^2 = 2.38$, is similarly close to the Breslow–Day statistic $BD = 2.37$. Note that it could be argued that the appropriate degrees of freedom for this test is 5, rather than 6, as was mentioned on page 318. The sufficient marginals for the homogeneous association model (MS, MA, AS) include a zero sum in the MA marginal, since no people over the age of 75 at the time of the original survey survived. If indicator variables are added to the model identifying the cells that lead to the zero marginal (treating them as structural zeroes), the fit (and therefore G^2) remains the same, but one more parameter is estimated, dropping the degrees of freedom of any model that includes the MA effect by one (only one of the indicator variables is actually needed to identify the two cells, since they are in the same column in the last partial table).

Model	Subset test	G^2	df	p	AIC_C
(MS, MA, AS)		2.4	6	.88	0.00
(MS, MA)		92.6	12	$< .001$	77.89
	$AS = 0 \mid MS, MA, AS$	90.3	6	$< .001$	
(MS, AS)		632.3	12	$< .001$	617.56
	$MA = 0 \mid MS, MA, AS$	629.9	6	$< .001$	
(MA, AS)		8.3	7	.30	3.88
	$MS = 0 \mid MS, MA, AS$	6.0	1	.015	
(S, MA)		101.8	13	$< .001$	85.04
	$AS = 0 \mid MA, AS$	93.5	6	$< .001$	
(M, AS)		641.5	13	$< .001$	624.71
	$MA = 0 \mid MA, AS$	633.2	6	$< .001$	

TABLE 8.2. Loglinear model fitting for smoking and mortality data.

The chosen model implies that the association between smoking and mortality is constant, given age. Since this association is represented by a 2×2 table, it can be summarized by the estimated odds ratio implied by the fits in any of the partial subtables, which is $\hat{\theta} = .652$ (compare this to the Mantel–Haenszel estimate $\hat{\theta}_{MH} = .655$). As we know, this pattern is very different from the marginal association, but this shouldn't be a surprise, since the homogeneous association model is not collapsible over any of the variables.

Note that in the model fitting summarized in Table 8.2 we allowed for the possibility of an unneeded AS effect (it turned out to be highly statistically significant, of course). If the mortality variable was considered the specific response variable in this context, with age and smoking status predictors, it would be sensible to always include the AS effect, even if there was no relationship between the variables. This would be consistent with allowing an arbitrary association pattern among the predictors (with the estimated counts being consistent with the actual relationship in the table).

We conclude this section with the analysis of a $5 \times 4 \times 2$ table.

Characterizing Types of Death in the U.S. Military

The table on the next page summarizes the manner of death of U.S. active duty personnel who died worldwide in 1999. The data are classified by the manner of death, branch of the service, and gender of the decedent, as reported by the Statistical Information Analysis Division, Directorate for Information Operations and Reports, of the Department of Defense. Is there information in the patterns of these deaths that would be useful for the leaders of the military? Note, by the way, that there were no hostile deaths of U.S. military personnel in 1999.

Table 8.3 summarizes model fitting for this table, where S represents branch of service, M represents manner of death, and G represents gender. The models (SM, MG) and (G, SM) are favored by AIC_C. The former model implies conditional independence between gender and service branch given manner of death, but the latter model is even simpler, implying joint independence of gender with both service and manner of death. That is, while there is a relationship between the branch of the service and the manner by which the person died, gender is completely unrelated to either variable.

The main effect of gender simply reflects that far more men than women died in the military in 1999 — 692 versus 69. This is reflected in the loglinear parameter estimates for gender, which are $\hat{\lambda}_{\text{Male}} = 1.153$ and $\hat{\lambda}_{\text{Female}} =$

−1.153, implying exp(2.306) = 10.03 times higher count for men than women.

		Branch of service			
Gender	Manner of death	Army	Navy	Air Force	Marine Corps
Males	Accident	158	96	54	69
	Illness	56	30	21	6
	Homicide	11	7	2	6
	Self-inflicted	51	32	13	7
	Unknown	24	9	13	27
Females	Accident	15	9	7	3
	Illness	5	4	4	0
	Homicide	5	2	0	1
	Self-inflicted	3	3	1	0
	Unknown	1	2	2	2

The three-dimensional table is collapsible over gender, so the association between manner of death and service branch can be found from the marginal table:

	Branch of service			
Manner of death	Army	Navy	Air Force	Marine Corps
Accident	173	105	61	72
Illness	61	34	25	6
Homicide	16	9	2	7
Self-inflicted	54	35	14	7
Unknown	25	11	15	29

The Army had the most deaths in 1999, followed by the Navy; the Air Force and Marine Corps totals are similar, and noticeably less. More than half of all deaths were accidental, while less than 5% were homicides. The association between manner of death and branch of service is driven to a large degree by the very different pattern in the Marine Corps, which has a higher than expected number of accidental and unknown deaths, and a lower than expected number of illness-related and self-inflicted deaths. Deaths, where the manner was unknown are also different from the other

Model	Subset test	G^2	df	p	AIC_C
(SM, SG, MG)		5.5	12	.94	1.93
(SM, SG)		14.2	16	.58	1.99
	$MG = 0\|SM, SG, MG$	8.7	4	.07	
(SM, MG)		10.7	15	.77	0.63
	$SG = 0\|SM, SG, MG$	5.2	3	.16	
(SG, MG)		61.4	24	$< .001$	32.25
	$SM = 0\|SM, SG, MG$	55.9	12	$< .001$	
(M, SG)		69.2	28	$< .001$	31.83
	$SM = 0\|SM, SG$	55.0	12	$< .001$	
(G, SM)		18.6	19	.48	0.00
	$SG = 0\|SM, SG$	4.4	3	.22	
	$MG = 0\|SM, MG$	7.9	4	.10	
(S, MG)		65.7	27	$< .001$	30.40
	$SM = 0\|SM, MG$	55.0	12	$< .001$	

TABLE 8.3. Loglinear model fitting for military death data.

types, occurring less often in the Army and Navy. Figure 8.3 demonstrates the usefulness of the fit, as the mosaic plots for the observed table and fitted counts are very similar.

This analysis ignores a key point: that the different branches of the military are not the same size, and do not have the same proportions of men and women. We can correct for this by including the logged number of people at risk for death as an offset, converting the analysis to one of death rates, rather than deaths. The number of men and women in each branch of the service in 1999 were as follows: 407,398 (Army men), 321,424 (Navy men), 294,117 (Air Force men), 162,477 (Marine Corps men), 72,028 (Army women), 51,622 (Navy women), 66,473 (Air Force women), and 10,164 (Marine Corps women). Any model that includes the SG interaction will have the exact same fit as before, since the eight distinct values of people at risk correspond to a marginal subtable that is fit exactly by any such model (that is, the number of people at risk doesn't provide any new information over the interaction effect itself). Details of all of the model fits are not given here, but the two models that we might consider are (SM, MG), with $G^2 = 6.7$ on 15 degrees of freedom, and (G, SM), with $G^2 = 15.5$ on 19 degrees of freedom. These two models have virtually identical values of AIC_C, so we will focus on the simpler one. Once again, gender is related

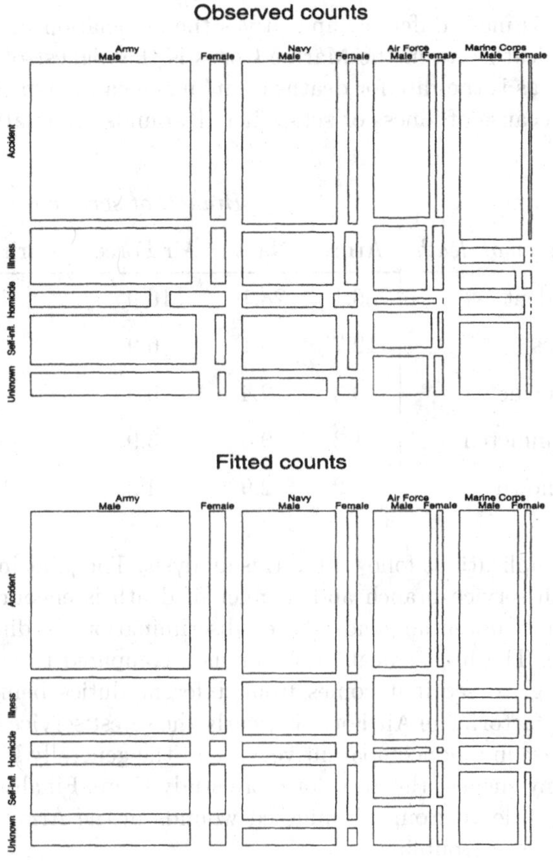

Observed counts

Fitted counts

FIGURE 8.3. Mosaic plots of military death data. Observed counts (top plot) and fitted counts based on (G, SM) model (bottom plot).

to death as a main effect, although now it is rates, rather than total count, that is being modeled. The parameter estimates for gender are $\hat{\lambda}_{\text{Male}} = .239$ and $\hat{\lambda}_{\text{Female}} = -.239$, implying that the death rate for men is 61% higher than that for women ($\exp(.478) = 1.61$). This is far lower than the ratio of deaths, since there are far more men than women in the military. Still, it is interesting to see that even when there were no combat-related deaths, the death rate is significantly higher for men.

The interaction between branch of service and manner of death says that death rates for the different services differ by the manner in which the person died. The death rates per 100,000 people are given on the next page. The death rates for the Army and Navy are similar, although they are lower for the Navy in all categories. Death rates are generally lower in the Air Force than in the other services. Once again it is the Marine

Corps that is most different, and drives the association in the table. The accidental death rate in the Marine Corps is the highest of any branch of the service, as is the rate for deaths of unknown cause. On the other hand, the rates because of illness or self-inflicted wounds are relatively low.

| | *Branch of service* | | | |
Manner of death	Army	Navy	Air Force	Marine Corps
Accident	36.1	28.1	16.9	41.7
Illness	12.7	9.1	6.9	3.5
Homicide	3.3	2.4	0.6	4.1
Self-inflicted	11.3	9.4	3.9	4.1
Unknown	5.2	2.9	4.2	16.8

Several implications follow from this analysis. The joint independence of gender with service branch and manner of death is encouraging, since it argues against potential gender-based discrimination (leading to death) in any service. The higher death rate for men compared to women is worth investigating, to see if it comes from different duties being assigned, or some other factor. The Air Force is clearly the safest service based on these data (at least in a noncombatant year), and the generally high death rates for the Army suggest the need for more study there. Finally, the relatively high rates of death from self-inflicted wounds in the Army and Navy are also particularly troubling.

8.3   Models for Tables with Ordered Categories

When some (or all) of the factors in a three-dimensional table are ordinal, it is advantageous to consider models that incorporate that ordinality, for the same reasons as are true for two-dimensional tables: the ability to fit simpler models with intuitive interpretations, and the ability to incorporate a highest-order interaction effect (here a three-way interaction) that is unsaturated and also easily interpretable. Just as was true for two-dimensional tables, these models can be interpreted in terms of odds ratios. As was noted earlier, the odds ratios $\{\theta_{ij(k)}\}$, $\{\theta_{i(j)k}\}$, and $\{\theta_{(i)jk}\}$ describe the local conditional associations between two variables within a fixed level of the third variable. The ratio of odds ratios

$$\theta_{ijk} = \frac{\theta_{ij(k+1)}}{\theta_{ij(k)}} = \frac{\theta_{i(j+1)k}}{\theta_{i(j)k}} = \frac{\theta_{(i+1)jk}}{\theta_{(i)jk}}$$

describes the local three-factor interaction, referring to the interaction in a $2 \times 2 \times 2$ section of the table consisting of adjacent values of X, Y, and Z. There are $(I-1)(J-1)(K-1)$ values of θ_{ijk}, and the lack of any three-way interaction occurs if all equal 1 (this is equivalent to all values of λ_{ijk}^{XYZ} equaling zero in a standard loglinear model).

The association models described earlier for two-dimensional tables generalize in different ways for three-dimensional tables. Assume first that all three variables are ordinal, with prespecified scores $\{u_i\}$, $\{v_j\}$, and $\{w_k\}$, respectively. The *homogeneous uniform association* model is a special case of the homogeneous association model (XY, XZ, YZ), and has the form

$$\log e_{ijk} = \lambda + \lambda_i^X + \lambda_j^Y + \lambda_k^Z + \beta^{XY}(u_i - \overline{u})(v_j - \overline{v})$$
$$+ \beta^{XZ}(u_i - \overline{u})(w_k - \overline{w}) + \beta^{YZ}(v_j - \overline{v})(w_k - \overline{w}). \quad (8.4)$$

The three β parameters describe the pairwise partial associations given the level of the third variable in a particularly simple way. For this model the local conditional odds ratios for unit-spaced scores correspond to the β's; that is,

$$\log \theta_{ij(k)} = \beta^{XY}$$
$$\log \theta_{i(j)k} = \beta^{XZ}$$
$$\log \theta_{(i)jk} = \beta^{YZ}$$
$$\log \theta_{ijk} = 0.$$

This model implies that, given the level of the third variable, the pairwise association of the other two variables is the same throughout the subtable and follows a uniform association pattern. Note that if a particular β parameter equals zero, then there is conditional independence between those two variables given the level of the third, so the likelihood ratio and Wald tests for each of the association parameters provide focused tests of conditional independence given homogeneous uniform association.

Models including row effects (X), column effects (Y), and layer effects (Z) are also possible, particularly if one or more of these variables are nominal (or if the appropriate ordinal scores are unclear or unknown). For example, if Z is nominal, a model that includes a uniform association between the ordinal variables X and Y and layer effects for the associations between X and Z and Y and Z, respectively, is

$$\log e_{ijk} = \lambda + \lambda_i^X + \lambda_j^Y + \lambda_k^Z + \beta^{XY}(u_i - \overline{u})(v_j - \overline{v})$$
$$+ \tau_k^{XZ}(u_i - \overline{u}) + \tau_k^{YZ}(v_j - \overline{v}) \quad (8.5)$$

with $\sum_k \tau_k^{XZ} = \sum_k \tau_k^{YZ} = 0$. In this model the estimated parameters τ^{XZ} and τ^{YZ} take the place of the layer scores $\mathbf{w} - \overline{w}$ in the XZ and YZ subtables, respectively. For this model the local conditional odds ratios for unit-spaced $\mathbf{u}$ and $\mathbf{v}$ scores satisfy

$$\log \theta_{ij(k)} = \beta^{XY}$$
$$\log \theta_{i(j)k} = \tau_{k+1}^{XZ} - \tau_k^{XZ}$$
$$\log \theta_{(i)jk} = \tau_{k+1}^{YZ} - \tau_k^{YZ}$$
$$\log \theta_{ijk} = 0.$$

Model (8.5) can be modified in the obvious ways to accommodate two or three nominal variables as well.

Models (8.4) and (8.5) (and related models) all assume that the local conditional odds ratios of any two variables are the same for each level of the third variable; that is, $\log \theta_{ijk} = 0$. For arbitrary tables, a three-way interaction makes the model saturated, but for ordinal tables local conditional odds ratios can be modeled parsimoniously. For example, the homogeneous association model can be augmented with a constant (but nonzero) local conditional odds ratio, yielding

$$\log e_{ijk} = \lambda + \lambda_i^X + \lambda_j^Y + \lambda_k^Z + \lambda_{ij}^{XY} + \lambda_{ik}^{XZ} + \lambda_{jk}^{YZ}$$
$$+ \beta^{XYZ}(u_i - \overline{u})(v_j - \overline{v})(w_k - \overline{w}). \quad (8.6)$$

For unit-spaced scores, $\log \theta_{ijk} = \beta^{XYZ}$, making this a *constant interaction* model. A special case of (8.6) replaces the arbitrary partial associations λ_{ij}^{XY}, λ_{ik}^{XZ}, and λ_{jk}^{YZ} with uniform associations $\beta^{XY}(u_i - \overline{u})(v_j - \overline{v})$, $\beta^{XZ}(u_i - \overline{u})(w_k - \overline{w})$, and $\beta^{YZ}(v_j - \overline{v})(w_k - \overline{w})$, respectively. In this model the partial association of any two variables is constant within levels of the third variable, and changes linearly across the levels of the third variable (that is, *uniform interaction*).

What is Important to Undergraduate Business Students?

Consider again the survey of business school undergraduates undertaken by LaBarbera and Simonoff (1999), discussed on page 212. In addition to asking when they decided on a major, the respondents also were asked the importance to them in choosing a major field of the number of anticipated employment opportunities and of the anticipated size of their salary (measured on a five-point Likert scale from very unimportant (VU) to very important (VI)). The resultant table is given on the next page.

When major chosen	Importance of employment opportunities	Importance of size of salary				
		VU	U	N	I	VI
High school	Very unimp.	2	0	0	0	1
	Unimportant	0	0	3	0	1
	Neutral	0	1	3	0	0
	Important	0	0	1	8	8
	Very imp.	1	0	2	10	25
Freshman	Very unimp.	0	0	0	0	0
	Unimportant	0	0	0	0	0
	Neutral	0	0	2	1	1
	Important	0	1	4	6	6
	Very imp.	0	0	2	15	21
Sophomore	Very unimp.	2	0	0	0	0
	Unimportant	0	1	0	1	1
	Neutral	0	1	4	3	1
	Important	0	0	7	19	10
	Very imp.	0	0	5	22	38
Junior	Very unimp.	0	0	0	0	0
	Unimportant	0	1	0	1	0
	Neutral	0	0	1	1	3
	Important	0	1	4	15	3
	Very imp.	0	1	2	14	24
Senior	Very unimp.	2	0	1	0	1
	Unimportant	0	1	0	0	0
	Neutral	0	0	1	2	2
	Important	0	0	0	8	2
	Very imp.	0	0	0	6	4

Model	Subset test	G^2	df	p	AIC_C
(SE, SM, EM)		42.6	64	.98	42.49
(SE, SM)		66.3	80	.86	21.13
	$EM = 0\|SE, SM, EM$	23.7	16	.10	
(SE, EM)		58.4	80	.97	13.26
	$SM = 0\|SE, SM, EM$	15.8	16	.46	
(SM, EM)		152.2	80	$< .001$	107.07
	$SE = 0\|SE, SM, EM$	109.7	16	$< .001$	
(S, EM)		171.5	96	$< .001$	85.94
	$SE = 0\|SE, EM$	113.1	16	$< .001$	
(SE, M)		85.6	96	.77	0.00
	$EM = 0\|SE, EM$	27.2	16	.04	
(S, E, M)		198.7	112	0	76.63

TABLE 8.4. Loglinear model fitting for undergraduate business data, ignoring special characteristics of the table.

Table 8.4 summarizes loglinear model fitting for this table, ignoring its ordinal nature. Here S refers to the salary question, E refers to the educational opportunities question, and M refers to the timing of major choice question. Note that the sufficient marginals for some of the models in the table contain zero entries; the models have been fit using Poisson regression with parameter estimates very large in absolute value, which yield very small estimated fitted counts for those marginal cells. The model (SE, EM), which posits conditional independence of importance of size of salary and time of choice of major given importance of employment opportunities, fits best among the models using two factor interactions. The simpler model in which time of major choice is jointly independent of both importance of salary and importance of employment opportunities (SE, M) is the AIC_C choice, although the subset test of $EM = 0$ given (SE, EM) is marginally significant. This model hypothesizes that timing choice of the major is not relevant to the importance questions, but there is a connection between the two importance questions. The table is collapsible over the salary and employment opportunities pair, implying a main effect of the timing of choice of major represents that roughly one-third of the students chose their majors as sophomores, less than 10% chose it as seniors, and the rest fairly evenly distributed over the high school, freshman, and junior categories. The table is also collapsible over the timing of major choice, which defines the SE interaction.

Model	Subset test	G^2	df	p	AIC_C
(SE, M)		85.6	96	.77	9.81
(U, M)		121.5	111	.23	11.41
	$U|SE, M$	35.9	15	.002	
(R, M)		119.6	108	.21	16.12
	$R|SE, M$	34.0	12	$< .001$	
	$U|R$	1.9	3	.59	
(C, M)		119.4	108	.21	15.95
	$C|SE, M$	33.8	12	$< .001$	
	$U|C$	2.1	3	.55	
$(R + C, M)$		114.6	105	.25	17.84
$(\text{Symmetry}, M)$		107.8	106	.43	8.83
$(\text{Quasi-symmetry}, M)$		89.9	102	.80	0.00
	Marginal homogeneity	17.9	4	.001	
$(\text{Ord. quasi-symm}, M)$		100.7	105	.60	3.98
	Ord. QS|QS	10.8	3	.01	

TABLE 8.5. Loglinear model fitting for undergraduate business data, taking into account special characteristics of the table.

Models that take the form of the table into account can help focus this SE relationship further. The natural association models to consider are ones that put association structure on the SE term plus a nominal M effect (it seems unlikely that effects relating to the timing of major choice are necessarily monotonically related to that timing, given the earlier analysis). Table 8.5 summarizes fitting of such models. The (SE, M) nominal effects model is also given in the table as a reference. In this table the employment opportunities variable E is referred to as the row variable, while the salary variable S is referred to as the column variable, so U refers to uniform association between employment opportunities and salary, R refers to a row effects model based on an interaction of a nominal employment opportunities variable and unit-based scores for salary, and so on.

Uniform association between S and E is slightly inferior to the nominal (SE, M) model according to AIC_C, and the subset test rejects it fairly strongly. Despite this, the association model based on only β^{SE} has the advantage of ease of interpretation. For this model $\hat{\beta}^{SE} = .59$, so the estimated local odds ratio of contiguous 2×2 subtables of importance of salary by importance of employment opportunities is $\exp(.59) = 1.81$, a strong direct association. That is, students who view size of salary as important in choosing a major also view employment opportunities as important. The association of these two aspects of a career focus is hardly surprising, especially in an undergraduate business school. Models that include row and/or column effects don't provide significantly better fit than the uniform association model.

The salary/educational opportunities partial table, besides having ordinal rows and columns, is also symmetric, so tables that include symmetry conditions are also plausible. Table 8.5 also includes information for symmetry, quasi-symmetry, and ordinal quasi-symmetry for this pair of variables. Symmetry does not fit particularly well, but quasi-symmetry does (which is in turn a bit better than ordinal quasi-symmetry). The following table gives the estimated joint probability distribution for the two importance variables (note that since these variables are independent of when the major is chosen, this is the estimated probability matrix for students no matter when they chose their major).

Importance of employment opportunities	Importance of size of salary				
	VU	U	N	I	VI
Very unimportant	.0176	.0000	.0026	.0000	.0062
Unimportant	.0000	.0088	.0099	.0070	.0037
Neutral	.0004	.0048	.0322	.0280	.0138
Important	.0000	.0048	.0395	.1642	.0936
Very important	.0026	.0051	.0390	.1879	.3284

Marginal homogeneity is rejected (there is an estimated .56 probability that employment opportunities are viewed as very important, and probability .38 of rating them important or neutral, while size of salary is rated very important with probability .45 and important or neutral with probability .51), but given this, there is a symmetric association pattern in the table.

Simonoff (1998b) used generalizations of the smoothing techniques of Section 5.5.2 to three-dimensional tables to analyze this table. That analysis supported the presence of a three-way interaction (different salary/employment opportunities association patterns depending on when the student chose their major). It should be noted, however, that the smoothing-based analysis did not include any assessment of whether the patterns hypothesized were genuine, or might have been due to random chance.

8.4   Higher-Dimensional Tables

Contingency tables continue to generalize to four dimensions and beyond. As always, interaction effects can be interpreted as reflecting differences in lower-order effects given the level of a control variable. So, for example, a two-way interaction is saying that the main effect of rows (say) differs, depending on the level of the columns. A three-way interaction is saying that the association between rows and columns (say) differs, depending on the level of the layers. Higher-order interactions are defined in the same way. Consider, for example, a four-dimensional table. The saturated model for such a table is

$$
\log e_{ijk\ell} = \lambda + \lambda_i^W + \cdots + \lambda_\ell^Z + \lambda_{ij}^{WX} + \cdots + \lambda_{k\ell}^{YZ} \\
+ \lambda_{ijk}^{WXY} + \cdots + \lambda_{jk\ell}^{XYZ} + \lambda_{ijk\ell}^{WXYZ}, \quad (8.7)
$$

where all of the λ parameters sum to zero over each index (other than the overall level λ, of course). Models are once again assumed to be hierarchical.

Models for four-dimensional tables can be difficult to interpret, but association diagrams can help. Association diagrams are constructed the same way as for models for three-dimensional tables, with terms that include three effects represented by three-way connections. Otherwise, the interpretations of association diagrams in terms of marginal and conditional independence are the same as for three-dimensional tables. That is, two variables are independent if they are not connected at all in the diagram, and are conditionally independent given a set of other variables if they are only connected via a path that passes through that set of variables. As was the case for three-dimensional tables, different models that substitute variables for each other have equivalent interpretations.

An additional complication for models for four-dimensional tables is that different models, involving different interaction terms, lead to the same association diagram, and thus the same interpretation in terms of marginal and conditional independence. In fact, several models are equivalent to the saturated model in this sense. What this means is that for these models, there is no simplifying representation of the associations in the table in terms of marginal and conditional independence.

Association diagrams for the different types of hierarchical models are given in Figure 8.4, and the models consistent with each diagram with its interpretation are given in Table 8.6. Note that any models that involve variables that are independent of all of the other variables also include the special case of uniform distributions for those variables, but these have not been given separately in the table.

When more than one model leads to the same diagram, there exists a simplest model and a most complicated one. The simplest models are the last ones in the lists, and represent all lines in the diagram with two-way interactions. The most complicated model is called the *graphical* model. It is

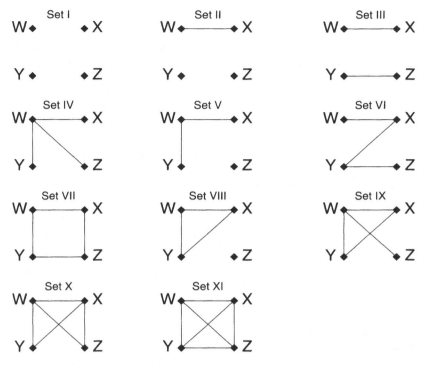

FIGURE 8.4. Association diagrams for four-dimensional loglinear models.

defined as follows. Define a *clique* to be the maximal subset of points in the diagram such that all are connected. So, for example, for Set X, the cliques are WXY and WXZ; for Set IX they are WXY and WZ; and for Set VIII they are WXY and Z. A model is graphical if all interactions corresponding to the cliques are included as sufficient marginals. The graphical models are given as the first ones in the lists.

Four-dimensional tables are not very easy to visualize, and for this reason it is a good idea to see if they can be thought of using a lower-dimensional structure. There are two different kinds of ways that this might happen. If a model always treats two variables (W and X, say) as associated, then the two-dimensional factor WX can be viewed as one factor with ij levels. Consider, for example, an example of Set X such as (WXY, WXZ). In this model W and X are always treated as a pair, in the sense that a variable has a possibly nonzero association with W if and only if it has a possibly nonzero association with X. The association diagram for (WXY, WXZ) shows that Y is connected to both W and X and Z is connected to both W and X. This means that the four-dimensional table could be considered three-dimensional, with factors $W \times X$, Y, and Z. The implication of this model is that Y and Z are conditionally independent, given the joint level of $W \times X$.

Set	Models	Description
I	(W, X, Y, Z)	$W \perp X \perp Y \perp Z$
II	(WX, Y, Z)	$W, X \perp Y, Z \;\&\; Y \perp Z$
III	(WX, YZ)	$W, X \perp Y, Z$
IV	(WX, WY, WZ)	$X \perp Y \perp Z \mid W$
V	(WX, WY, Z)	$Z \perp W, X, Y \;\&\; X \perp Y \mid W$
VI	(WX, XY, YZ)	$Z \perp W, X \mid Y \;\&\; W \perp Y, Z \mid X$
VII	(WX, WY, XZ, YZ)	$W \perp Z \mid X, Y \;\&\; X \perp Y \mid W, Z$
VIII	$(WXY, Z), (WX, WY, XY, Z)$	$Z \perp W, X, Y$
IX	$(WXY, WZ), (WX, WY, WZ, XY)$	$Y \perp Z \mid W, X \perp Z \mid W$
X	$(WXY, WXZ), (WXY, WZ, XZ),$	$Y \perp Z \mid W, X$
	(WX, WY, WZ, XY, XZ)	
XI	$(WXYZ), (WXY, WXZ, WYZ, XYZ),$	
	$(WXY, WXZ, WYZ), (WXY, WXZ, YZ),$	
	$(WXY, WZ, XZ, YZ), (WX, WY, WZ, XZ, YZ)$	

TABLE 8.6. Loglinear models for four-dimensional tables, with descriptions.

Models from Set IX can be viewed three-dimensionally in a similar way. Variables X and Y are jointly modeled for these models, since they are connected to each other, W is connected to each, and Z is connected to neither. This implies that the four-dimensional table can be considered a three-dimensional table with factors $X \times Y$, W, and Z, wherein Z is conditionally independent of $X \times Y$ given W.

Another way to think of a four-dimensional table using lower dimensions is to collapse the table over one or more margins. Naturally, care must be taken to only collapse over margins where the collapsed table has the same association structure as in the full table. Collapsibility conditions generalize the rule for three-dimensional tables. Suppose that the variables in a four- (or higher-) dimensional table can be separated into three mutually exclusive subsets A, B, and C, such that B separates A and C. Then, if the table is collapsed over the variables in C, effects related to variables in A, and to associations between variables in A and B, are unchanged. So, for example, for models from Set X, the factor WX is between Y and Z in the association diagram, so the table can be collapsed over either Y (to examine the $WX \times Z$ association) or Z (to examine the $WX \times Y$ association). For models from Set IX, the factor W is between XY and Z, so the table can be collapsed over either XY or Z. The model in Set VI has an association diagram that implies legitimate collapsing over W, WX,

WXY, Z, YZ, or XYZ. The diagrams for the models from Sets IV and V both imply collapsibility over X, Y, or Z, while the model from Set III allows collapsibility over WX or YZ.

These models and principles generalize to more than four dimensions, but at the price of increasing complexity. It is a challenge to understand the intuition behind three-way interactions without some simplifying structure (such as that provided by the association models of the previous section); it is extremely difficult to grasp the intuition behind a four-way interaction. It seems impossible to understand anything beyond that. The key to analyzing any higher-dimensional table is to try to find models that imply understandable and appropriate independence conditions, and allow for collapsing over margins in useful ways, so that the model's form can ultimately be reduced to a manageable state.

Stillbirth and Premature Birth Among Australian Aborigines

Stillbirth is defined as the death of a fetus at any time after the twentieth week of pregnancy. A child that is born alive after the twentieth week, but before the thirty-seventh week, is considered a premature birth. It is known that certain environmental factors can increase a mother's risk of stillbirth, including being more than 35 years old, malnutrition, inadequate prenatal care, smoking, and alcohol or drug abuse. Other causes include infectious disease and birth defects.

Coory (1998) reported the results of an examination of stillbirth in the Australian state of Queensland, based on data from the Queensland Perinatal Data Collection for the years 1987–1992. These data are intended to include every birth in Queensland during this time period. Besides the birth status of the child (stillbirth or live birth), the gestational age of the fetus (in weeks), the gender, and the race (white or Aborigine) were recorded, resulting in the following table (which is continued on the next page). Statistics for post-term births (gestational age more than 41 weeks) were not given in the paper, because only two post-term Aboriginal stillbirths were reported during this time period.

| | | Race | | | |
| | | Aborigine | | White | |
Gender	Gestational age	Stillbirth	Live birth	Stillbirth	Live birth
Male	≤ 24 weeks	22	16	171	121
	25–28	21	42	109	358
	29–32	12	73	95	944
	33–36	4	387	112	5155
	37–41	7	3934	169	101,776

Race

Gender	Gestational age	Aborigine		White	
		Stillbirth	Live birth	Stillbirth	Live birth
Female	≤ 24 weeks	17	16	167	107
	25–28	13	19	100	314
	29–32	10	76	78	727
	33–36	10	451	92	4224
	37–41	13	3729	209	96,077

Queensland has the largest number of Aboriginal births in Australia, which account for roughly 4% of all births in the state. Given that Australian Aborigines are an economically disadvantaged group, it is important to isolate health risks for this group, including perinatal risks. Table 8.7 describes the fitting of models designed to do this for this table. Here A represent gestational age, B represents birth status, R represent race, and G represents gender.

It is apparent from Table 8.7 that at least one three-way interaction is needed to fit this table. The subset test for ARG is highly significant, while that for ABR is marginally so. Indeed, the model (ABR, ARG, BG) fits very well, and has smallest AIC_C value. Unfortunately, this is a model from Set XI in Figure 8.4, which means that there is no independence interpretation of the model, and no collapsibility over any margin. The second-best model, (ARG, AB, BR, BG), is also from Set XI, but the model (almost) tied for third-best, (ARG, AB, BG) (whose AIC_C value is only 1.89 above that of the best model), is from Set X, implying independence and collapsibility possibilities. Since the test of $BR = 0$ is marginally insignificant, we will pursue this last model.

The association diagram for the model (Set X from Figure 8.4, substituting A and G for W and X, and B and R for Y and Z, respectively), implies conditional independence of birth status and race, given gestational age and gender. This is a very important result, since it says that any direct relationship between race and stillbirth occurrence is weak (if there at all), and that an observed difference between stillbirth rates for the white and Aboriginal population is being driven primarily by the other two variables.

The association diagram tells us that we can pursue these indirect effects, since we can examine the ARG effect by collapsing over B, and can examine the ABG effect by collapsing over R. The first table is given on the top of page 343. The table does not relate to stillbirth at all, but it does describe the connection between the other three variables. For full-term births (37–41 weeks), there is no association between the race and gender of the child. Among premature births, the odds of a male child being Aborigine versus white compared to those for a female child were higher for lower gestational

Model	Subset test	G^2	df	p	AIC
(A, B, R, G)		6848.9	32	$< .001$	6795.7
(AB, AR, AG, BR, BG, RG)		48.5	17	$< .001$	25.3
(ABR, ABG, ARG, BRG)		3.2	4	.52	6.0
(ABG, ARG, BRG)		12.5	8	.13	7.3
	$ABR = 0	ABR, ABG, ARG, BRG$	8.7	4	.07	
(ABR, ARG, BRG)		7.0	8	.54	1.8
	$ABG = 0	ABR, ABG, ARG, BRG$	3.8	4	.43	
(ABR, ABG, BRG)		35.8	8	$< .001$	30.6
	$ARG = 0	ABR, ABG, ARG, BRG$	32.6	4	$< .001$	
(ABR, ABG, ARG)		3.6	5	.61	4.4
	$BRG = 0	ABR, ABG, ARG, BRG$	0.4	1	.53	
(ABR, ARG, BG)		7.1	9	.63	0.0
(ARG, AB, BR, BG)		16.4	13	.23	1.2
	$ABR = 0	ABR, ARG, BR, BG$	9.3	4	.05	
(ABR, ARG)		12.4	10	.26	3.2
	$BG = 0	ABR, ARG, BG$	5.3	1	.02	
(ARG, BR, BG)		6091.6	17	.23	6068.4
	$AB = 0	ARG, AB, BR, BG$	6075.2	4	$< .001$	
(ARG, AB, BG)		19.0	14	.17	1.8
	$BR = 0	ARG, AB, BR, BG$	2.6	1	.11	

TABLE 8.7. Loglinear model fitting for Australian stillbirth data.

age (more premature) and lower for higher gestational age (less premature); that is, for Aborigines being male was associated with being more premature, while for whites it was associated with being less premature. Not surprisingly, the total number of births increases with gestational age. For both male and female children, there is a very strong association between race and gestational age at birth, with Aboriginal births progressively more likely as the birth is more premature. That is, Aborigine mothers have a higher risk for premature delivery than white mothers, with that relative risk declining as the birth approaches full-term status. Since many of the causes of premature delivery are environmental, this suggests the need for a thorough investigation as to potential health initiatives that could reduce this risk for Aboriginal mothers.

Gestational age	Gender	Race	
		Aborigine	White
≤ 24 weeks	Male	38	292
	Female	33	274
25–28	Male	63	467
	Female	32	414
29–32	Male	85	1039
	Female	86	805
33–36	Male	391	5267
	Female	461	4316
37–41	Male	3941	101,945
	Female	3742	96,286

The tables that follow give two views of the ABG table, allowing the AB and BG effects to be seen more easily.

Gender	Gestational age	Birth status	
		Stillbirth	Live birth
Male	≤ 24 weeks	193	137
	25–28	130	400
	29–32	107	1017
	33–36	116	5542
	37–41	176	105,710
Female	≤ 24 weeks	184	123
	25–28	113	333
	29–32	88	803
	33–36	102	4675
	37–41	222	99,806

| Gestational | | Birth status | |
age	Gender	Stillbirth	Live birth
≤ 24 weeks	Male	193	137
	Female	184	123
25–28	Male	130	400
	Female	113	333
29–32	Male	107	1017
	Female	88	803
33–36	Male	116	5542
	Female	102	4675
37–41	Male	176	105710
	Female	222	99806

As would be expected, the chance of a stillbirth decreases with increasing gestational age. At each gestational age, the odds of a male child being stillborn versus born alive are lower than those for a female child; that is, being female is associated with stillbirth. This is quite striking, since, according to the World Health Organization, "[i]n all societies significantly more male foetuses are spontaneously aborted or stillborn" (World Health Organization, 1998, page 12). It would be worthwhile to see if this is just a one-time anomaly or a persistent Australian pattern worth investigating further.

8.5 Conditional Analyses

Conditional (exact) analysis of $2 \times 2 \times K$ tables is based on treating each 2×2 partial table separately and then combining the results based on independence of the observations across partial tables. The test for homogeneity of association (corresponding to the asymptotic Breslow–Day test) is based on a null distribution that is a product of separate hypergeometric distributions for each partial table, conditional on the observed sum $S = \sum_k n_{11k}$.

The probability of the observed table occurring, based on the product hypergeometric distribution, then can be compared to the distribution of all tables with sum S, with the tail probability of the test being the sum of all probabilities that are no greater than that for the observed table.

An exact confidence interval for θ, the common odds ratio in a $2 \times 2 \times K$ table (assuming a common odds ratio is reasonable) is based on a product of (now) noncentral hypergeometric distributions for the partial table counts, conditioning again on S. The interval can be constructed by inverting an exact test for the hypothesis $\theta = \theta_0$ (that is, the interval includes those values of θ such that the observed $\hat{\theta}$ is not statistically significantly different from them at level α).

Exact inference based on loglinear models for multidimensional tables is based, as always, on the sufficient statistics for the model, the sufficient marginals (page 320). Many tests for these tables can be simplified by considering the sufficient marginals and collapsibility conditions. For example, consider a three-dimensional table, and the comparison of model (XY, Z) to (XY, YZ). The sufficient marginals for the former model are $\{n_{ij.}, n_{..k}\}$, while those for the latter model are $\{n_{ij.}, n_{.jk}\}$. A test of the presence of the YZ interaction, given the XY interaction, is based on the distribution of $\{n_{.jk}\}$ given $\{n_{ij.}, n_{..k}\}$. Since X and Z are (conditionally) independent under both models, the table is collapsible over X, so the distribution of $\{n_{.jk}\}$ given $\{n_{ij.}, n_{..k}\}$ (conditioning on XY) is identical to that of $\{n_{.jk}\}$ given $\{n_{.j.}, n_{..k}\}$ (conditioning on X). That is, exact inference for $YZ = 0$ can be based on the $Y \times Z$ two-dimensional table. Hypotheses that cannot be stated in terms of collapsed (lower-dimensional) tables require enumeration of products of multiple hypergeometric probabilities (or, failing that, Monte Carlo approximation of such products).

Is There an Association Between Inbreeding and Schizophrenia?

Schizophrenia is defined as any mental disorder that is characterized by withdrawal from reality (such as from delusions or hallucinations) and social contact, and disturbances of thought and language, often accompanied by other behavioral, emotional, or intellectual difficulties. It is well-known that heredity plays a strong role in the pathology of schizophrenia, but the actual genes involved are unknown. Indirect evidence regarding the genetic mechanism can help to further research into the disease. For example, for a recessive disease (one that does not manifest itself if the dominant (nondisease) allele is present), inbreeding is positively associated with the chances of being affected. Thus, investigation into schizophrenia occurrence in isolated populations (where inbreeding is more likely to occur) can provide evidence of the genetic character of schizophrenia.

Zhang (1999) reported on the results of such an investigation. Daghestan, the largest of the North Caucasian autonomous republics of the

Russian Federation, is home to roughly two million people of more than 35 nationalities. The western part of the country is comprised of Caucasus Mountains divided by four major rivers, creating canyons thousands of feet deep. This has resulted in many of the communities in the region being very isolated. There is a tendency in Daghestan to retain property in families by encouraging marriage within ethnic groups and villages, and the Muslim religion that was brought to the region 800 years ago permits marriages within the same family line. The resultant inbreeding has resulted in large genetic differences between different communities, and small differences within communities.

The following table gives the available data. The data are separated by membership in one of 12 extended families (or pedigrees). The units of observations are sets of siblings (sibships), and the level of inbreeding of the sibship is rated as high or low based on a family history. If at least one member of the sibship is schizophrenic, the sibship is considered positive for schizophrenia.

Counts

Pedigree	Low inbreeding		High inbreeding	
	Not schiz.	Schizophrenic	Not schiz.	Schizophrenic
1	2	2	1	1
2	3	1	0	2
3	2	0	0	2
4	1	0	2	1
5	0	1	1	0
6	2	1	2	2
7	2	0	0	1
8	3	2	2	3
9	1	0	1	1
10	1	0	0	1
11	11	3	1	2
12	2	0	0	1

Conditional independence between inbreeding and schizophrenia can be tested using the Cochran–Mantel–Haenszel test, which is $CMH = 8.0$ ($p = .005$), strongly rejecting conditional independence. The corrected Breslow–Day statistic $BD^* = 12.2$ on 11 degrees of freedom ($p = .35$), implying that a constant odds ratio is reasonable ($BD = 12.6$, implying the same). The Mantel–Haenszel estimate of the common odds ratio is $\hat{\theta}_{MH} = 4.43$, with 95% confidence interval $(1.48, 13.23)$.

The small counts in the table suggest exploring a conditional analysis. The exact test of conditional independence has p-value $p = .007$, which is similar to the asymptotic value. The exact test for homogeneous association has p-value $p = .68$; while this is noticeably larger than the asymptotic value, the implications are of course the same. The exact 95% confidence interval for θ is $(1.38, 15.45)$, which is almost 20% wider than the asymptotic interval, but this is in large part because of conservativeness, since the interval based on a mid-p test is $(1.52, 13.54)$.

Thus, there is strong evidence of a direct association between inbreeding and schizophrenia in a Daghestan pedigree, with the odds of a schizophrenic member of the pedigree more than four times higher in a high inbreeding pedigree than in a low one. This suggests that recessively acting mutations contribute towards tendency to schizophrenia.

8.6   Background Material

Yule (1903) noted the possibility of Simpson's paradox, although this reversal of conditional and marginal associations gets its name from Simpson (1951). The graphical representation of Simpson's paradox given in Figure 8.1 was described in Jeon et al. (1987) and Baker and Kramer (2001); see also Wainer (2002). Fleiss (1981) discusses extensively the analysis of $2 \times 2 \times K$ tables. Tarone (1985) noted the need for the corrected version BD^* of the Breslow–Day statistic BD.

Bishop et al. (1975) introduced the theory and practice of loglinear modeling for multidimensional contingency tables to a wide audience. Darroch et al. (1980) used the theory of Markov fields to suggest representing loglinear models using association diagrams, and defined the class of models called graphical models. Andersen (1997) gives a thorough discussion of loglinear modeling and association diagrams for multidimensional tables, and is noteworthy in distinguishing a uniform effect (lack of a main effect) from the presence of a main effect in association diagrams. An alternative approach to modeling multidimensional contingency tables that incorporates assumptions about the causalities behind associations (and their directions) are graphical chain models, which were introduced by Wermuth and Lauritzen (1990); see Edwards (2000) for further discussion and Tunaru (2001) for a comparison of analyses based on graphical association models and graphical chain models for a four-dimensional contingency table related to road accidents.

Theus and Lauer (1999) discussed how mosaic plots can be used to guide loglinear model selection. Friendly (1999) described extensions to the mosaic plot for multidimensional tables.

Lloyd(1999, Chapter 7) discusses applications of conditional inference to three-dimensional tables, and also provides computer code for some of these problems. Kim and Agresti (1997) discussed extending (nearly) exact inference to tests of conditional independence and marginal homogeneity for $I \times J \times K$ tables.

8.7 Exercises

1. Consider the following major league baseball comparisons.

   (a) In 1983, Ken Oberkfell of the St. Louis Cardinals had 143 hits in 488 times at bat, for an average of .293. In 1984 (during which he spent part of the year with the Atlanta Braves) he had 87 hits in 324 at bats, for an average of .269. This yields an overall average of .283. Mike Scioscia of the Los Angeles Dodgers had 11 hits in 35 times at bat in 1983, for an average of .314, 93 hits in 341 times at bat in 1984, for an average of .273, with an overall average of .277. Who was the better hitter (in terms of average) during these two years?

   (b) In 1989, David Justice of the Atlanta Braves had 12 hits in 51 times at bat, for an average of .235. In 1990, he had 124 hits in 439 at bats, for an average of .282. This yields an overall average of .278. Andy Vanslyke of the Pittsburgh Pirates had 113 hits in 476 times at bat in 1989, for an average of .237, 140 hits in 493 times at bat in 1990, for an average of .284, with an overall average of .261. Who was the better hitter (in terms of average) during these two years?

   (c) In 1995, Derek Jeter of the New York Yankees had 12 hits in 48 times at bat, for an average of .250. In 1996, he had 183 hits in 582 at bats, for an average of .314. In 1997, he had 190 hits in 654 at bats, for an average of .291. This yields an overall average of .300. David Justice, then of the Atlanta Braves and Cleveland Indians, had 104 hits in 411 times at bat in 1995, for an average of .253, 45 hits in 140 times at bat in 1996, for an average of .321, 163 hits in 495 times at bat in 1997, for an average of .329, with an overall average of .298. Who was the better hitter (in terms of average) during these three years? (Ross, 2001)

   (d) Do you think E.H. Simpson or G.U. Yule would enjoy the intricacies of baseball averages?

2. The Birth to Ten study is a cohort study conducted in the Johannesburg/Soweto metropolitan area of South Africa starting in 1990 designed to follow children born between April and June 1990 in hopes of identifying risk factors for cardiovascular disease. After five years the children and caregivers were invited to participate in interviews that included questions about living conditions, exposure to tobacco smoke, and other health-related issues. Many children did not participate in these interviews, leaving open the possibility of biases if inferences were made based on those who participated. Morrell (1999) gave data comparing the children who participated to those who did not with respect to whether the mother had medical aid at the time of the birth.

(a) According to the available information, 195 of 1174 children who did not participate in the five-year interviews had mothers with medical aid, while the mothers of 46 of 416 children who did participate in the interviews had medical aid. Is there evidence of a relationship between medical aid status and participation in the interviews?

(b) The study classified children by their racial group, with the following results:

White children

	No interview	Interview
Had medical aid	104	10
No medical aid	22	2

Black children

	No interview	Interview
Had medical aid	91	36
No medical aid	957	368

Is there evidence of a relationship between medical aid status and participation in the interviews for white children? What about for black children?

(c) How do you account for the difference in your answers to parts (a) and (b)? Does a picture like that in Figure 8.1 help explain the difference?

3. Verify the following relationships between different types of independence using the definitions on page 321.

(a) If X is jointly independent of Y and Z, and Y is independent of Z, then X, Y, and Z are completely independent.

(b) If X and Y are conditionally independent given Z, and Y is independent of Z, then Y is jointly independent of X and Z.

(c) If X is jointly independent of Y and Z, and Z is jointly independent of X and Y, then X, Y, and Z are completely independent.

4. Moore (1999) investigated outcomes when teaching finance using case methods and lectures. Sections of undergraduate business classes were taught lessons on two topics, working capital management and capital budgeting, using either case-based methods or traditional lectures. After the lessons, students were evaluated regarding performance, and were questioned regarding their opinions of the experience.

(a) The following table summarizes responses to a question regarding the usefulness of the topic in a possible future career.

| | | Teaching method | |
Topic	Usefulness	Lecture	Case
Working capital	Positive	33	13
	Negative	8	12
Capital budgeting	Positive	18	28
	Negative	3	9

Is there a relationship between teaching method and the perception of usefulness of the lessons, given the topic of the lesson? Use the CMH statistic to answer this question. What form does the observed association take? Are the association patterns similar for the two types of topics?

(b) The table on the next page summarizes responses to a question regarding whether the class experience resulted in a different appreciation of the subject. Is there a relationship between teaching method and the change of appreciation of the topic, given that topic? What form does the observed association take? Are the association patterns similar for the two types of topics? Are the results consistent with recent trends in business education to deemphasize lectures in favor of more active learning techniques?

Topic	Different appreciation	Teaching method Lecture	Case
Working capital	Yes	30	17
	No	11	8
Capital budgeting	Yes	15	21
	No	6	16

5. The possibility of racial discrimination in the application of the death penalty in the United States is an extremely controversial issue, and various studies of the patterns of death penalty application have been undertaken to investigate it. The following tables are two examples of such studies. The first table cross-classifies blacks who had been convicted of murder in Georgia by whether they received the death penalty, the race of the victim, and the aggravation level of the crime from 1 (least aggravated) to 6 (most aggravated).

Aggravation level	Race of victim	Death penalty Yes	No
1	White	2	60
	Black	1	181
2	White	2	15
	Black	1	21
3	White	6	7
	Black	2	9
4	White	9	3
	Black	2	4
5	White	9	0
	Black	4	3
6	White	17	0
	Black	4	0

(Woodworth, 1988; see also Ramsey and Schafer, 1997, page 540). The second cross-classifies homicide cases in North Carolina where

the death penalty was possible by whether the defendant received
the death penalty, the race of the defendant, and the race of the
victim.

Race of victim	Race of defendant	Death penalty Yes	No
Nonwhite	Nonwhite	29	587
	White	4	76
White	Nonwhite	33	251
	White	33	541

(Butterfield, 2001). The policy question of interest is whether the
death penalty is applied in an unfair way. For the first study, "fairness" would presumably imply (conditional) independence between
death penalty application and the race of the victim, while in the second it would presumably imply independence between death penalty
application and the race of the victim and the race of the defendant.
Are these forms of independence consistent with the observed tables?
Each table includes the race of the victim as a variable; are association patterns related to race of the victim similar in the two tables?
What do you think these studies say about the fairness of application
of the death penalty in Georgia and North Carolina, respectively?

6. The following table cross-classifies cases between July 1, 1978 and
   December 31, 1999 in Maryland where the death penalty could have
   been pursued by race of the victim, race of the defendant, and whether
   the prosecutor initially filed notice of intent to seek the death penalty
   (Paternoster et al., 2003).

Race of victim	Race of defendant	Death penalty initially sought Yes	No
Black	Black	97	455
	White	11	11
White	Black	120	145
	White	105	148

(a) Is there evidence of unfairness in the process by which prosecutors initially seek the death penalty? (See the previous question.)

(b) Paternoster et al. (2003) found that when the prosecutorial jurisdiction was taken into account, the effect of the victim's race greatly diminishes. Scheidegger (2003) argued that this pattern supports use of the death penalty, since it shows that apparent disparities in seeking the death penalty actually reflect local differences in attitudes towards the death penalty (that is, more conservative areas, which tend to have higher percentages of white residents, elect prosecutors who are more likely to seek the death penalty, while more liberal areas, which tend to have higher percentages of black residents, elect prosecutors who are less likely to seek the death penalty). Do you agree with this argument? What does the observed association between the race of the defendant and initial seeking of the death penalty for black victims, and how it differs from that for white victims, say about this argument?

7. Fleiss (1981), Section 10.1, described an alternative to the CMH and BD statistics for the analysis of conditional independence and homogeneous association in $2 \times 2 \times K$ tables. Let T^2 be the sum of squares of the components of lack of fit from each partial table; that is,

$$T^2 = \sum_{k=1}^{K} \frac{(n_{11k} - e_{11k})^2}{V(n_{11k})},$$

where e_{11k} is the expected value of n_{11k} under conditional independence. This is a test of the overall fit of conditional independence to the table. A one-degree-of-freedom χ^2 test for conditional independence assuming homogeneous association is

$$A^2 = \frac{[\sum_k (n_{11k} - e_{11k})/V(n_{11k})]^2}{\sum_k [1/V(n_{11k})]}.$$

This is an alternative to the CMH statistic. An easily constructed χ^2_{K-1} test for homogeneous association is then

$$H^2 = T^2 - A^2,$$

which is comparable to the Breslow–Day statistics.

(a) Show that if the marginal totals $\{n_{1 \cdot k}, n_{2 \cdot k}, n_{\cdot 1k}, n_{\cdot 2k}\}$ for each partial table are the same for all k then $A^2 = CMH$.

(b) Analyze the association patterns in the Whickham smoking mortality data using A^2 and H^2. Are the results different from those when using CMH and BD, respectively?

(c) As part of the approval process for a new drug, pharmaceutical companies submit the results of clinical investigations into the effectiveness of the drug, which are then reviewed by the Division of Biostatistics and Epidemiology of the Food and Drug Administration. The following data come from such a review for the drug Palivizumab, which was approved for use on June 19, 1998 (Neeman, 1998). Palivizumab is used to prevent infection caused by respiratory syncytial virus (RSV), by giving the patient's body the antibodies it needs to protect against the infection. A clinical trial involving 1502 premature infants, or infants with bronchopulmonary dysplasia (BPD, a lung disease often associated with premature birth), was conducted starting in late 1996. In this trial, subjects were given either Palivizumab or a placebo, and were followed for five months (over the winter season) to see if they were hospitalized for RSV. The trial was conducted at sites in the United States, the United Kingdom, and Canada, with the following results:

| | | Treatment | |
Location	RSV hospitalization	Placebo	Palivizumab
United States	Yes	44	39
	No	382	812
United Kingdom	Yes	4	3
	No	36	80
Canada	Yes	5	6
	No	29	62

Is there evidence that Palivizumab is associated with a lower RSV hospitalization rate? That is, are the treatment and hospitalization result conditionally independent given location of the trial? Do CMH and A^2 give different answers to this question? Is the observed association pattern homogeneous across locations? Do BD (or BD^*) and H^2 give different answers to this question? What about a loglinear modeling approach to these questions? Construct a 95% confidence interval for the common odds ratio relating hospitalization and treatment. Does this interval support the decision to approve the drug?

(d) The following table separates the results by initial condition (premature or BPD).

Initial condition	RSV hospitalization	Treatment Placebo	Palivizumab
BPD	Yes	34	39
	No	232	457
Premature	Yes	19	9
	No	215	497

Repeat the questions of part (b), now controlling for initial condition. How does the apparent effectiveness of Palivizumab differ between premature infants and those with BPD?

8. Kass et al. (1991) analyzed the question of whether homosexual men with AIDS were more likely to lose their private health insurance coverage than those without AIDS. Questionnaires were given to participants in the Multicenter AIDS Cohort Study, and patients with AIDS and leukemia at the Johns Hopkins Hospital, and respondents who had private health insurance were asked if they had lost that coverage, with the following results.

Education level	AIDS	Lost coverage Yes	No
High school	Yes	1	23
	No	8	116
College	Yes	3	41
	No	24	699
Postgraduate	Yes	3	25
	No	6	591

Is there evidence that an AIDS diagnosis is related to losing insurance coverage? Is education level related to the association between diagnosis and insurance loss?

9. Sleep apnea is a medical condition wherein people stop breathing repeatedly during sleep, sometimes hundreds of times during the night.

Untreated sleep apnea can cause cardiovascular disease (including high blood pressure), memory loss, weight gain, headaches, and impotency. Diagnosis of sleep apnea is typically performed in a sleep laboratory using polysomnography (PSG), the measurement and recording (and subsequent interpretation) of signals used to analyze sleep and breathing. Steltner et al. (2002) developed and investigated diagnostic software for detection and identification of respiratory events in subjects based on an analysis of recordings of nasal pressure and mechanical respiratory input impedance. Nineteen subjects who had been referred to a German sleep laboratory were enrolled into a study. PSG was performed on each, and the resultant signals were scored by two sleep experts into the five categories of normal sleep (notated N), hypopnea (abnormally slow, shallow breathing, notated H), obstructive apnea (caused by a blockage of the airway, such as when the soft tissue in the rear of the throat collapses and closes during sleep, notated OA), central sleep apnea (where there is no blockage of the airway, but the brain fails to signal the muscles to breathe, notated CA), and mixed apnea (a combination of the two, notated MA). An automatic algorithm was also used to classify time periods into the same five categories using nasal pressure and mechanical respiratory input impedance. I am indebted to Holger Steltner for making these data available here.

(a) The file apnea1 contains the results of comparisons of second-by-second evaluations by the two scorers for each of the 19 subjects. What can you say about the association between the two scorers? Since these are agreement data, the methods of Section 7.2.6 would be reasonable tools to use. Can you find a loglinear model that fits these data adequately? Is there a homogeneous association pattern across subjects?

(b) The files apnea2 and apnea3 give results comparing the automatic algorithm to scorer 1 and scorer 2, respectively. (The scorer values are not directly comparable to those in apnea1, because they were generated based on cross-validated test sets. Also, there are data for only 18 subjects in apnea3.) How do the algorithm's classifications agree with the scorers' classifications? Is the performance similar across subjects?

(c) If the strength of association between the algorithm and the scorers is comparable to that between scorers, that would be an argument for the use of the algorithm, which is less expensive and easier to use. Do the results here support the use of the algorithm?

10. Firearms are implicated in 65% of deaths among persons less than 19 years old. Strategies for reducing access to firearms in the home

include trigger locks, gun safes and gun lock boxes. If adults in residential settings are the primary sources for firearms ultimately being used in suicides and unintentional firearm injury, such strategies may be effective preventive measures, but they are less plausible if firearms are being obtained from peers, adult friends, or illegal sources. The location of a firearms-related incident can help to identify patterns in procurement of such weapons. Grossman et al. (1999) examined data for 124 youths up to 19 years old who sought treatment at a trauma center in Seattle, Washington, or presented to the county medical examiner, for a self-inflicted or unintentional firearm injury between 1990 and 1995. The results are as follows:

Injury location	Injury type	Fatal Yes	No
Victim's home	Suicide	35	10
	Unintentional	5	19
Relative/friend's home	Suicide	3	2
	Unintentional	4	17
Other or unknown	Suicide	8	1
	Unintentional	1	19

What can you say about the association pattern in this table? Is there a relationship between injury type and whether or not it is fatal? Is this relationship affected by the location of the incident? Can you summarize the observed relationship in terms of any independence relationships? What do these results say about the potential usefulness of prevention strategies such as those mentioned earlier?

11. As computer usage in the classroom has increased, so has interest in conducting educational assessments using computers. If testing using paper-and-pencil results in different assessments than when using the computer, the results of tests are potentially invalidated. If differences arise because of familiarity of the students with computers (or the lack thereof), biases related to that access (which could be related to socioeconomic status) becomes a potentially serious problem. Russell and Haney (1997) investigated this question using writing assessment tests given to middle school students in Massachusetts. Those students who completed both the multiple choice and writing

assessment parts of the test are cross-classified by their score, the test medium, and the gender of the student in the following table.

Test medium

Gender	Score	Computer	Paper-and-pencil
Female	1–5	1	1
	6–10	5	1
	11–15	10	7
	16–20	7	0
	≥ 21	2	0
Male	1–5	0	1
	6–10	1	5
	11–15	9	5
	16–20	2	0
	≥ 21	1	0

(a) Find an appropriate loglinear model for this table, ignoring the ordering in the score. What does the model imply about a connection between the medium of the test and performance on it? Can you summarize this relationship in a two-dimensional form?

(b) Now, fit a model that takes into account the ordering of test scores. Are the implications similar to those in part (a)?

(c) Convert this $2 \times 5 \times 2$ table into a $2 \times 4 \times 2$ table by combining the last two score categories. Does this lead to different implications than those in parts (a) and (b)?

12. The file chlamydia gives the reported cases of chlamydia (the most common bacterial sexually transmitted disease in the United States) in 1996, separated by age, gender, and race, as given by the Centers for Disease Control and Prevention. Describe the pattern of counts for these data. Be sure to take into account the ordering of the age variable (including the possibility of scores that are not simply linear in the category), and the fact that the table contains more than 350,000 observations. How might you adjust your analysis to take into account the age and ethnicity distributions in the United States?

13. Market researchers have long believed that market leaders can (and do) maintain market leadership over many years, and even decades. Golder (2000) explored this question by following up on a book written in 1923 by G.B. Hotchkiss and R.B. Franken entitled *The Leadership of Advertised Brands*. This book listed market leaders in 100 different categories based on a survey of over 1000 people, and Golder determined 1997 market share for the market leaders in 95 of the categories (five no longer seemed relevant). The results were as follows, separated by whether the product was a durable or nondurable good:

Type of good	1923 position	1	2	3	4–5	6–10	> 10	Failed
					1997 position			
Durable	No. 1	7	2	6	3	1	8	19
	No. 2	3	2	0	0	3	7	13
	No. 3	1	1	0	1	1	1	12
	No. 4	1	0	1	1	1	3	1
	No. 5	0	0	2	0	1	0	1
Nondurable	No. 1	15	5	3	5	6	9	8
	No. 2	4	4	2	3	3	11	14
	No. 3	1	2	1	0	3	5	13
	No. 4	0	1	0	0	1	8	7
	No. 5	0	0	1	0	0	5	1

(a) Does it appear that market leadership has endured over this 50-year time period? That is, are 1923 market position and 1997 market position related to each other? Does the type of product (durable or nondurable) matter? Does any loglinear model provide a good fit to these data?

(b) Can the ordering in the two market position variables be exploited to provide a simpler representation of the data?

(c) The "Failed" category is clearly different from the others. What happens if you remove this category from the table? Do your answers to parts (a) and (b) change?

(d) Consider now collapsing all of the 1997 market position categories together, so that the 1997 variable only has two categories ("Did not fail" and "Failed"). Is there a relationship between 1923 position and long-term success or failure of the brand? Does the type of good matter?

14. Legionnaires' disease is a sometimes fatal airborne bacterial respiratory disease that is characterized by severe pneumonia, headache, and a dry cough. Since it was first recognized after an outbreak at a convention of the American Legion in Philadelphia in 1976, many nations have implemented reporting mechanisms whereby causes of the infection can be identified and controlled. Infuso et al. (1998) described cases of Legionnaires' disease in France that were identified in 1995 via at least one of three sources: a notification system to public health officials, the Centre National de Référence des *Legionella* (the national reference laboratory), and a survey of hospital laboratories. They then categorized the cases by the source, resulting in the following table:

Notification	*Hospital*	*Reference laboratory*	
system	*laboratories*	Yes	No
Yes	Yes	14	6
	No	10	7
No	Yes	100	73
	No	46	—

Note that the cell corresponding to no report from any of the three sources is a structural zero, and is the value of interest here, since it represents unreported cases of Legionnaires' disease.

(a) What is the "best" model for this table? What does this model imply about the independence relationships of the different notification methods?

(b) Estimate the number of cases of Legionnaires' disease that were not reported. (*Hint*: If you fit the model using a Poisson generalized linear model based on indicators for the variables, with 1 = Yes and 0 = No, then the desired value is simply $\exp(\hat{\lambda})$.)

(c) Estimate the sensitivity of each of the reporting mechanisms (recall that the sensitivity of a test is the proportion of people who actually have the disease who are correctly identified).

(d) Construct an approximate 95% confidence interval for the total number of people who contracted Legionnaires' disease in France in 1995. (*Hint*: Given the known reported cases, the only estimated value is the number of unreported cases determined in part (b).)

15. The file **assessment** contains the original data from the study of student reactions to assessment methods in a multivariate statistics course discussed on page 254. I am indebted to Ann O'Connell for making these data available here. The file consists of the responses of the 14 students to 16 questions regarding the assessment techniques used in the class: structured data analysis assignments, open-ended data analysis assignments, article reviews, and annotating computer output (difficulty, appropriateness, level of learning, and rating from most to least preferred for each).

(a) Explore these data for interesting structure. Possibilities include the four-dimensional relationships within each assessment strategy, the four-dimensional relationships within type of evaluation (difficulty, appropriateness, or level of learning), or the four-dimensional ranking relationship. Take advantage of the ordering in the levels of the variables. Can you extract any interesting structure?

(b) With only 14 observations, any table you construct will necessarily be very sparse. How does that affect your analysis? Are any inferential or modeling techniques useful for data like these?

16. Wermuth and Cox (1998) presented data from two general social surveys in Germany in 1991 and 1992, including responses on how well the political system is perceived to function, age group, year of the survey, the type of formal schooling, and location of the survey (East or West Germany). The data are given in the file **germangss**.

(a) Particularly interesting questions for these data are whether opinions on the political system are different in the two regions of Germany, and different at the two time periods. Is that the case? That is, are association patterns between opinions on the political system, schooling, and age homogeneous over region and/or time? Does the natural ordering of the opinion, schooling, and age variables allow for simplifications in the analysis? What can you say about trends in the feelings about the political system in the two parts of Germany?

(b) Wermuth and Cox (1998) noted several forms of prior knowledge that could be exploited in the analysis of the table. Specifically, since different school systems were established in East and West Germany after 1949, a three-way interaction between schooling, age, and location would be expected; since the surveys were designed to be representative, schooling and age should be conditionally independent given year, and possibly jointly independent of year given region; and main effects of schooling, age, and year on political opinion within region might be expected, but not interaction effects. Are these expectations consistent with your model choice in part (a)?

(c) Wermuth and Cox (1998) combined the two lowest levels of the schooling variable and the two highest levels of the age variable in their analysis. Is this decision supported by AIC_C?

17. Construct the sufficient marginals for the business student data (page 332) for all three two-way effects. Do any have zero marginal sums? Given this, incorporate appropriate indicator variables into a log-linear model to fit cells as structural zeroes so that the relevant fitted marginals are also zero. Does this change the choice of best loglin-ear model? (*Hint*: Remember to include the appropriate number of parameters in any AIC_C values you calculate.)

18. On February 12, 1999, the U.S. Senate voted whether to remove President Bill Clinton from office, based on two impeachment articles passed by the House of Representatives (article I: perjury, and article II: obstruction of justice). Reifman (1999) collected information regarding the 100 senators that could be related to their votes on the impeachment articles, including the senator's political party, degree of political conservatism (as judged by the American Conservative Union), percentage of the vote Clinton received in the 1996 presidential election in the senator's state, the year in which the senator's seat is up for reelection, and whether or not the senator was a first-term senator. The data are given in the file **impeach**. Treating the joint distribution of votes on the two impeachment articles as the response of interest, what variables provide predictive power for this response? Treat the conservatism and vote percentage variables as categorical by discretizing them in a sensible way. Does the conservatism variable add anything given the senator's political party?

19. Margolis et al. (2000) analyzed data for 1991–1996 from the Fatality Analysis Reporting System, a nationwide registry of motor vehicle deaths in the United States. They reported deaths of children (younger than 16) cross-classified by type (passenger or pedestrian/bicyclist), gender, age, and whether the accident was alcohol-related. The data are given in the file **vehicle**. Summarize the association pattern in this table. Are there noteworthy patterns related to gender or the child's age in the table?

20. Kvam (2000) studied the effects of active learning on performance in an introductory engineering statistics class. Students were taught using either traditional or active learning techniques, and were then tested twice (immediately after the course, and eight months later). The results were as follows:

Learning style	Test 1 score	Test 2 score	
		Low	high
Traditional	Low	7	1
	High	2	13
Active	Low	2	3
	High	2	8

(a) It was anticipated before the study began that poor students from the active learning group would have better retention of the material than poor students from the traditional learning group. This would be reflected in different odds ratios in the partial tables separated by learning method. Is there sufficient evidence to reject homogeneity of odds ratios here?

(b) Since these data are sparse, a conditional analysis is reasonable. Does such an analysis lead to different inferences than does the asymptotic analysis in part (a)?

21. On April 16, 1846, nine covered wagons left Springfield, Illinois, and headed west to California. The group consisted of the families of George and Jacob Donner and James Frazier Reed, along with two servants and seven teamsters who drove the wagons. Eventually they joined other families, becoming the so-called Donner Party. The party took a purported shortcut to California starting in Wyoming in July that was far longer and more difficult than they had believed, were attacked by Paiute Indians, and eventually became stranded while crossing the Sierra Nevada mountains near Lake Tahoe in November because of snow. Of the 89 people in the Donner Party, only 48 survived. The file **donner** cross-classifies information on the Donner Party by age group, gender, and survival into a $7 \times 2 \times 2$ table, based on data from the web site "New Light on the Donner Party" (http://www.utahcrossroads.org/DonnerParty), authored by Kristin Johnson, a librarian at Salt Lake Community College; see also McCurdy (1994).

(a) Do the data support an association between gender and survival that is homogeneous across age groups? If not, where does the heterogeneity arise?

(b) Does a conditional (exact) analysis lead to the same conclusions? Assuming a homogeneous association, how does an exact confidence interval for the common odds ratio compare to an asymptotic one?

22. A lawsuit was filed in 1975 against the State of Mississippi Department of Justice accusing it of discriminating against black lawyers and black support personnel who applied for positions in the state. The following table gives the number of people hired and number of applicants for the years 1973–1982 (Finkelstein and Levin, 2001, page 246).

Year	White	Black	Year	White	Black
1973	10/21	0/2	1978	8/18	0/1
1974	6/18	0/7	1979	6/15	1/4
1975	2/15	0/7	1980	7/27	3/12
1976	3/16	2/5	1981	4/21	1/5
1977	13/19	1/4	1982	1/4	3/13

Is race conditionally independent of hiring status given year? What would that imply about the merits of the lawsuit? Is there any evidence of a change in hiring pattern over time? Does a conditional (exact) analysis give different results from an asymptotic one?

9
Regression Models for Binary Data

9.1 The Logistic Regression Model

9.1.1 Why Logistic Regression?

The loglinear models of the previous four chapters are designed for count data, where a Poisson or multinomial distribution is appropriate. The most basic form of categorical data, however, is binary — 0 or 1. It is often of great interest to try to model the probability of success (the outcome coded 1) or failure (the outcome coded 0) as a function of other predictors. Consider a study designed to investigate risk factors for cancer. Attributes of people are recorded, including age, gender, packs of cigarettes smoked, and so on. The target variable is whether or not the person has lung cancer (a 0/1 variable, with 0 for no lung cancer and 1 for the presence of lung cancer). A natural question is then "What factors can be used to predict whether or not a person will have lung cancer?" Substitute businesses for people, financial characteristics for medical risk factors, and whether a company went bankrupt for whether a person has cancer, and this becomes an investigation of the question "What financial characteristics can be used to predict whether or not a business will go bankrupt?" The models of this chapter are designed to address such questions.

Let $p|\mathbf{x}$ be the probability of success given a set of k predictor values $\mathbf{x}$ (we use k instead of p for the number of predictors in this chapter and the next chapter, to avoid confusion with the probability p). The first question to consider is the form of $p|\mathbf{x}$. We might propose a linear model for p, consistent with

$$p|\mathbf{x} = \beta_0 + \beta_1 x_1 + \cdots + \beta_k x_k,$$

but is this reasonable? Consider the following situation. We wish to model the probability of an applicant for a \$100,000 mortgage having their application accepted as a function of their family income. There are three distinct kinds of situations:

1. For applicants with a large income, there comes a point where the probability of being approved is so high that additional income adds little to the probability of being approved. In other words, the probability curve as a function of income levels off for high incomes.

2. The situation is similar for low incomes. At some point the probability of being approved is so low that \$1000 less income subtracts little from the probability of being approved. In other words, the probability curve as a function of income levels off for low values of income.

3. For people with a moderate family income (say around \$50,000 annually), we might believe that each additional thousand dollars of income is associated with a fixed (constant) increase in the probability of being approved. That is, a linear relationship in this range is reasonable.

These three patterns suggest that the probability curve is likely to have an S-shape (a sigmoid). Figure 9.1 provides empirical support for this idea. It is a plot of the proportion of girls menstruating by age, from a sample of 3918 Warsaw girls in 1965 (Milicer and Szczotka, 1966), and the S-shape of the empirical proportions is apparent.

There are many different functions that can lead to such an S-shaped curve, so how can one be chosen? This is where some of the results of previous chapters can be helpful. Let y_i be the result for the ith observation in a sample of size N, where $y_i = 1$ for a success and $y_i = 0$ for a failure. A reasonable model for data of this type is that $(y_i|\mathbf{x}_i) \sim \text{Bin}(1, p_i|\mathbf{x}_i)$. A generalization of this model is to allow for multiple trials n_i for each set of $\mathbf{x}_i$ values, and record the number of successes y_i in the n_i trials. In that case, $(y_i|n_i, \mathbf{x}_i) \sim \text{Bin}(n_i, p_i|\mathbf{x}_i)$.

The log-likelihood for the sample is

$$\sum_{i=1}^{N} \left[\log \binom{n_i}{y_i} + y_i \log p_i + (n_i - y_i) \log(1 - p_i) \right]. \tag{9.1}$$

Rewriting this as

$$\sum_{i=1}^{N} \left[y_i \log \left(\frac{p_i}{1 - p_i} \right) + n_i \log(1 - p_i) + \log \binom{n_i}{y_i} \right],$$

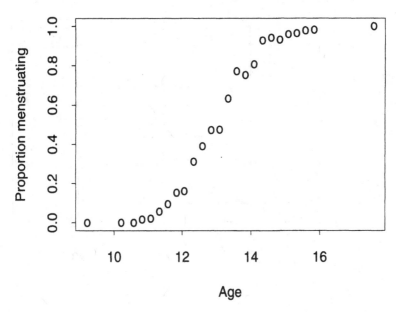

FIGURE 9.1. Proportion of Warsaw girls menstruating at different ages.

and referring to (5.1), shows that the binomial is a member of the exponential family, with canonical parameter the log-odds, or logit,

$$\theta = \log\left(\frac{p}{1-p}\right).$$

If it is assumed that the predictors are related to p_i through the linear predictor

$$g(p_i) = \beta_0 + \beta_1 x_{1i} + \cdots + \beta_k x_{ki},$$

this is a generalized linear model, and the link function g determines the shape of the relationship between p and $\mathbf{x}$ (more correctly, the inverse of the link function determines the shape).

Recall that the canonical link function equates the linear predictor to the canonical parameter, which leads to the *logistic linear regression model*,

$$\log\left(\frac{p_i}{1-p_i}\right) = \beta_0 + \beta_1 x_{1i} + \cdots + \beta_k x_{ki}.$$

The inverse of the logit link,

$$p_i = \frac{\exp(\beta_0 + \beta_1 x_{1i} + \cdots + \beta_k x_{ki})}{1 + \exp(\beta_0 + \beta_1 x_{1i} + \cdots + \beta_k x_{ki})} \tag{9.2}$$

has the desired S-shape. Figure 9.2 illustrates the logistic regression model with one predictor, and can be compared to Figures 2.2 and 5.1. The shaded

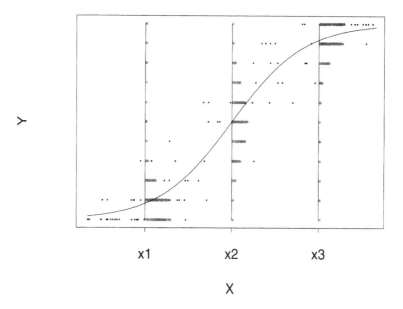

FIGURE 9.2. The logistic regression model

bars represent the binomial probability distributions based on $n_i = 10$ trials at three different values of x. The relationship between p and x is S-shaped, and when p is close to zero or one the distribution of $y|x$ is quite asymmetric. It is also apparent that variability is highest when p is close to .5, and decreases as p gets closer to zero or one. Different choices of β give curves that are steeper (larger $|\beta_1|$) or flatter (smaller $|\beta_1|$), or shaped as an inverted (backwards) S (negative β_1).

The logistic regression model is a generalized linear model, and as such, all of the methods for inference and model checking discussed in previous chapters apply. Estimation of β is based on maximum likelihood. Being based on the canonical link function for the binomial random variable, logistic regression has the advantages that go with that link, as described in Section 5.1.1. The logistic model implies that the log of the odds follows a linear model. Equivalently, β_1 has the following interpretation: a one unit increase in x_1 is associated with multiplying the odds of success by $\exp(\beta_1)$, holding all else fixed; that is, $\exp(\beta_1)$ is the odds ratio of success relative to failure for predictor values $\mathbf{x}_0^*$ versus $\mathbf{x}_0$, where the two sets of predictor values differ only in that x_1 is one unit higher in $\mathbf{x}_0^*$ than in $\mathbf{x}_0$. The odds of success when all predictor values equal zero is obtained from the constant term as $\exp(\beta_0)$.

The logit link also has an important advantage that comes from the odds ratio interpretation of the slope coefficients. As was noted in Section 6.1.3, the odds ratio from a population is unchanged whether observations are sampled prospectively or retrospectively. So, for example, if an investigation into a medical treatment is based on a case-control (retrospective) study, the estimated slope coefficients in a logistic regression estimate the relationships between the variables and success (in terms of odds ratios) from a prospective point of view (as if the study had been a cohort study). Given the added time and expense that often go with a cohort study, this is a major advantage for logistic regression over other choices of link function. If the probability of the outcome of interest is small, the odds ratio for a specific variable x_j given the other predictors, $\exp(\hat{\beta}_j)$, is close to the relative risk, a number that is otherwise not estimable from a retrospective study without more information (see Section 6.1.3).

9.1.2   Inference for the Logistic Regression Model

The usual inferential techniques for generalized linear models naturally apply to the logistic regression model. A test of the overall strength of the regression tests the hypotheses

$$H_0 : \beta_1 = \cdots = \beta_k = 0$$

versus

$$H_a : \beta_j \neq 0 \text{ for at least one } j.$$

As usual, the likelihood ratio test is based on comparing the quality of the fit without any predictors to the quality of the fit using predictors, as measured by difference in the deviance for the models with and without predictors. Let $y_i \sim \text{Bin}(n_i, p_i)$, where p_i satisfies the logistic regression model. The N distinct values of the predictors are called the *covariate patterns* in the data. The likelihood ratio test for the overall significance of the regression is

$$LR = 2 \sum_{i=1}^{N} \left[y_i \log \left(\frac{\hat{p}_i^a}{\hat{p}_i^0} \right) + (n_i - y_i) \log \left(\frac{1 - \hat{p}_i^a}{1 - \hat{p}_i^0} \right) \right],$$

where $\hat{\mathbf{p}}^a$ are the estimated probabilities based on the fitted logistic regression model, and $\hat{\mathbf{p}}^0$ are the estimated probabilities under the null hypothesis. This can be compared to a χ^2 distribution on k degrees of freedom, as long as either N is large, or the n_i values are reasonably large. Any set of hypotheses where the null is a (linear) special case of the alternative also can be tested using the likelihood ratio test as a difference of deviance values, with the appropriate degrees of freedom being the difference in the number of parameters estimated in the two models. As is true in general for such tests for generalized linear models, they usually follow assumed χ^2 distributions reasonably well, even if the underlying deviances do not.

Two tests of the additional predictive power provided by an individual predictor are easily available. These are tests of the hypotheses

$$H_0 : \beta_j = 0$$

versus

$$H_a : \beta_j \neq 0.$$

These hypotheses can be tested using the likelihood ratio test form described above, where $\hat{\mathbf{p}}^0$ is calculated based on all predictors except the jth predictor and $\hat{\mathbf{p}}^a$ is calculated based on all predictors, but this requires fitting $k + 1$ different models. The alternative (and standard) approach is to use the Wald test statistic,

$$z_j = \frac{\hat{\beta}_j}{\widehat{\text{s.e.}}(\hat{\beta}_j)},$$

where $\widehat{\text{s.e.}}(\hat{\beta}_j)$ is calculated based on the iteratively reweighted least squares approximation to the maximum likelihood estimate. These statistics are then compared to a Gaussian distribution to determine significance. Similarly, confidence intervals for individual coefficients take the form $\hat{\beta}_j \pm z_{\alpha/2}\widehat{\text{s.e.}}(\hat{\beta}_j)$.

Logistic regression also can be used for prediction of group membership, or *classification*. A fitted model (with estimated $\hat{\boldsymbol{\beta}}$) can be applied to new data to give an estimate of the probability of success for that observation using (9.2). If a simple success/failure prediction is desired, this can be done by calling an estimated $\hat{p}$ less than .5 a predicted failure, and an estimated $\hat{p}$ greater than .5 a success, although in specific situations other choices might be preferable. If this process is applied to the same data used to fit the logistic regression model, a *classification table* results. Here is a hypothetical example of such a table:

Predicted result

		Success	Failure	
Actual	Success	6	1	7
result	Failure	2	6	8
		8	7	

In this example 80% of the observations (12 out of 15) were correctly classified to the appropriate group. Is that more than we would expect just by random chance? The answer to this question is not straightforward, because we have used the same data to both build the model (getting $\hat{\boldsymbol{\beta}}$) and evaluate its ability to do predictions (that is, we have used the data twice). For this reason, the observed proportion correctly classified can be expected to be biased upwards compared to the situation where the model is applied to completely new data.

What can be done about this? The best solution is to do what is implied by the last line in the previous paragraph — get some new data, and apply the fitted model to it to see how well it does (that is, validate the model on new data). If no new data are available, two diagnostics have been suggested that can be helpful. One approach is to determine the proportion that would be right if every observation had been predicted to have come from the larger group (termed C_{max}), since performance using such a simplistic strategy would have to be considered a lower bound for performance overall. For these data, this is

$$C_{max} = \max\left(\frac{7}{15}, \frac{8}{15}\right) = 53.3\%.$$

The observed 80% correctly classified is considerably larger than C_{max}, supporting the usefulness of the (hypothetical) logistic regression.

A more sophisticated argument is as follows. If the logistic regression had no power to make predictions, the actual result would be independent of the predicted result. That is, for example,

P(Actual result a success and Predicted result a success) =

$\quad P$(Actual result a success) $\times$ P(Predicted result a success).

The right side of this equation can be estimated using the marginal probabilities from the classification table, yielding

P(Actual result a success and Predicted result a success)

$$= \left(\frac{7}{15}\right)\left(\frac{8}{15}\right) = 24.9\%.$$

That is, we would expect to get 24.9% of the observations correctly classified as successes just by random chance. A similar calculation for the failures gives

P(Actual result a failure and Predicted result a failure)

$$= \left(\frac{8}{15}\right)\left(\frac{7}{15}\right) = 24.9\%.$$

Thus, assuming that the logistic regression had no predictive power for the actual result (success or failure), we would expect to correctly classify $24.9\% + 24.9\% = 49.8\%$ of the observations. This still doesn't take into account that we've used the data twice, so this number is typically inflated by 25% before being reported, resulting the C_{pro} measure; that is,

$$C_{pro} = 1.25 \times 49.8\% = 62.2\%.$$

The observed 80% is considerably higher than C_{pro}, which would support the usefulness of the logistic regression as a classifier.

9.1.3   Model Fit and Model Selection

As is the case in any regression model fitting, it is important to check assumptions and assess the adequacy of a logistic regression fit. In principle, this is a straightforward application of the principle described in Section 5.1.6, basing residuals, diagnostics, and so on, on the iteratively reweighted least squares form of the maximum likelihood estimate. So, for example, the Pearson residuals have the form

$$r_i^P = \frac{y_i - n_i \hat{p}_i}{\sqrt{n_i \hat{p}_i (1 - \hat{p}_i)}},$$

and can be used to identify outlying observations.

There is, however, a subtlety in these constructions that we have not seen before. Consider again the log-likelihood for the data, (9.1). Rather than thinking of the data as a collection of N observations, each reflecting y_i successes and $n_i - y_i$ failures, an alternative way to think of the data is as a collection $\sum_i n_i$ decomposed observations, each being either a success or a failure (that is, each is Bernoulli distributed). Since a binomial random variable can be represented as a sum of independent and identically distributed Bernoulli random variables, decomposing each binomial y_i into n_i Bernoullis this way is not unreasonable. It is easy to see that, except for the constant term that is not a function of **p**, the log-likelihoods for these two representations are identical. Thus, the maximum likelihood estimate for β is the same, as are confidence intervals and likelihood ratio and Wald tests of hypotheses.

Regression diagnostics, on the other hand, change completely from one representation to the other. Consider, for example, an observation in the original formulation where there were three successes in ten trials, and $\hat{p} = .15$. This observation has a Pearson residual equal to

$$r^P = \frac{3 - (10)(.15)}{\sqrt{(10)(.15)(.85)}} = 1.33,$$

indicating a reasonable fit to the observation. If the observation is decomposed into its ten constituent Bernoulli parts, however, the result is ten residuals, three of which equal

$$r^P = \frac{1 - (1)(.15)}{\sqrt{(1)(.15)(.85)}} = 2.38$$

and seven of which equal

$$r^P = \frac{0 - (1)(.15)}{\sqrt{(1)(.15)(.85)}} = -0.42,$$

with clearly different fit implications.

The usual measures of goodness-of-fit, X^2 and G^2, are similarly affected by the way the data are represented. The Pearson statistic X^2 is the sum of squares of the Pearson residuals, which obviously changes with the change in representation. The deviance G^2 depends on the saturated model, which assigns a separate parameter to each observation. In the original formulation, there are N observations, and the saturated model has fitted values $\hat{p}_i = \bar{p}_i = y_i/n_i$, the observed proportion of successes in the ith observation. This leads to deviance

$$G^2 = 2 \sum_i \left[y_i \log \left(\frac{\bar{p}_i}{\hat{p}_i} \right) + (n_i - y_i) \log \left(\frac{1 - \bar{p}_i}{1 - \hat{p}_i} \right) \right]. \qquad (9.3)$$

The decomposed observation form of the data is based on $\sum_i n_i$ observations, and for each the fitted value under the saturated model is the observed 0 or 1, yielding a deviance that can be written

$$G^2 = -2 \sum_i [y_i \log \hat{p}_i + (n_i - y_i) \log(1 - \hat{p}_i)].$$

Once again, this is completely different from (9.3).

The issue comes down to how one defines covariate patterns. Three possibilities have been suggested, and are used in common statistical software:

1. Determine the covariate patterns based on the underlying process that generated the data. So, if each y_i was physically generated as a distinct binomial random variable, these constitute the covariate patterns.

2. Determine the covariate patterns based on the distinct values of all of the potential predictors in the regression model.

3. Determine the covariate patterns based on the distinct values of only the predictors used in the regression model. This is similar to method (2), except that the set of covariate patterns changes depending on what predictors are used in the model. Note that this implies that different models fit to the same data would not necessarily be comparable regarding fit, since the structure of the measures of fit changes with the changing covariate patterns.

It is easy to argue in favor of approach (1), assuming the underlying process that generated the data is known. Consider the following representation of the data in this situation.

Predictors	Successes	Failures
$x_{11}, \ldots, x_{1k}$	y_1	$n_1 - y_1$
$x_{21}, \ldots, x_{2k}$	y_2	$n_2 - y_2$
$\vdots$		
$x_{N1}, \ldots, x_{Nk}$	y_N	$n_N - y_N$

This represents the response data as an $N \times 2$ contingency table, with the covariate patterns determining the rows, and success or failure determining the columns. First, consider a situation with multiple replications at each covariate pattern; that is, the $\{n_i\}$ are reasonably large. In this situation, the usual goodness-of-fit tests for contingency tables apply. In particular, the likelihood ratio test for goodness of fit compares the $\{y_i\}$ success values to $n_i \hat{p}_i$ and the $\{n_i - y_i\}$ failure values to $n_i(1 - \hat{p}_i)$, and yields the statistic (9.3). There are $k + 1$ parameters fit in the logistic regression model, including the intercept; in addition, once the fitted successes are calculated the fitted failures are known (since they sum to n_i for the ith covariate pattern), so an additional N parameters (corresponding to the $n_i(1 - \hat{p}_i)$ values) are fit. Thus, the G^2 statistic is compared to a χ^2 distribution on $2N - N - k - 1 = N - k - 1$ degrees of freedom, and with large values of $\{n_i\}$ the χ^2 approximation is reasonable.

The connection between logistic regression and the contingency table form given above suggests that approach (1) to goodness-of-fit is best, while approach (3), where the covariate patterns are determined by the variables actually in the model, is inappropriate. In all of the models for contingency tables considered in the previous three chapters, the form of the table (and hence the form of the saturated model) did not change depending on the terms actually included in the model, so it does not seem reasonable for it to change in the logistic regression context. Further connections between logistic regression and contingency tables will be explored in Section 9.3. The second approach, where covariate patterns are determined by the set of potential predictors in the model, is a reasonable compromise when the underlying probability mechanism is unclear, or where the initial set of predictors is very large, and it is desirable to preliminarily find a subset of the predictors that are "potential predictors."

What if there are few (or no) replications at individual covariate patterns? An obvious example of this is when there is only a single binary outcome for each set of predictors (for example, an individual patient living or dying given his or her medical condition), implying that $n_i = 1$ and y_i equals either zero or one. The contingency table form of the data shows that standard χ^2 goodness-of-fit tests are inappropriate in this situation, since the cell fitted values $\{n_i \hat{p}_i\}$ and $\{n_i(1 - \hat{p}_i)\}$ are too small. In fact, it can be shown that the distribution of G^2 is degenerate given $\hat{\beta}$ in this situation, which implies that G^2 provides no information at all about goodness-of-fit.

An alternative goodness-of-fit test has been proposed for this circumstance. The *Hosmer–Lemeshow* statistic has the same form as X^2, but the underlying cells on which the test is based are chosen in an adaptive way. In this test, the observations are ordered by the fitted probabilities $\hat{p}_i$. Then, the data are formed into g groups (typically $g = 10$) of roughly equal size based on the smallest $1/g$ of the fitted probabilities, the next $1/g$, and so on. For each group, the expected number of successes in the group is the sum of the fitted probabilities for that group, and a Pearson-type statistic can be constructed that compares the observed number of successes in each group to the expected number. This is then compared to a χ^2 distribution on $g - 2$ degrees of freedom (if the number of expected successes in a group is (essentially) zero, the group is not used in determining the statistic). Different statistical packages use different methods of resolving ties, and some set g less than ten if there are a small number of distinct covariate patterns in the data.

The same issues that arise in selecting appropriate models for Poisson regression and contingency table data arise in the logistic regression context, and the same approaches are available. Models can be compared using AIC or AIC_C, as in (5.12) and (5.13), respectively, and subset tests can be used to compare models if one is a special case of the other.

The Flight of the Space Shuttle Challenger

On January 28, 1986, the space shuttle *Challenger* took off on the 25th flight in the space shuttle program of the National Aeronautics and Space Administration (NASA). Less than 2 minutes into the flight, the spacecraft exploded, killing all on board. A Presidential Commission headed by former Secretary of State William Rogers was formed to explore the reasons for this disaster.

First, a little background information: the space shuttle uses two booster rockets to help lift it into orbit. Each booster rocket consists of four pieces whose joints (called field joints) are sealed with rubber O-rings, which are designed to prevent the release of hot gases produced during combustion. Each booster contains three primary O-rings (for a total of six for the orbiter). In the 23 previous flights for which there were data (the hardware for one flight was lost at sea), the O-rings were examined for damage.

One interesting question is the relationship of O-ring damage to temperature. It was (forecasted to be) cold — 31 °F — on the morning of January 28, 1986, which would have been 22° colder than for any previous flight. It is well known that O-rings are sensitive to temperature. A warm O-ring will quickly recover its shape when compression is removed, but a cold one will not. This can lead to joints not being sealed, and can result in a fuel leak. The Rogers Commission determined that it was the combustion of this leaking fuel that led to *Challenger*'s destruction. There was a good deal of discussion among people at Morton Thiokol (the manufacturers of the solid rocket motor), NASA's Marshall Space Flight Center, and the Kennedy Space Center the night before the flight as to whether the flight should go on as planned or not because of the cold weather and the possibility of its effect on O-ring reliability. A simplified version of one of the arguments made is as follows. There were 7 previous flights where there was damage to at least one primary O-ring. Consider the following table.

Ambient temperature	Proportion of O-rings damaged
53°	.333
57°	.167
58°	.167
63°	.167
70°	.167
70°	.167
75°	.333

Based on this table, there is no apparent relationship between temperature and the probability of damage, since higher damage occurred at both lower and higher temperatures. Thus, the fact that it was going to be cold on the day of the flight wouldn't seem to imply that the flight should be scrubbed. (In fact, this table was not actually constructed the night of January 27th, but was rather given later by two Thiokol staff members as an example of the reasoning in the prelaunch debate. The actual charts faxed from the Thiokol engineers to NASA that night were considerably less informative than even this seriously flawed table. See Chapter 2 of Tufte (1997) for more details on the charts actually used.)

Unfortunately, this analysis is completely inappropriate. The problem is that it is ignoring the 16 flights where there was no O-ring damage, acting as if there is no information in those flights, which is not reasonable. If flights with high temperatures **never** had O-ring damage, for example, that would certainly tell us a lot about the relationship between temperature and O-ring damage! The top plot in Figure 9.3 gives a scatter plot of the proportion of primary O-rings damaged versus the ambient temperature for all of the launches. The plot is based on data given in Dalal et al. (1989). It is apparent that, except for one flight, there is an inverse relationship between temperature and O-ring damage. A plot of this kind would certainly have raised some alarms as to the advisability of launching the shuttle (recall

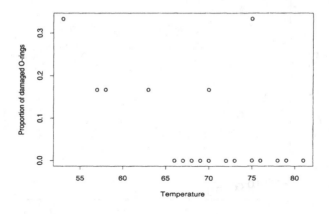

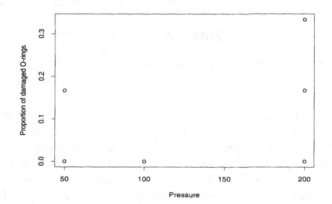

FIGURE 9.3. Scatter plots of *Challenger* data. Proportion of O-rings damaged versus temperature (top plot) and proportion of O-rings damaged versus pressure (bottom plot).

that this plot is based only on data from previous flights). Unfortunately, such a plot was never constructed.

The second plot in Figure 9.3 relates the proportion of O-rings damaged to the pressure setting (in pounds per square inch) used in a pressure test performed after assembly. While it is possible that using a higher pressure in the test could cause damage to the O-ring (or the putty that surrounds it), there is no evidence of this in the plot. It should be remembered, of course, that (as is true in any regression model) these marginal relationships might not reflect the predictive power (and relationships) in a joint prediction of O-ring damage from both variables.

The natural way to view the underlying process of damaged O-rings is as a set of 23 binomial random variables, each based on six trials (the

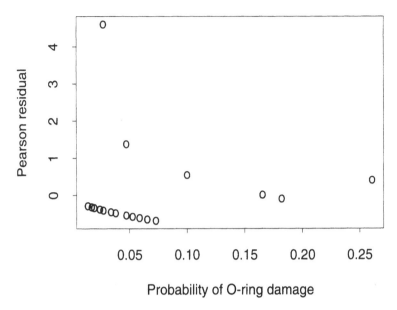

Probability of O-ring damage

FIGURE 9.4. Plot of Pearson residuals versus estimated probabilities of O-ring damage.

six primary O-rings). A logistic regression fit using both predictors fits the data well ($G^2 = 16.5$ on 20 degrees of freedom, $p = .69$). Note that a χ^2 approximation for the deviance is marginally reasonable here, with six replications per covariate pattern. The Hosmer–Lemeshow statistic also indicates a reasonable fit ($HL = 11.0$ on 8 degrees of freedom, $p = .20$). The Wald statistic for the pressure variable is insignificant ($z = 1.12$). The model using only ambient temperature is preferred to the model using both ($G^2 = 18.1$ on 21 degrees of freedom, and the value of AIC_C for the latter model is 0.55 lower).

Figure 9.4 is a plot of the Pearson residuals for this model versus the estimated probabilities of O-ring failure. It is apparent that the unusual flight noted earlier is an outlier (the stripes of observations in the plot comes from the fact that the number of damaged O-rings only takes on three values in the data; see Section 9.6). This observation corresponds to the flight of the *Challenger* from October 30 through November 6, 1985. In fact, the O-ring damage on this flight was different from that in the other flights in that the two O-rings damaged in that flight suffered only "blow-by" damage (where hot gases rush past the O-ring), while in all of the other flights damaged O-rings suffered "erosion" (where the O-rings burn up), (possibly) as well as blow-by.

It is therefore reasonable to address this flight by either changing the number of damaged O-rings from two to zero (and then note that by "damage" we mean only erosion damage), or by omitting it from the data set. We

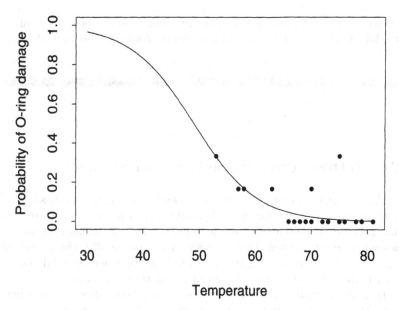

FIGURE 9.5. Plot of *Challenger* data with the fitted logistic regression curve overlaid.

will do the latter here. Once again the pressure variable does not add any significant predictive power ($z = 0.35$), while the model using only temperature fits the data well ($G^2 = 9.4$ on 20 degrees of freedom, $HL = 5.3$ on 7 degrees of freedom). Examination of residuals shows nothing unusual. The likelihood ratio test of the significance of the relationship between O-ring damage and temperature is highly significant ($LR = 10.7$ on one degree of freedom), as is the Wald test for the slope of the regression ($z = 3.0$).

The fitted relationship between O-ring damage probability and ambient temperature is

$$\log\left(\widehat{\frac{p}{1-p}}\right) = 8.662 - .177 \times \text{Temperature}.$$

The estimated slope coefficient implies that each increase in temperature of one degree Fahrenheit is associated with an estimated multiplication of the relative odds of an O-ring damage by $\exp(-.177) = 0.838$, or a 16% decrease.

And what about the morning of January 28, 1986? Substituting into the logistic function gives a probability estimate of O-ring damage for a temperature of 31° of .96. Figure 9.5 gives the O-ring damage/Temperature portion of the data set, with the estimated logistic curve overlaid. The fit for 31° is an extrapolation, but the implications are obvious. Indeed, with the benefit of hindsight, it can be seen that the *Challenger* disaster was not at all surprising, given data that were available at the time of the flight. As a result of its investigations, one of the recommendations of the commission was that a statistician be part of the ground control team from that time on.

<hr>

9.2   Retrospective (Case-Control) Studies

Recall again the discussion of prospective and retrospective studies of bankruptcy discussed in Section 6.1.3. It might seem that a retrospective sampling scheme is a decidedly inferior choice, since we cannot answer the question of most interest; that is, what is the probability that a business will go bankrupt given its initial debt level? More generally, while (as was noted earlier) the slope coefficients in a logistic regression are equivalent whether the sampling scheme is prospective or retrospective, and hence inferences about predictors are meaningful in either situation, it isn't possible to construct an estimate of the probability of success from a logistic regression model based on a retrospective study, using only the observed data.

There is, however, a way that we can answer this question, if we have additional information. Let π_Y and π_N be the true (unconditional) probabilities that a randomly chosen business goes bankrupt or does not go bankrupt, respectively. These numbers are called *prior probabilities.* Consider the probability of bankruptcy given a low initial debt level, $p_{Y|L}$. By the definition of conditional probabilities,

$$
\begin{aligned}
p_{Y|L} &= p_{LY}/p_L \\
&= p_{L|Y}\pi_Y/p_L \\
&= p_{L|Y}\pi_Y/(p_{LY} + p_{LN}) \\
&= p_{L|Y}\pi_Y/(p_{L|Y}\pi_Y + p_{L|N}\pi_N).
\end{aligned}
$$

The probabilities $p_{L|Y}$ and $p_{L|N}$ are estimable in a retrospective study, so if we have a set of prior probabilities (π_Y, π_N) we can estimate the probability of interest.

In order to estimate the probability of success from a retrospective study, the constant term in the fitted logistic regression is adjusted. The common approach to this problem is as follows. If the prior probabilities are π_Y (for success) and π_N (for failure), the adjusted constant term has the form

$$
\tilde{\beta}_0 = \hat{\beta}_0 + \log\left(\frac{\pi_Y n_N}{\pi_N n_Y}\right), \tag{9.4}
$$

where n_Y (n_N) is the total number of successes (failures). Thus, probabilities of success are estimated for a retrospective study using (9.2), with β_0 estimated using $\tilde{\beta}_0$ (and the rest of the coefficients using the usual maximum likelihood estimates).

Predicting Bankruptcy in the Telecommunications Industry

Understanding and predicting bankruptcy has always been important — now more than ever, given the failure in recent years of multibillion-dollar enterprises including Adelphia Communications Corporation, Enron Corporation, and WorldCom, Inc. Effective bankruptcy prediction is useful for investors and analysts, allowing for accurate evaluation of a firm's prospects. Historically, discriminant analysis has been used for this purpose, but this is a natural logistic regression problem.

The following analysis is based on a retrospective sample of 25 telecommunications firms that declared bankruptcy between May 2000 and January 2002 that had issued financial statements for at least two years, and information from the December 2000 financial statements of 25 telecommunications that did not declare bankruptcy (data given in the file bankruptcy). I am indebted to Jeffrey Lui for gathering these data. The nonbankrupt firms were chosen to try to match asset sizes with the bankrupt firms, although no formal matching has been done here. Five financial ratios were chosen as potential predictors of bankruptcy:

1. Working capital as a percentage of total assets (WC/TA, expressed as a percentage). Working capital is the difference between current assets and liabilities, and is thus a measure of liquidity as it relates to total capitalization. Firms on the road to bankruptcy would be expected to have less liquidity.

2. Retained earnings as a percentage of total assets (RE/TA, expressed as a percentage). This is a measure of cumulative profitability over time, and is thus an indicator of profitability, and also of age. A younger firm is less likely to be able to retain earnings, since it would reinvest most, if not all, of its earnings in order to stimulate growth. Both youth and less profitability would be expected to be associated with an increased risk of insolvency.

3. Earnings before interest and taxes as a percentage of total assets (EBIT/TA, expressed as a percentage). This is a measure of the productivity of a firm's assets, with higher productivity expected to be associated with a healthy firm.

4. Sales as a percentage of total assets (S/TA, expressed as a percentage). This is the standard capital turnover ratio, indicating the ability of

a firm's assets to generate sales; lower sales would be expected to be associated with unhealthy prospects for a firm.

5. Book value of equity divided by book value of total liabilities (BVE/ BVL). This ratio measures financial leverage, being the inverse of the debt to equity ratio. A smaller value is indicative of the decline of a firm's assets relative to its liabilities, presumably an indicator of unhealthiness. While it is typical in bankruptcy studies to use the market value of equity in this ratio, the "Internet bubble" of the late 1990s makes this problematic. It was not at all unusual during this time period for so-called dot-coms to have very high stock prices that collapsed within a matter of months, making the market value of equity unrealistically high before the collapse.

A good way to get a feeling for the predictive power of individual variables in this type of situation, where the target variable only takes on the values 0 (not bankrupt) or 1 (bankrupt), is to construct side-by-side boxplots, to see if there is separation between the two groups on the variables. This does not take into account the variables having joint effects, and doesn't necessarily imply that a linear logistic model is appropriate, but is still helpful.

Figure 9.6 gives side-by-side boxplots for these data. The working capital, retained earnings, and earnings variables all show clear separation between bankrupt and nonbankrupt firms, in the ways that would have been expected. The sales variable shows less predictive power, with the bankrupt firms actually having the highest values of sales as a percentage of assets. Nonbankrupt firms have a generally higher equity to liabilities ratio (lower debt to equity), although the long right tail of the variable make this a little harder to see. Note that while there is no assumption being made here on the distributions of the predictors, the long tail in this variable does suggest that logging this variable might be helpful. In fact, for these data the unlogged variable works just as well, so we won't pursue this further.

A logistic regression model fit to these data strongly supports the predictive power of the financial ratios ($LR = 57.5$ on 5 degrees of freedom), and the model fits the data ($HL = 8.4$ on 8 degrees of freedom; note that since there is only one replication per covariate pattern, the deviance and Pearson goodness-of-fit statistics are not meaningful here). Each of the Wald statistics is less than 1.5 in absolute value, however, indicating that this model is overspecified. The S/TA variable is clearly not needed, as the model omitting it preserves virtually all of the predictive power ($LR = 56.4$) using one less variable. Once this variable is omitted, things get complicated, since omitting any other variable reduces the fit in at least a marginally significant way.

There is a more fundamental problem here anyway. Figure 9.7 is an index plot of the Pearson residuals for this model. It is clear that the first

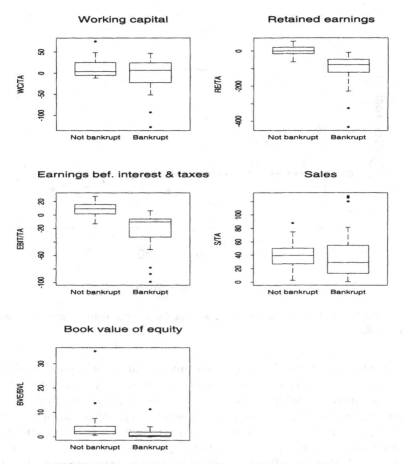

FIGURE 9.6. Side-by-side boxplots of bankruptcy data.

observation is an outlier. This observation corresponds to the firm 360Networks. This firm was in the business of building computer networks, and was one of only two firms that ultimately went bankrupt that had positive earnings the year before insolvency. Its retained earnings as a proportion of total assets were also not very negative (as would be expected for a bankrupt firm), but part of this could be from the nature of its business; the thousands of miles of cable that it owned resulted in the firm having $6.3 billion in total assets only three months before it declared bankruptcy, making RE/TA less negative.

If this observation is removed, a new problem arises. The model using all five predictors fits perfectly — that is, the deviance is zero. This might seem like a good thing, but consider again the form of the data. With only one replication for each covariate pattern, the only way that the model can fit perfectly is if all observations have fitted probabilities of 0 (for nonbankrupt

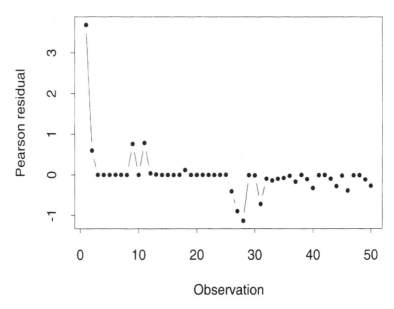

FIGURE 9.7. Index plot of Pearson residuals for logistic regression fit to bankruptcy data.

companies) and 1 (for bankrupt companies). This is only possible if the fitted logit is $\pm\infty$. That is, the zero deviance is actually highlighting that the maximum likelihood estimate of at least one of the slope parameters is infinite in absolute value. Naturally, a computer algorithm cannot yield an infinite parameter estimate. Rather, what typically occurs in this situation is that one or more of the estimated slopes are very large (in absolute value), and all of the estimated standard errors are very large, resulting in Wald statistics that are close to zero (as Mantel, 1987, noted, the problem is related to the Wald statistic's use of parameter estimates, rather than null values, to estimate the variances of the slope estimates, when an alternative hypothesis far from the null is actually true).

A solution to this problem is to find a simpler model that fits almost as well, but where the maximum likelihood estimates are finite. The model dropping the sales variable S/TA still has zero deviance, so we are left with models with three or fewer predictors. We will focus on the three variable model that includes RE/TA, EBIT/TA, and BVE/BVL. This model fits the data very well ($HL = 0.5$ on 8 degrees of freedom), and accounts for a highly significant amount of predictive power ($LR = 58.5$ on 3 degrees of freedom). The fitted model is outlined in the output on the next page. The predictive power of the overall model is actually quite strong, but each of the Wald statistics is only weakly significant. The asymptotically equivalent likelihood ratio tests for each of the three predictors are noticeably different from the Wald tests here, and much more consistent with the observed

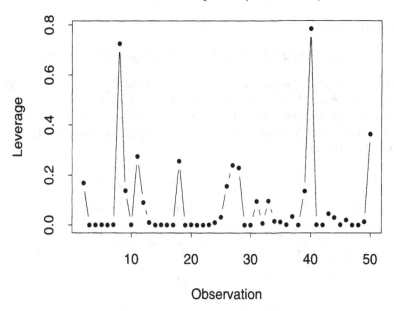

FIGURE 9.8. Index plot of leverage values for logistic regression fit to bankruptcy data omitting 360Networks. Index numbers refer to full sample numbers.

strength of the regression (the test statistics are 11.5, 5.5, and 10.8, each χ_1^2 distributed, and each highly statistically significant).

Predictor	Coef	s.e.	Wald	p
Constant	-0.092	1.471	-0.06	0.950
RE/TA	-0.082	0.042	-1.95	0.052
EBIT/TA	-0.268	0.159	-1.69	0.091
BVE/BVL	-1.218	0.765	-1.59	0.111

All three coefficients are negative, as would be expected. Exponentiating the coefficients gives the partial odds ratios. For example, $\exp(-.082) = .92$, implying that a one percentage point increase in retained earnings as a proportion of total assets is associated with an 8% decrease in the odds of a firm going bankrupt, given earnings before interest and taxes (as a proportion of total assets) and book value of equity (as a proportion of book value of liabilities). Regression diagnostics do not indicate any outliers, although there are two leverage points, IDT Corporation and eGlobe (see Figure 9.8).

It is hard to imagine that omitting these observations could lead to a much better fit. For example, if the model is used to classify the observations as bankrupt or not bankrupt on the basis of the estimated probabilities being above or below .5, the classification table on the following table results. Forty-seven of 49, or 95.9%, of the firms were correctly classified, far higher than

$$C_{\text{pro}} = (1.25)[(.4898)(.4898) + (.5102)(.5102)] = 62.5\%$$

and $C_{\text{max}} = 51.0\%$, reinforcing the strength of the logistic regression. Thus, the three financial ratios do an excellent job of classifying the firms into the bankrupt and nonbankrupt groups (even considering that 360Networks would have been misclassified, 94% of the original firms would have been correctly classified).

Predicted result

		Bankrupt	Not bankrupt	
Actual	Bankrupt	23	1	24
result	Not bankrupt	1	24	25
		24	25	

This fitted logistic regression can be used to estimate prospective probabilities of bankruptcy by adjusting the constant term of the regression using prior probabilities of bankruptcy. The first step is to choose this prior probability; we can use a 10% bankruptcy probability, which is roughly consistent with what would be expected for firms with a corporate bond rating of B. This yields the adjusted intercept

$$\tilde{\beta}_0 = \hat{\beta}_0 + \log\left[\frac{(.10)(25)}{(.90)(24)}\right] = \hat{\beta}_0 - 2.1564.$$

The plots in Figure 9.9 show how the estimated probabilities are related to each of the predictors for the observed firms; the strong marginal discriminating power of RE/TA and EBIT/TA is particularly clear. That is, earnings figures seem to be the best marginal discriminators between bankrupt and non-bankrupt telecommunications firms, at least in this time period. The estimated probability of 360Networks going bankrupt is only .012, reinforcing how surprising it was that it did so.

Nonexistence of maximum likelihood estimates results from *separation*, the situation where there is no overlap in the distribution of the covariates between successes and failures (complete separation has occurred if there is no overlap; quasicomplete separation has occurred if the overlap is at a few tied values). The MLEs will exist if there is no separation, but they can be very unstable if the covariates are almost separated.

Exact (conditional) estimation can be a viable alternative in this situation. This is discussed in more detail in Section 9.7.

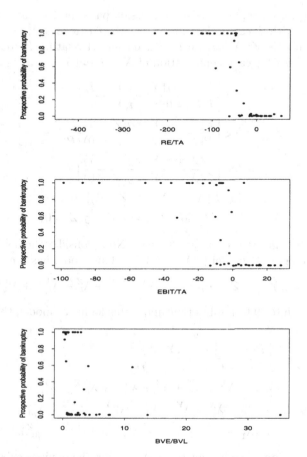

FIGURE 9.9. Plots of prospective probability estimates against each predictor for bankruptcy data omitting 360Networks.

9.3 Categorical Predictors

Categorical (nominal) predictors can be incorporated into logistic regression models in the same way that they are into least squares regression models (Section 3.2) and count regression models (Section 5.2). As was noted in Section 9.1.3, however, there is an intimate connection between logistic regression models and loglinear models, which is particularly apparent for models with only categorical predictors.

Consider, for example, a logistic regression based on two categorical predictors, X (with I categories) and Y (with J categories), and a binary response variable Z. A logistic regression model relating Z to X and Y defines the logit for each combination of $X = i$ and $Y = j$ as

$$\beta_0 + \beta_i^X + \beta_j^Y = \log\left[\frac{P(Z = 1 | X = i, Y = j)}{P(Z = 0 | X = i, Y = j)}\right] \tag{9.5}$$

$$= \log\left[\frac{P(X = i, Y = j, Z = 1)/P(X = i, Y = j)}{P(X = i, Y = j, Z = 0)/P(X = i, Y = j)}\right]$$

$$= \log\left[\frac{P(X = i, Y = j, Z = 1)}{P(X = i, Y = j, Z = 0)}\right]$$

$$= \log[P(X = i, Y = j, Z = 1)]$$

$$- \log[P(X = i, Y = j, Z = 0)] \tag{9.6}$$

where the $\boldsymbol{\beta}^X$ and $\boldsymbol{\beta}^Y$ terms sum to zero. Now, consider the homogeneous association loglinear model (XY, XZ, YZ). This model has the form

$$\log[nP(X = i, Y = j, Z = k)] = \lambda + \lambda_i^X + \lambda_j^Y + \lambda_k^Z + \lambda_{ij}^{XY} + \lambda_{ik}^{XZ} + \lambda_{jk}^{YZ}.$$

Substituting into (9.6) implies that under this loglinear model, the logit for Z satisfies

$$\log\left[\frac{P(Z = 1 | X = i, Y = j)}{P(Z = 0 | X = i, Y = j)}\right]$$

$$= \lambda + \lambda_i^X + \lambda_j^Y + \lambda_1^Z + \lambda_{ij}^{XY} + \lambda_{i1}^{XZ} + \lambda_{j1}^{YZ}$$

$$- \lambda - \lambda_i^X - \lambda_j^Y - \lambda_0^Z - \lambda_{ij}^{XY} - \lambda_{i0}^{XZ} - \lambda_{j0}^{YZ}$$

$$= (\lambda_1^Z - \lambda_0^Z) + (\lambda_{i1}^{XZ} - \lambda_{i0}^{XZ}) + (\lambda_{j1}^{YZ} - \lambda_{j0}^{YZ}).$$

The last equation is equivalent to (9.5), making the obvious substitutions. That is, the homogeneous association model is equivalent to the additive logistic regression model.

It might seem that the loglinear model (XZ, YZ) is also equivalent to the logistic regression model, since the λ_{ij}^{XY} term drops out of the logit equation. While it is true that the logit for Z in the (XZ, YZ) model has the same form as that of (XY, XZ, YZ), it is the latter model that is equivalent to the logistic regression. This is because the logistic regression model makes no assumptions about the association between the predictors X and Y, while (XZ, YZ) implies that X and Y are conditionally independent given Z. More generally, the equivalent loglinear model to a given logistic regression model includes the highest-order interaction involving all of the potential predictors in the data, along with interactions of the predictors with the target variable corresponding to effects in the model. So, for example, if an interaction between two predictors W and X is believed to be predictive for the binary target Z, the equivalent loglinear model includes the WXZ three-way interaction as part of the model.

Nonexistence of the MLEs also can occur when using categorical predictors in a logistic regression. In this situation separation corresponds to all observations that take a particular value of the predictor being either successes or failures. The solution to this problem is to remove all of those observations from the data. Indeed, this is not really a problem at all, since the omitted observations are perfectly predicted as successes or failures anyway.

The Sinking of the Titanic

One of the most famous maritime disasters occurred during the maiden voyage of the ocean liner *Titanic*, which struck an iceberg in the North Atlantic and sank on April 15, 1912. Many articles, books, and movies have told the story of this disaster, but a relatively straightforward statistical analysis tells the story in a remarkably evocative way. The following table summarizes the mortality experiences of the 2201 people on board the ocean liner, given as survival percentages of the number of people of certain subgroups at risk (Dawson, 1995; Simonoff, 1997). People are separated by gender, age group (child or adult) and economic status (first class, second class, third class [steerage], or crew), or G, A, and E, respectively.

Gender		**Age**	
Male		*Adult*	*Child*
	First class	32.6% of 175	100% of 5
Economic status	*Second class*	8.3% of 168	100% of 11
	Third class	16.2% of 462	27.1% of 48
	Crew	22.3% of 862	—
Female			
	First class	97.2% of 144	100% of 1
Economic status	*Second class*	86.0% of 93	100% of 13
	Third class	46.1% of 165	45.2% of 31
	Crew	87.0% of 23	—

This table is actually a re-expression of a four-dimensional contingency table (with survival being the fourth dimension), put into a form that highlights the characterization of survival as the target variable. Note that there were no children among the crew; that is, there are four structural zeroes in the underlying contingency table corresponding to the cells (male, child, crew, survived), (male, child, crew, did not survive), (female, child, crew, survived), and (female, child, crew, did not survive). There are also random zeroes corresponding to children from first and second class, all of whom survived.

As is always the case, it's a good idea to look at the data to try to see what patterns emerge. Since the predictors are all categorical, tables summarize marginal relationships with survival in this case (that is, main effects). Note that just as is true in regression models with numerical predictors, these marginal relationships (which correspond to two-way contingency tables) don't necessarily correspond to the partial relationships implied by a logistic regression model, given the other predictors; still, they are likely to be informative to at least some extent.

Economic status	Percent survived	**Age**	Percent survived
First class	62.5% of 325	*Child*	52.3% of 109
Second class	41.4% of 285	*Adult*	31.3% of 2092
Third class	25.2% of 706		
Crew	24.0% of 885		

Gender	Percent survived
Female	73.2% of 470
Male	21.2% of 1731

The chance of survival was apparently related to all three of these factors. Mortality was much higher for men than for women, and higher for adults than for children. This is, of course, consistent with the "rule of the sea," which says that women and children should be saved first in a disaster. The observed survival percentage is directly related to economic status, with higher status associated with higher survival probability, and the crew having survival rates similar to those in steerage.

We can use logistic regression to try to decide which effects are useful to build a model predicting survival probability accurately, including interaction effects. Note that the observed data summarize the mortality experience of all of the people on the *Titanic*, so these data could be viewed as the entire population of interest, rendering useless any notions of statistical inference. This view can be countered in (at least) two ways. First, while the particular sinking of the *Titanic* was a unique event, it was hardly the only maritime disaster of the 20th (or 21st) century; what is learned from the sinking of this ship has implications for other maritime incidents. Second, even if formal inference (and hence test statistics and p-values) is viewed as inappropriate, the logistic regression model still provides an efficient way of summarizing the patterns in the data, and AIC_C is still a useful measure for determining the best summary.

Model	G^2	df	p	AIC_C
G	237.5	12	$< .001$	217.69
E	491.1	10	$< .001$	475.27
A	652.4	12	$< .001$	632.60
E, G	131.4	9	$< .001$	117.64
A, G	231.6	11	$< .001$	213.80
E, A	465.5	9	$< .001$	451.70
E, A, G	112.6	8	$< .001$	100.79
EG	66.2	6	$< .001$	58.49
AG	215.3	10	$< .001$	199.49
EA	436.3	7	$< .001$	426.53
A, EG	45.9	5	$< .001$	40.17
G, EA	76.9	6	$< .001$	69.16
E, AG	94.6	7	$< .001$	84.79
EA, EG	1.7	3	$.64$	0.00
EG, AG	37.3	4	$< .001$	33.55
EA, AG	65.0	5	$< .001$	59.29
EA, EG, AG	0.0	2	1.00	0.33

TABLE 9.1. Logistic regression model fitting for *Titanic* survival data.

Table 9.1 summarizes model fitting. All of the models fit are hierarchical, in that the presence of an interaction effect in the model implies that the associated main effects are also present. Since there were no children in the crew, the interaction between economic status and age (EA) is fit using only two of the effect codings corresponding to pairwise products of those for the main effects, rather than three. The models are given ordered from smallest to largest AIC_C value within model class (one main effect, two main effects, three main effects, and so on) to make model comparison easier (subset tests are not given, although of course they also can be used to guide model fitting).

The only two models that fit the table include the two interactions EA and EG, or all three pairwise interactions EA, EG, and AG, with AIC_C preferring the simpler model. The simpler model fits the table very well, with fitted survival counts within 2 or 3 of the actual counts in every category. It also has a clear advantage over the (EA, EG, AG) model related to its equivalence to a loglinear model. The simpler model's equivalent loglinear model is (EAS, EGS), where S represents survival. This model is a member of Set X from Figure 8.4. As was noted on page 339, this model is therefore collapsible over both G (to isolate the EAS effect) and A (to isolate the EGS effect). Since the EAS effect is the EA interaction in the logistic regression, and the EGS effect is the EG effect, this means that the survival pattern on the *Titanic* can be studied by examining the following two tables of survival percentages.

The first table refers to the EG effect.

	Gender	
Economic status	*Female*	*Male*
First class	97.2% of 145	34.4% of 180
Second class	87.7% of 106	14.0% of 179
Third class	45.9% of 196	17.3% of 510
Crew	87.0% of 23	22.3% of 862

First, we see that the gender effect of females having much higher survival rates than males is very strong, at every economic status. Given this, the interaction effect shows that this was much less true for females in third class than in any other category, with less than half of those females surviving. That is, steerage female survival percentage is considerably lower than would be expected from the main effects alone. There is considerably less evidence of difference in survival by economic status among males, although first class survival is somewhat higher than among the other three groups.

The following table summarizes the interaction of economic status and age.

	Age	
Economic status	*Child*	*Adult*
First class	100.0% of 6	61.8% of 319
Second class	100.0% of 24	36.0% of 261
Third class	34.2% of 79	24.1% of 627
Crew	—	24.0% of 885

The age main effect is evident, as noted earlier, but the interaction also makes clear the different nature of third class compared with the others.

While no children of the other classes died, almost two-thirds of those in steerage did. Thus, the story of the *Titanic* seems to be that while the "rule of the sea" did seem to hold, it was less applicable to those in third class than those in the other categories. There has been speculation that this was because of unfair treatment of the passengers by the crew, although there is little or no evidence to support this allegation. The official report on the disaster, *Lord Mersey's Report*, blamed the accident on excessive speed, and explained the lower survival rates of third class women and children as being due to their greater reluctance to leave the ship (and quite possibly all of their belongings) and the greater difficulty in getting them from their quarters to the lifeboats.

9.4   Other Link Functions

The logit link is the canonical link for binomial data, and thus shares the favorable properties of canonical links, including a simpler form for the sufficient statistic for the linear predictor and equality of the observed and expected Fisher information. Being based on odds ratios, it also has the great advantage of yielding regression coefficients that are directly interpretable in both prospective and retrospective studies. Despite this, there are situations where using a noncanonical link can be useful.

9.4.1   Dose Response Modeling and Logistic Regression

The following scenario illustrates the possibilities. A toxicology *dose response study* (also called a quantile response study) examines the relationship between dosage level of a toxic substance and animal mortality, the ultimate goal being to model the probability of death p as a function of dosage x. Let y be the result for a randomly selected animal, with $y = 1$ corresponding to the animal dying. A reasonable model for this process is that the animal has a certain tolerance T to the toxic substance, with mortality occurring if the dosage is above that tolerance. If $f(t)$ is the density function for the population distribution of tolerance values, then the probability that an animal dies when given dosage x is

$$P(y = 1) = P(x \geq T) = P(T \leq x) = \int_{-\infty}^{x} f(t)dt.$$

That is, the probability of death at dosage x equals the cumulative distribution function of the tolerance distribution, $F(x)$. From the point of view of a generalized linear model, the dose response model implies a link function of the form $F^{-1}(p)$. Since the tolerance distribution is not directly observed, but determines success or failure, this formulation of regression for binary data is sometimes more generally called a *latent variable* formulation of the problem.

Say the tolerance distribution has a *logistic distribution*, with

$$f(t) = \frac{\exp[-(t-\alpha)/\gamma]}{\gamma\{1 + \exp[-(t-\alpha)/\gamma]\}^2}. \tag{9.7}$$

This distribution has mean α and standard deviation $\gamma\pi/\sqrt{3}$. Its density function (based on zero mean and unit standard deviation) is given in the top plot of Figure 9.10 as the solid line, and its distribution function,

$$F(x) = \frac{\exp[(x-\alpha)/\gamma]}{1 + \exp[(x-\alpha)/\gamma]},$$

is given in the bottom plot. The latter, of course, is the logit probability curve (9.2), making the substitution $\beta_0 = -\alpha/\gamma$ and $\beta_1 = 1/\gamma$. That is, logistic regression can be justified for dose response modeling if the tolerance distribution is logistic.

9.4.2   Probit Regression

The logistic density function is symmetric and bell-shaped, which suggests a Gaussian density $\phi(\cdot)$ as an alternative. This is the idea behind *probit regression*, as first suggested in Bliss (1934, 1935). In this model the tolerance values follow a Gaussian distribution with mean μ and variance σ^2. The standard normal density (given in Figure 9.10 as the dotted line) is shorter tailed than the logistic density. Since the probability of success as a function of x is the cumulative distribution function, this means that the logistic probability function approaches zero and one less steeply than does the probit. As the curves in the bottom plot show, however, the difference between the two probability curves is very small, even in the tails, and typically probit and logit fits to a given data set are very similar to each other.

The probability curve for probit regression is the cumulative normal $\Phi[(x-\mu)/\sigma]$, which is equivalent to a curve for a linear predictor $\beta_0 + \beta_1 x$ making the substitutions $\beta_0 = -\mu/\sigma$ and $\beta_1 = 1/\sigma$. The slope coefficient in a probit regression refers to the relationship between changes in x and the standard normal score (z-score) of a subject. That is, for a probit regression with fitted parameters $\hat{\beta}_0$ and $\hat{\beta}_1$, the estimated z-score when $x = 0$ is $\hat{\beta}_0$, and each additional unit value of x is associated with an estimated change in the z-score of $\hat{\beta}_1$.

As we would expect from the similarity of the probit and logit probability functions, there is a connection between the parameters in the two models.

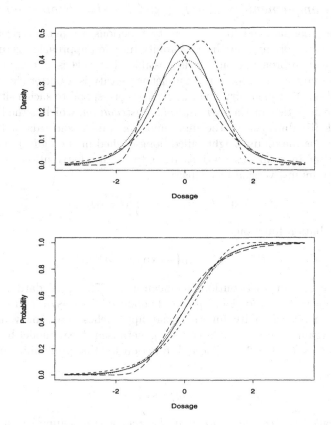

FIGURE 9.10. Density (top plot) and distribution (bottom plot) functions for dose response formulation of binomial regression. Solid line refers to logit link, dotted line to probit link, short dashed line to complementary log–log link, and long dashed line to log–log link.

The logit tolerance distribution has the same standard deviation as the probit tolerance distribution if $\sigma/\gamma = \pi/\sqrt{3} \approx 1.8$, and the logit probability function has the same slope when $p = .5$ (when $\beta_0 + \beta_1 x = 0$, the so-called *median lethal dose*, LD_{50}, since it corresponds to the point where half of the subjects die) as the probit probability function if $\sigma/\gamma = 4\phi(0) \approx 1.6$. Let β_j^L be a logistic regression parameter, and let β_j^P be a probit regression parameter. Then $\beta_j^L/\beta_j^P = \sigma/\gamma$. That is, if both models fit well, we should expect the logistic regression parameters to be roughly 1.6 to 1.8 times larger than the probit regression parameters.

9.4.3 Complementary Log–Log and Log–Log Regression

Both the Gaussian and logistic density functions are symmetric around their means, implying that the probability function approaches zero at the same rate it approaches one. It is possible that this is not appropriate in some circumstances, and an asymmetric density is a better choice. The long- and short-dashed curves in Figure 9.10 correspond to such a situation. These are densities for the *extreme value distribution*, which is the limiting family of distributions for the minimum (left-tailed, short-dashed in the figure) or the maximum (right-tailed, long-dashed in the figure) of a large collection of random observations from the same distribution. The left-tailed (minimum) version has density

$$f(t) = \exp\{(t - \alpha)/\gamma - \exp[(t - \alpha)/\gamma]\}/\gamma$$

and distribution function

$$F(x) = 1 - \exp\{-\exp[(x - \alpha)/\gamma]\}. \tag{9.8}$$

This extreme value distribution has mean $\alpha - .57722\gamma$, standard deviation $\pi\gamma/\sqrt{6}$, median $\alpha + \log[\log(2)]]\gamma$, and mode α. The asymmetry in f is reflected in a probability function that approaches zero slowly and one quickly. Again, this is formulated as a generalized linear model by taking $\beta_0 = -\alpha/\gamma$ and $\beta_1 = 1/\gamma$. The link function for this probability function satisfies

$$\beta_0 + \beta_1 x = \log\{-\log[1 - p(x)]\}.$$

This is called the *complementary log–log* (cloglog) link function, as it is a log–log transformation of the complement of p.

The right-tailed (maximum) version of the extreme value distribution has density

$$f(t) = \exp\{-(t - \alpha)/\gamma - \exp[-(t - \alpha)/\gamma]\}/\gamma$$

and distribution function

$$F(x) = \exp\{-\exp[-(x - \alpha)/\gamma]\}.$$

This probability function approaches zero quickly and one slowly. The link function for this probability function satisfies

$$\beta_0 + \beta_1 x = \log\{-\log[p(x)]\},$$

and is therefore called the *log–log* (loglog) link function. The complementary log–log and log–log models are a natural pair, in that if one holds for the probability of a success, the other holds for the probability of failure (see page 420).

The slope coefficients for the complementary log–log link can be interpreted based on the appropriate probability function. From equation (9.8), and making the substitution $(x - \alpha)/\gamma = \beta_0 + \beta_1 x$, we can see that increasing x by 1 is associated with changing the probability of failure $1 - p$ from

$$\exp[-\exp(\beta_0 + \beta_1 x)]$$

to

$$
\begin{aligned}
\exp\{-\exp[\beta_0 + \beta_1(x + 1)]\} &= \exp[-\exp(\beta_0 + \beta_1 x + \beta_1)] \\
&= \exp[-\exp(\beta_0 + \beta_1 x)\exp(\beta_1)] \\
&= \exp[-\exp(\beta_0 + \beta_1 x)]^{\exp(\beta_1)}.
\end{aligned}
$$

That is, adding 1 to x is associated with raising the complement probability $1 - p$ to the $\exp(\beta_1)$ power. Since a log–log model for p is equivalent to a complementary log–log model for $1 - p$, the slope coefficient from the log–log model has a corresponding interpretation.

A Classic Dose-Response Study

Bliss (1935) first laid out the theory of tolerance models, and illustrated them with data from the following experiment. Sets of beetles were exposed to different concentrations of gaseous carbon disulphide CS_2 for five hours (roughly 60 beetles at each concentration), and the proportions of insects that died were recorded. The data are presented in Figure 9.11 and given in the file **beetles**, as given on page 247 of Agresti (2002).

Logistic regression (solid line) and probit regression (dotted line) fits are also given in the figure. The logit and probit fits are very similar, as would be expected, but neither fits very well ($G^2 = 11.1$ for the logit model, and $G^2 = 10.0$ for the probit model, both on 6 degrees of freedom, each yielding tail probabilities of roughly .1). The complementary log–log model provides a much better fit, with $G^2 = 3.5$, and is represented by the dashed line in Figure 9.11. The fitted regression line,

$$\log[-\log(1 - p)] = -39.52 + 22.01 \times \text{Log dose}$$

implies an extreme value tolerance distribution for the beetles with $\alpha = 1.795$ and $\gamma = .045$. A graph of the density function for this tolerance distribution is given in Figure 9.12, and the asymmetry implied by the complementary log–log fit is evident. The median lethal dose (the dose where half of the beetles are expected to die) based on this density is 1.779, slightly larger than the mean tolerance (1.769) and slightly smaller than the modal value (1.795).

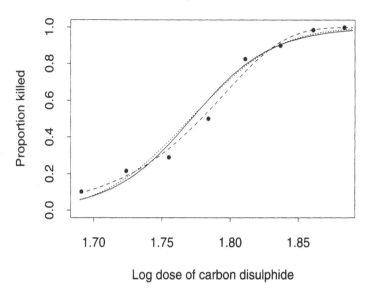

Log dose of carbon disulphide

FIGURE 9.11. Observed mortality proportions for beetle dose-response data, with logistic (solid line), probit (dotted line), and complementary log–log (dashed line) fits superimposed.

9.5 Overdispersion

The binomial random variable, like the Poisson, has the strong restriction that the mean of the random variable np exactly determines the variance $np(1 - p)$. In many data sets this assumption is violated, typically in that the variance is larger than would be expected. As was discussed in Section 4.5.3, this overdispersion can arise from unmodeled heterogeneity in the population or positively correlated underlying Bernoulli trials. Many of the techniques to address this problem in the Poisson regression context, discussed in Section 5.3, have corresponding approaches in the binary regression situation. Recall that overdispersion cannot occur for ungrouped binary data ($n_i = 1$ for all i), so these methods are only relevant for data with replications at the distinct covariate patterns.

9.5.1 Nonparametric Approaches

A simple quasi-likelihood approach takes the variance of y_i to satisfy

$$V(y_i) = \phi n_i p_i (1 - p_i),\qquad\qquad (9.9)$$

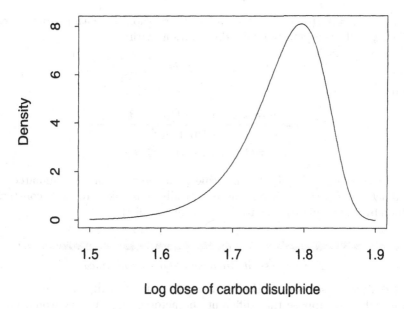

Log dose of carbon disulphide

FIGURE 9.12. Tolerance distribution of beetles implied by fitted complementary log–log regression.

a constant inflation over the binomial variance function. Just as was true for Poisson regression, the variance inflation can be estimated using the Pearson χ^2 goodness-of-fit statistic, with $\hat{\phi} = X^2/(N - k - 1)$. Estimated parameters remain the same, while Wald statistics are deflated by the constant factor $\sqrt{\hat{\phi}}$.

The overdispersion-resistant sandwich variance estimate also can be derived for binary regression models. The estimate has a similar form to that for overdispersion in Poisson regression, as in (5.19); since $V(y_i) = n_i p_i(1 - p_i)$ under the binomial distribution, the estimate of this variance replaces $\hat{\mu}_i$, yielding

$$\hat{V}(\hat{\beta}) = (X'\text{diag}(n_i\hat{p}_i[1 - \hat{p}_i])X)^{-1}$$
$$\times (X'\text{diag}[(y_i - n_i\hat{p}_i)^2]X)(X'\text{diag}(n_i\hat{p}_i[1 - \hat{p}_i])X)^{-1}.$$

9.5.2   Beta-Binomial Regression

Section 4.5.3 described the beta-binomial distribution, an alternative parametric model that exhibits overdispersion relative to the binomial distribution. The model generalizes to regression models in a straightforward way, although computation is a challenge, since the beta-binomial distribution is not part of the exponential family. Assume that given p_i, $y_i \sim \text{Bin}(n_i, p_i)$, and p_i is beta distributed with parameters (α_i, β_i). Let

$E(p_i) = \pi_i = \alpha_i/(\alpha_i + \beta_i)$ satisfy a logistic relationship with predictors $\mathbf{x}_i$. Then y_i follows a beta-binomial distribution, with

$$E(y_i) = n_i\pi_i$$

and

$$V(y_i) = \frac{n_i\alpha_i\beta_i[1 + (n_i - 1)(\alpha_i + \beta_i + 1)^{-1}]}{(\alpha_i + \beta_i)^2}$$
$$\equiv n_i\pi_i(1 - \pi_i)[1 + (n_i - 1)\theta_i],$$

where $\theta_i = (\alpha_i + \beta_i + 1)^{-1}$. The scale parameter θ_i can be formulated as a function of predictors, but more typically it is taken to be a constant (thereby restricting $\alpha_i + \beta_i$ to be constant for all i).

The Success of Broadway Shows (continued)

Recall the discussion of Broadway shows data starting on page 96. It is reasonable to suppose that different characteristics of a show would have predictive power for the critical success of the show, as measured by Tony nominations and awards. The file **broadway** contains information on 92 Broadway shows including the type of show (musical, musical revue, or play), whether or not the show was a revival of an earlier show, a rating of the critical reviews for the show in *The New York Times* and the *Daily News*, respectively, on a scale from 0 to 5, and the number of Tony nominations and awards earned by the show in the six major categories (for more details see Simonoff and Ma, 2003). Note that $n_i = 6$ for all observations for both the analysis of earning a nomination and that of winning an award. In all of the regressions performed the status of a show as a revival had no predictive power, so that variable is dropped from the analysis. The show type effect is fit using effect codings for musical and musical revue.

Consider first the probability of earning a major Tony nomination. There appears to be a show type effect, as musicals averaged 2.3 nominations in the major categories, musical reviews averaged 1.1, and plays averaged 1.5. Figure 9.13 doesn't provide much indication of any effects related to reviews in either newspaper.

Table 9.2 summarizes the results of a logistic regression fit to these data in part (a). As expected, musicals are associated with a higher probability of earning a Tony nomination, and both newspaper reviews are highly significant positive predictors of nomination (a better review being associated with a higher probability of nomination). Unfortunately, the model does not fit the data very well, with $G^2 = 205.9$ on 87 degrees of freedom.

As was noted earlier, unmodeled heterogeneity is a reasonable hypothesis here, suggesting correcting for overdispersion. The quasi-likelihood estimate $\hat{\phi} = X^2/(N - p) = 2.04$, implying dividing all of the Wald statistics by 1.43;

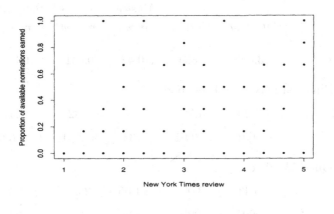

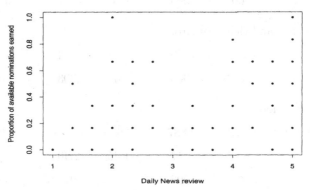

FIGURE 9.13. Scatter plots of proportion of available Tony nominations earned by *Times* (top plot) and *Daily News* (bottom plot) reviews.

the resultant statistics are given in part (b) of the table. The Wald statistics obtained when using the sandwich variance estimate are given in part (c) of the table. The results here are similar, suggesting that the variance form (9.9) is probably reasonable for these data. The entries in part (d) of the table refer to a beta-binomial fit. Despite the fact that this model does not fit the data very well ($G^2 = 171.6$), the regression coefficients are similar to the logistic regression coefficients, and the implications are similar. That is, a show being a musical more than doubles the odds of earning a Tony nomination, and the reviews in both the *Times* and *Daily News* are predictive for Tony nominations, with the effects of similar magnitude.

What about winning the Tony award? Once again there is an apparent show type effect, with musicals averaging 0.6 awards per show, musical revues 0.1, and plays 0.4. Figure 9.14 suggests stronger effects relating to newspaper reviews than were seen for nominations.

	Intercept	Musical	Musical revue	Times review	Daily News review
Coefficient	−3.505	0.804	−0.425	0.401	0.396

(a) Binomial maximum likelihood

	Intercept	Musical	Musical revue	Times review	Daily News review
Wald	−9.20	4.29	−1.50	3.82	4.41
p	< .001	< .001	.133	< .001	< .001

(b) Quasi-likelihood

	Intercept	Musical	Musical revue	Times review	Daily News review
Wald	−6.43	3.00	−1.05	2.67	3.08
p	< .001	.005	.294	.007	.002

(c) Robust sandwich estimator

	Intercept	Musical	Musical revue	Times review	Daily News review
Wald	−6.69	2.79	−0.88	2.94	3.17
p	< .001	.005	.381	.003	.002

(d) Beta-binomial

	Intercept	Musical	Musical revue	Times review	Daily News review
Coefficient	−3.378	0.887	−0.656	0.345	0.388
Wald	−6.47	3.17	−1.45	2.40	3.12
p	< .001	.002	.147	.017	.002

TABLE 9.2. Regression estimation for Tony nominations data. Coefficients are based on the logistic regression model, with Wald statistics based on binomial maximum likelihood, the robust sandwich estimator, and quasi-likelihood, and beta-binomial regression.

Table 9.3 summarizes model fitting for these data. The logistic regression model (part (a)) is a better fit than it was to the nominations data, but overdispersion is still suggested ($G^2 = 112.0$ on 87 degrees of freedom, $p = .04$). The effects are similar to those for Tony nominations, except that the *Daily News* review is only marginally predictive for Tony awards given the other variables. The quasi-likelihood Wald statistics are based on $\hat{\phi} = 1.54$, which has the effect of making the *Daily News* variable quite insignificant. The Wald statistics based on the sandwich estimate are similar. The beta-binomial model has deviance $G^2 = 92.9$, indicating a good fit to the data.

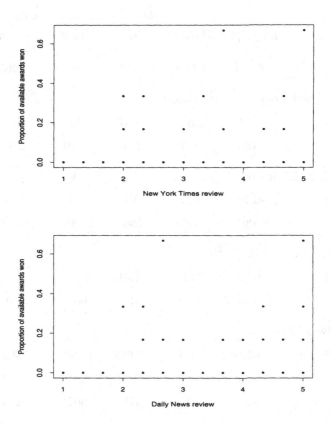

FIGURE 9.14. Scatter plots of proportion of available Tony awards won by *Times* (top plot) and *Daily News* (bottom plot) reviews.

9.6 Smoothing Binomial Data

The assumption of a linear relationship between the predictors and the binomial target (through the link function) can be relaxed to be an unspecified smooth curve in the same way as was described for Poisson data in Section 5.5. Local likelihood estimation proceeds in exactly the same way as before, based on the local log-likelihood (5.23), only now using the binomial log-likelihood rather than the Poisson log-likelihood. This has the effect of smoothing in the (canonical) logit space, which is then transformed back to the probability space. Multiple predictors can be addressed by smoothing in a multidimensional space, or by fitting an additive model

	Intercept	Musical	Musical revue	Times review	Daily News review
Coefficient	−6.653	1.115	−0.729	0.866	0.263
(a) Binomial maximum likelihood					
Wald	−8.01	2.83	−1.07	4.37	1.71
p	< .001	.005	.284	< .001	.087
(b) Quasi-likelihood					
Wald	−6.40	2.25	−0.85	3.51	1.37
p	< .001	.025	.396	< .001	.169
(c) Robust sandwich estimator					
Wald	−6.50	2.51	−1.01	3.41	1.16
p	< .001	.012	.312	.001	.246
(d) Beta-binomial					
Coefficient	−8.320	1.351	−0.685	1.050	0.314
Wald	−5.64	2.20	−0.72	3.04	1.19
p	< .001	.028	.471	.002	.235

TABLE 9.3. Regression estimation for Tony awards data. Coefficients are based on the logistic regression model, with Wald statistics based on binomial maximum likelihood, the robust sandwich estimator, quasi-likelihood, and beta-binomial regression.

that replaces some or all of the linear terms with smooth curves. Just as was true in the count regression context, the specific smoothing technique used is generally less important than the amount of smoothing done, the goal being to avoid both undersmoothing and oversmoothing.

Another way that smoothing can be useful in the binomial context is when there are no replications in the data — that is, the target variable is binary (0/1). In this situation residual plots are very difficult to interpret, since if there is only one predictor all of the observations line up in the form of two curved stripes whether the model fits or not (see Figure 9.4 for a corresponding example with three stripes). Smoothing the Pearson residuals (using a least squares–based smoother) can help identify undesirable structure in the residuals (since all should have mean roughly equal to zero), and potentially the source of lack of fit.

Smoking and Mortality in Whickham (continued)

Recall the loglinear analysis of the cohort study relating smoking and age to survival in Whickham (page 325). The model of homogeneous association between smoking, age, and mortality was the preferred model for these data, implying 35% lower odds of a smoker surviving to the followup survey than a nonsmoker, given age. As was noted in Section 9.3, this model is equivalent to the logistic regression model that relates the probability of survival to smoking status and age, when age is treated as a nominal (factor) variable.

The age variable is, of course, numerical (and hence ordinal), which suggests simplifying the model by using that numerical character. Table 9.4 summarizes the properties of several such models. Since age is only given in ranges, we treat it as numerical with values $\{21, 30, 40, 50, 60, 70, 80\}$. The coefficient for age in a linear logistic model is not significantly different from zero ($p = .12$), and implies less of a risk to smokers (23% lower odds of survival) than did the homogeneous association model. This model, however, clearly does not fit well, with $G^2 = 33.7$ on 11 degrees of freedom.

This suggests that a nonlinear function of age might be an appropriate middle ground between the linear model and the model treating age as nominal. Figure 9.15 gives the smooth function of age estimated in a semiparametric logistic model that also includes a smoking status indicator variable. The nonlinearity of the age effect is clear, with the log-odds of survival becoming more negative as age increases. The smoothing parameter for the local linear smoother was chosen so as to minimize AIC_C, which results in a model with AIC_C virtually equal to that of the homogeneous association model. The estimated odds of survival for a smoker are 33% lower than those for a nonsmoker given the smooth term, very similar to those in the homogeneous association model.

The curve in Figure 9.15 looks roughly quadratic, which suggests fitting a model that includes Age and Age^2. This model fits reasonably well ($G^2 = 12.2$ on 10 degrees of freedom), although not as well as the homogeneous association model (AIC_C is 1.69 higher). The assessment of the smoking

Model	G^2	df	p	AIC_C	OR
Smoker, factor(Age)	2.4	6	.88	0.00	0.65
Smoker, Age	33.7	11	< .001	21.23	0.77
Smoker, smooth(Age)	8.6	9.1	.49	0.00	0.67
Smoker, Age, Age^2	12.2	10	.28	1.69	0.65
Smoker, Age × Age $<\geq 55$	8.6	9	.47	0.17	0.64

TABLE 9.4. Logistic linear and semiparametric model fitting for smoking and mortality data. The last column gives the estimated odds ratio of survival for smokers versus nonsmokers.

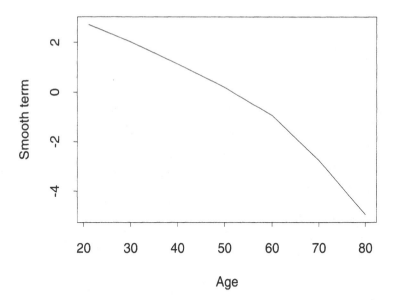

FIGURE 9.15. Smooth term for age in additive logistic model of survival on smoking status and age.

effect is very similar to the latter model, with 35% lower odds of survival given age.

A final parametric attempt is to recognize that the curve in Figure 9.15 looks most like two separate straight lines, separated at (roughly) age 60. The final line in Table 9.4 refers to an interaction model, with separate lines for age groups less than 55 and more than 55. This model fits very well ($G^2 = 8.6$ on 9 degrees of freedom), with AIC_C virtually identical to that of the homogeneous association model. For the younger age groups, the estimated model is

$$\log\left(\frac{p}{1-p}\right) = 6.111 - .447 \times \text{Smoker} - .087 \times \text{Age},$$

while for the older age groups it is

$$\log\left(\frac{p}{1-p}\right) = 13.413 - .447 \times \text{Smoker} - .211 \times \text{Age}.$$

Thus, given age, being a smoker is associated with 36% lower odds of survival 20 years later. Further, for initial age less than 55, given smoking status each additional year older is associated with 8.3% lower odds of survival 20 years later, but for initial age at least 55, each additional year older is associated with 19% lower odds of survival. Given general life expectancies in the 70s, these patterns are eminently reasonable.

9.7   Conditional Analysis

The standard tests and intervals for logistic regression models are based on Gaussian and χ^2 approximations that are known to be inappropriate when the sample size is small, the data are sparse in certain regions of the space of predictors, or when the overall percentage of observations that are successes is close to zero or one. Recall also from the discussion of the telecommunications bankruptcy data (page 386) that certain maximum likelihood estimates are undefined if there is complete separation in the data, and unstable if there is near separation.

As we have seen in earlier sections, a solution to this problem is to use exact inference, proceeding conditionally on the sufficient statistics for parameters not of direct interest. The resultant conditional likelihood is independent of the other parameters, and inference proceeds based on it. The conditional maximum likelihood estimate maximizes the conditional likelihood, and tests and intervals can be constructed based on the permutation distributions from the conditional likelihood. The properties of conditional maximum likelihood estimators can be considerably different from those of the standard (unconditional) MLE; indeed, if the number of nuisance parameters is large relative to the sample size, the unconditional MLE can be inconsistent. Note also that since by definition the unconditional MLE minimizes the deviance, conditional estimates can result in a model with noticeably higher deviance than do unconditional ones.

Conditional inference is possible if a generalized linear model is based on the canonical link (logit for binomial regression). Recall from Section 5.1.1 that in this case the sufficient statistic for the regression coefficients is a linear function of the data, $X'y$. In fact, the sufficient statistic for any subset of β is the corresponding subset of $X'y$; that is, $S_j = \sum_{i=1}^{n} x_{ij} y_i$ is sufficient for β_j. Say the coefficient β_1 was being studied. The distribution of S_1 given $S_2, \ldots, S_p$ depends on only β_1, and it is this distribution that is used to estimate β_1 and construct intervals and tests for it. As always, conditional tests can be conservative because of the discreteness of the conditional distribution, but use of the mid-p-value can help alleviate this.

Determination of the ESR

The following analysis is based on data originally from Collett and Jemain (1985); see also Collett (2003), page 8, and King and Ryan (2002). The erythrocyte sedimentation rate (ESR) is the rate at which red blood cells settle out of suspension in blood plasma when measured under standard conditions. The ESR level is associated with the occurrence of certain rheumatic and malignant diseases and chronic infections, and is thus routinely screened in samples of blood. A healthy person should have ESR < 20mm/hour. A study in Kuala Lumpur, Malaysia, explored whether

	Fibrinogen		Gamma-globulin	
	Unconditional	Conditional	Unconditional	Conditional
Full data set				
$\hat{\beta}$	1.910	1.727	0.156	0.105
CI	$(.007, 3.814)$	$(.165, 3.893)$	$(-.079, .390)$	$(-.095, .364)$
p	.049 (Wald)	.027	.193 (Wald)	.340
	.015 (LR)		.171 (LR)	
Data set with outliers removed, both predictors				
$\hat{\beta}$	24.877	2.549^*	-0.139	0.000^*
CI	$(-38.091, 87.845)$	$(.816, \infty)$	$(-1.006, .728)$	$(-1.832, \infty)$
p	.439 (Wald)	.002	.754 (Wald)	1.000
	$< .001$ (LR)		.715 (LR)	
Data set with outliers removed, one predictor				
$\hat{\beta}$	17.459	14.479		
CI	$(-8.905, 43.822)$	$(1.818, 64.104)$		
p	.194 (Wald)	$< .001$		
	$< .001$ (LR)			

TABLE 9.5. Unconditional and conditional logistic regression results for ESR data. Entries marked * are median unbiased estimates rather than conditional MLEs.

a healthy ESR in an individual is related to the level of two plasma proteins, fibrinogen and gamma-globulin, given in grams/liter. These proteins are known to occur at generally higher levels in the presence of inflammatory disease, and the goal of the study was to see if ESR could be useful as a diagnostic for the presence of elevated levels of this protein. The data are given in the file esr. Table 9.5 summarizes the results of model fitting for these data.

A logistic regression fit for these data results in a model that apparently fits reasonably well ($HL = 6.6$, $df = 8$), and provides significant predictive power ($LR = 7.9$, $df = 2$, $p = .019$). Fibrinogen level is seen as a significant predictor of ESR status, while gamma-globulin level is not. Note that while the Wald statistic p-value for fibrinogen is only marginally significant ($p = .049$), the likelihood ratio test is strongly significant ($p = .015$). The

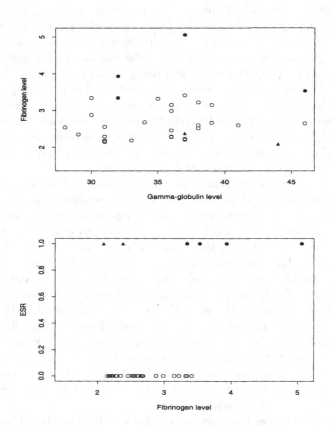

FIGURE 9.16. Scatter plots of ESR data. Top plot: plot of fibrinogen level versus gamma-globulin level. Bottom plot: plot of ESR health status versus fibrinogen level. In both plots the open symbols refer to healthy individuals, while the solid symbols refer to unhealthy individuals.

conditional tail probabilities are similar to the unconditional ones, as are the 95% confidence intervals for the slope parameters.

The top plot of Figure 9.16 gives a graphical representation of the data. The open symbols refer to healthy individuals ($ESR = 0$), while the solid symbols refer to unhealthy individuals ($ESR = 1$). It is apparent that, except for two observations, lower fibrinogen level is associated with healthy ESR status, which accounts for the positive slope coefficient for fibrinogen (recall that it is an unhealthy ESR status that is being modeled as the "success" here). These two individuals, which are marked as solid triangles in the plot, are marginal outliers because of the low fibrinogen/unhealthy ESR combination, which suggests refitting the model with them omitted, to see the effect on the fit.

This effect is, in fact, dramatic. The model fits very well ($HL = 0.3$, $df = 8$), and provides highly significant predictive power ($LR = 18.1$, $df = 2$, $p < .001$). The MLE of the coefficient for fibrinogen becomes much larger; since we have omitted two observations with low fibrinogen and unhealthy status, this is not surprising. What is surprising, however, is that the Wald statistic for fibrinogen is not at all significant ($p = .439$), which is hardly what would be expected after omitting two observations that *weaken* the relationship! Similarly, the usual confidence interval for the fibrinogen slope coefficient includes values considerably below zero. The likelihood ratio test is more sensible here ($p < .001$) but is, of course, more difficult to use to construct a confidence interval.

The top plot in Figure 9.16 shows what is going on here. Once the two outliers (solid triangles) are removed, the data are almost completely separable (that is, a straight line can be drawn on the figure that almost completely separates the solid circles from the open ones). This leads to instability in the MLE, and the strange Wald statistic and confidence interval results. The conditional slope estimate is hardly changed from the fit before omitting the outliers, although it should be pointed out that the numbers given for the two slopes are not conditional MLEs, but rather median unbiased estimates, the value for which the conditional likelihood for that parameter equals .5 (this is because the conditional log-likelihood cannot be maximized, due to each of the sufficient statistics for the nuisance parameters being at the limit of its range; see Cytel Software Corporation, 1999, page 257). The conditional p-value is small ($p = .002$), as would be expected, and the confidence interval for the slope for fibrinogen is distinctly away from zero (and extends to $+\infty$).

Both analyses agree that gamma-globulin level has little predictive power given fibrinogen level, which suggests fitting the model just on fibrinogen level. This corresponds to the bottom plot of Figure 9.16, again omitting the two observations coded as solid triangles. The strong effect of fibrinogen is clear, but so is the near-separability of the data. The model fits well ($HL = 0.5$ on 7 degrees of freedom), and provides highly significant predictive power ($LR = 18.0$ on 1 degree of freedom, $p < .001$), but the Wald confidence interval and p-value are again misleading. The conditional estimates and tests highlight the (correct) strong association between fibrinogen level and ESR health status, and the confidence interval now has a large, but finite, upper limit. It can be noted, however, that while the conditional tests are better-behaved than the Wald tests for these data, the advantage of the conditional parameter estimates is less clear. The data's near-separability suggests that the conditional coefficient for fibrinogen is too small in the two-variable regression with the outliers removed, an impression reinforced by the much larger slope in the one-variable regression. See Webber (2003) for further discussion.

9.8   Background Material

Most texts on generalized linear models discuss logistic regression as a special case. Hosmer and Lemeshow (2000) is a more thorough and detailed reference on many aspects of the method. Kagan (2001) showed that the logistic is the only link function for generalized linear models of binary data where the retrospective and prospective likelihoods differ only by an intercept; thus, logistic regression is the only generalized linear model for binomial data for which the slope coefficients have the same interpretation under both prospective and retrospective sampling.

Simonoff (1998c) discussed in detail how different definitions of covariate patterns affect logistic regression goodness-of-fit, and described how the appropriate tests can be obtained from statistical packages that use inappropriate covariate pattern rules for calculating the tests. The dependence of the Hosmer–Lemeshow goodness-of-fit statistic on the way that the data are grouped has led to several proposals for alternative tests; see, for example, Hosmer et al. (1997) and Pigeon and Heyse (1999). Cai and Tsai (1999) described eight plots for identifying nonlinearity in generalized linear models, and showed how they can be used in both the binary and binomial situations. Menard (2000) describes several alternatives for a "pseudo-R^2" for logistic regression models, ultimately preferring (5.10), based on its intuitive form and relative insensitivity to the base rate (the overall proportion of successes in the sample). Section 4.5 of Hosmer and Lemeshow (2000) discusses the numerical problems associated with complete and quasi-complete separation.

Hinde and Demétrio (1998) provide an overview of models and techniques for overdispersed binomial data, including quasi-likelihood, pseudo-likelihood, and fully parametric approaches. An important early quasi-likelihood approach based on the beta-binomial variance form was proposed by Williams (1982). Dean (1992) proposed several tests for overdispersion in binomial regression models. As is true in the Poisson regression context, unmodeled correlation between trials can lead to overdispersion relative to the binomial random variable. Kupper and Haseman (1978) proposed a parametric model for data where Bernoulli trials within groups (for example, survival of a litter of animals from the same mother) are expected to be correlated within one another. Zero-inflated versions of binomial and beta-binomial regression models are also possible; see Hall (2000), Vieira, Hinde, and Demétrio (2000), and Hall and Berenhaut (2002) for further discussion.

Chapter 7 of Lloyd (1999) discusses conditional inference in the logistic regression context. Brazzale (1999) describes how saddlepoint approximations (sometimes called *small-sample asymptotics*) provide a mechanism to approximate the conditional distributions needed for conditional inference. Using a double saddlepoint approximation to the conditional log-likelihood, an adjusted form of the profile log-likelihood is an approximate substitute for the conditional log-likelihood. Brazzale (1999) describes how the necessary calculations can be performed in the statistical package S-PLUS. See also Lloyd (1999, Section 7.9).

9.9   Exercises

1. Consider a prospective (cohort) study with a binary outcome, where a predicting variable x_j is binary, such as exposed/not exposed to a potential risk factor. Recall from Section 6.1.3 that the odds ratio OR and the relative risk RR have the relationship

$$OR = \frac{p_1/(1-p_1)}{p_0/(1-p_0)} = RR \times \left(\frac{1-p_0}{1-p_1}\right),$$

where p_0 is the incidence of the outcome in the nonexposed group and p_1 is the incidence of the outcome in the exposed group.

(a) Show that

$$RR = \frac{OR}{1 - p_0 + p_0 \times OR}. \qquad (9.10)$$

(b) Now consider a logistic regression model for the outcome of interest as a function of x_j (as well as other variables). Of course, exponentiating the estimated slope coefficient $\hat{\beta}_j$ provides an estimate of the conditional odds ratio given the other predictors, and if the incidence of the outcome of interest is small in both groups, the estimated odds ratio is a good estimate of the true relative risk. Zhang and Yu (1998) proposed using (9.10) to convert the estimated odds ratio from a logistic regression into an estimated relative risk. Note that this is only valid in a prospective study, since otherwise p_0 is not estimable from the data. Use this method to estimate the relative risk of dying for smokers versus nonsmokers for the Whickham smoking cohort study for the models considered in Section 9.6. Is the estimated odds ratio very different from the estimated risk ratio in this case?

(c) Construct a 95% confidence interval for the true risk ratio by adjusting the limits of a confidence interval for the odds ratio. (Note, however, that this ignores the variability of the sample estimate of incidence in the nonexposed group, resulting in intervals that are too narrow; see McNutt et al., 1999.)

(d) Robbins et al. (2002) suggest estimating the relative risk directly by fitting a binomial regression model with log link, and exponentiating the coefficient of x_j. Apply this method to the

Whickham smoking data. What is the estimated relative risk? Does this estimate seem reasonable to you? Does the regression model with log link provide a good fit to the data?

2. Consider the situation where there are K items, and the task is to provide a ranking for them. Paired preference data for this problem corresponds to the situation where pairs of items are compared to each other multiple times, and a "winner" is determined in each comparison. Let p_{ij} be the probability that item i beats item j in a paired comparison, with ties not allowed (note that this implies that $p_{ji} = 1 - p_{ij}$). The *Bradley–Terry model* assumes that these probabilities satisfy

$$\log\left(\frac{p_{ij}}{p_{ji}}\right) = \beta_i - \beta_j.$$

This model can be fit using a logistic regression model without an intercept. The observations are the $K(K-1)/2$ pairs (n_{ij}, n_{ji}), with $n_{ij} \sim \text{Bin}(n_{ij} + n_{ji}, p_{ij})$. There are K predictor variables corresponding to the K items, the kth of which has the value 1 if $k = i$, -1 if $k = j$, and 0 otherwise (that is, for the (n_{ij}, n_{ji}) comparison, the variable for item i equals 1, the variable for item j equals -1, and the other variables equal 0). The model is overspecified, in that one of the parameters is redundant. The fitted coefficients then provide a ranking of the items from best to worst.

(a) Show that the fit from this model is identical to that if a quasi-symmetry model (Section 7.2.3) is fit to the $K \times K$ contingency table that cross-classifies the n_{ij} counts.

(b) The NCAA Division I men's basketball tournament, commonly referred to as "March Madness," is the tournament that ultimately leads to the crowning of the collegiate national basketball champion. The tournament is organized into four regional 16-team single-elimination tournaments, with the winners from each regional tournament advancing to the national semifinals (the "Final Four"). Within each region, the teams are seeded from number 1 (the team viewed as strongest in the region by the tournament selection committee) to number 16, and first-round games then match the highest seeded teams against the lowest (1 versus 16, 2 versus 15, and so on). The second round matches the winner of the $(1, 16)$ game against the winner of the $(8, 9)$ game, the winner of the $(2, 15)$ game against the winner of the $(7, 10)$ game, and so on. The file **ncaa** gives the results of all games played during the first 12 years of the current format of the tournament (1985–1996), with the observations being the 120 possible distinct seeding pairings, although many of the seeding pairings never occurred (Smith and Schwertman, 1999). Fit the Bradley–Terry model to these data. (*Note*: The file also contains the 16 constructed variables needed to fit the model as a logistic regression.) Does the model provide a good fit to these data?

(c) Order the seedings from best to worst on the basis of the model. Do the data support the quality assessments of the committee through the years? Are there any particularly surprising patterns?

(d) Two seeding levels i and j can be tested as being significantly different from each other by constructing a test (or confidence interval) for $\beta_i - \beta_j$. Are seedings 8 and 9 significantly different from each other? What about seedings 1 and 2? (*Hint*: The standard error of $\hat{\beta}_i - \hat{\beta}_j$ satisfies

$$\text{s.e.}(\hat{\beta}_i - \hat{\beta}_j) = [V(\hat{\beta}_i) + V(\hat{\beta}_j) + 2\text{Cov}(\hat{\beta}_i, \hat{\beta}_j)]^{1/2};$$

these values can be obtained from the covariance matrix of $\hat{\boldsymbol{\beta}}$, which then allows construction of a Wald test or confidence interval. A likelihood ratio test for $\beta_i = \beta_j$ can be constructed by comparing the deviance for the Bradley–Terry model to the model where the variables for seedings i and j are replaced by one that treats those seedings as identical.)

(e) An alternative model to consider is a logistic regression using the difference in seedings as the predictor variable. How does the fit of this model compare to that of the Bradley–Terry model? (Note that this model assumes that the seedings accurately reflect an ordering of the quality of the teams, while the Bradley–Terry model ignores this information.) What about a two-predictor model with each of the seeding variables as the predictors? How do these models compare to an ordinal quasi-symmetry fit?

3. The file japansolvent contains data from a retrospective study on 52 Japanese firms, 25 of which became bankrupt, including EBIT/TA, Net income divided by total capital (NI/TC), Sales/TA (S/TA), EBIT/S, NI/S, WC/TA, BVE/BVL, and BVE/TA (see page 381 for a description of some of these variables). I am indebted to Shin Ichiro Nishino and Korehiro Odate for gathering these data, and to Gary Simon for sharing them with me. Build a model for the bankruptcy status of Japanese firms as a function of (possibly a subset of) these variables. How does the model compare to the one derived for U.S. firms in Section 9.2?

4. Consider a retrospective study where the sampling rate for successes is τ_Y and that for failures is τ_N. The sampling rate of successes is defined for finite populations as the proportion of the population of successes (with population size N_Y) in the sample; the sampling rate of failures is defined similarly. Show that the correct adjustment to the intercept of a logistic regression fit to data of this type in order to produce correct prospective probability estimates is

$$\tilde{\beta}_0 = \hat{\beta}_0 - \log\left(\frac{\tau_Y}{\tau_N}\right).$$

That is, show that this is equivalent to (9.4).

5. The Cy Young Award is awarded annually to the best pitchers in the National and American Leagues, respectively, based on a vote of the members of the Baseball Writers Association of America. The file cyyoung8092 gives information for winners and losers of the award for the years 1980–1992. Data are given for the pitchers with the two highest wins totals and two highest saves totals in each league for each year, along with the actual Cy Young winner, if he was not in the former groups. Included in the file is the year, league, type of pitcher (starter or reliever), wins, winning percentage, saves, earned run average, strikeouts, innings pitched, and whether the pitcher won the Cy Young Award. Data for the 1981 season are not included, since a labor stoppage resulted in the cancelation of more than one-third of the games that season. I am indebted to Sean McCormick for gathering these data and sharing them with me.

   (a) Build a logistic regression model for the probability that a pitcher wins the Cy Young Award as a function of all of these predictors, as well as any variables you think might be useful to use (you might consider standardizing variables such as strikeouts by innings pitched, for example). Construct a classification table based on this model, classifying the pitcher from each league and year group with highest estimated probability of being Cy Young winner as the predicted winner. How well does it classify the winners versus the losers?

   (b) Now construct a (presumably simpler) "best" model based on the predictors that you think add to the predictive power of the model. How well does this model classify the winners versus the losers?

   (c) The file cyyoung9302 gives corresponding data for the years 1993–2002 (other than 1994, when a labor stoppage resulted in almost one-third of the season being canceled). Validate the models from parts (a) and (b) on the 1993–2002 data, constructing a predictive classification table of Cy Young winning and losing. Which model makes more accurate classifications?

(d) A reasonable strategy is to update the model after each season, and use the updated model to classify the next season. So, for example, the 1980–1992 data are used to predict the 1993 award, the 1980–1993 data are used to predict the 1995 award, and so on. Compare the predictive power of the approaches in parts (a) and (b) to this updating strategy. Does updating the model improve predictive accuracy?

(e) The logistic regression models assume that the observations are independent, but that is not the case here, since within a given year and league, knowing that one pitcher won the award guarantees that the other pitchers lost. How do you think this affects the properties of the model? How might you take this problem into account in your analysis?

6. An April 1997 lawsuit involved workers in a large multinational corporation, and concerned a nationwide downsizing in 1995 of certain repair technicians (the lawsuit dealt more specifically with layoffs in southern Manhattan). The downsizing was implemented based on a scoring system, with points being awarded to each technician on the basis of length of service (in years), a previous performance appraisal measure (normalized), and "Comparative Appraisal Points" (CAP), a rating by the employee's immediate supervisor from 0 to 60 on the employee's performance in technical, business, innovativeness, effectiveness, execution, and analysis categories. This latter measure, which was completely subjective, accounted for 60% of the overall score. After all of the employees were scored, they were ordered from highest to lowest, with the lowest scorers being laid off. Ultimately, 19 of 264 workers were laid off from Manhattan South, 15 of whom were black or Hispanic (there were 97 minority workers total). That is, the layoff rate was 15.5% for minority workers, and 2.4% for nonminority workers. Ten of the laid off minority workers filed a discrimination lawsuit. The data are given in the file lawsuit.

(a) The defendant's statistical expert attempted to argue against discrimination using the method of *reverse regression*. The principle is to build a regression model with the alleged discriminatory action as the target variable, and predicting variables that include uncontroversial job qualification measures (such as length of service, education qualifications, and so on) and a variable defining the discriminated group membership (here minority or nonminority). Then, if the group membership variable is statistically significant, this is taken as evidence that, given generally accepted factors, the action being examined is related to group membership. So, for example, if the target is salary, a significant gender effect given educational level, job description, etc., can be taken as *prima facie* evidence of possible discrimination. See Sections 13.3.1 and 14.7.1 of Finkelstein and Levin (2001) for further discussion of this type of analysis. In this case the discriminatory action is not payment of salary, but rather the act of laying off the workers or not, so the appropriate regression analysis is based on logistic regression. What are the results of a logistic regression fit modeling the probability of being laid off? Does minority status add significant predictive power to the model, once the other variables are included in the model?

(b) The plaintiff's statistical expert pointed out that this is not a valid application of reverse regression. When using regression to try to establish a pattern of discrimination (or refute one), only variables that are uncontroversially and objectively related to the action at hand can be included in the regression. Measures like length of service and normalized PA were accepted as fair measures of quality, but the CAP score was precisely what the plaintiffs claimed caused the discrimination (recall that it was a completely subjective measure determined by the employee's immediate supervisor). It is not surprising that once this (possibly discriminatory) measure is included in the regression, the minority effect is not significant; it has simply been absorbed into the CAP effect. How does the model change if CAP score is omitted from the logistic regression? What is the estimated odds ratio of being laid off versus not being laid off for minority versus nonminority workers, given the other variables? Is this odds ratio significantly different from 1 (indicating potential racial discrimination)?

7. Consider again the *Challenger* space shuttle data, given in the file `challenger`.

   (a) Build a logistic regression model for primary O-ring damage, where each flight is coded 1 if at least one O-ring is damaged, and 0 if there is no O-ring damage. Is the October 30, 1985 flight still an outlier?

   (b) The data set also includes an indicator variable as to whether there was O-ring damage in the nozzle joints of the shuttle for that flight. Fit a logistic regression model for nozzle joint O-ring damage. How does it compare to the model in part (a) for field joint O-ring damage?

8. The file `creditscore` (page 193), also contains information on whether the consumer's application for credit was accepted. Construct a

credit scoring model for acceptance of credit based on the available variables. Which variables have the strongest relationship to credit acceptance?

9. The file `halloffame` contains data for all of the modern position players eligible for membership in the Major League Baseball Hall of Fame. The original data source (Cochran, 2000) included various measures of player productivity that have been proposed by baseball researchers and fans, but the file given here only includes official statistics provided by major league baseball.

   (a) Build your "best" model for predicting Hall of Fame membership from the data given using career total variables (total hits, runs, home runs, runs batted in, etc.). How well does the model classify the data? Does position played matter? Why do you think that might be the case? (*Note*: The Hall of Fame membership variable is coded 0 [not a member], 1 [elected by vote of the members of the Baseball Writers' Association of America (BBWAA) during the first 20 years of eligibility), and 2 [selected by a vote of the Veterans Committee if not elected by the members of the BBWAA]. For this question, do not distinguish between the two different methods of being elected to the Hall.)

   (b) Build your "best" model for predicting Hall of Fame membership from variables based on average performance (batting average, slugging average, on base percentage, home runs per at bat, etc.) How well does the model classify the data? Does position played matter?

   (c) Plot the estimated probabilities of being a member of the Hall of Fame from each of your two models against each other. How closely do they agree? Are there any players for whom there are very different estimated probabilities? What accounts for the difference?

   See Berry (2000) for related discussion.

10. The following table gives the number of natural (spontaneous) abortions from pregnancies in three districts (rural, intermediate, and urban) of Shizuoka City, Japan, by the degree of consanguinity of the parents (no relation, second cousins, $1\frac{1}{2}$ cousins [the common ancestor of the parents is a grandparent of one and a great-grandparent of the other], first cousins) (McDonald and Diamond, 1983).

Consanguinity

District	None	2nd cousins	$1\frac{1}{2}$ cousins	1st cousins
Rural	27/958	1/160	3/65	12/293
Intermediate	67/2670	11/338	11/237	23/654
Urban	7/543	4/70	3/110	7/260

(a) Build a model for spontaneous abortion rate, treating district and consanguinity as nominal variables. Is type of district related to abortion rate?

(b) It is reasonable to suppose that the abortion rate would increase with closeness in relation of the parents. Based on the estimated regression coefficients, that is not the case here, with the first and $1\frac{1}{2}$ cousins groups in the incorrect order. If a desired ordering of the coefficients is violated, it can be enforced by pooling groups together. Fit the model with the two groups pooled. Does enforcing this restriction hurt the fit significantly?

(c) A simple way to enforce an ordering is to treat consanguinity as a numerical variable (district also can be treated this way). Does this lead to a simpler model that fits well?

11. The Minority Engineering Programs office of the Florida A&M University–Florida State University College of Engineering provides programs designed to encourage student development. Ohland et al. (2000) report the six-year graduation rates from the 1987–1993 cohorts of students who participated in the programs and those who did not, which are given in the file famufsu.

(a) The analysis in Ohland et al. (2000) did not take the time ordering of cohort year into account. Fit a logistic regression to these data consistent with treating cohort as nominal. Is there evidence that the programs are effective in increasing graduation rates? Is the effect apparently changing over time?

(b) Now fit a logistic regression model incorporating cohort as a numerical variable. Does this change the estimated effect of the programs? What does this model say about changing graduation rates over time?

(c) Fit an additive logistic model that brings in cohort as a smooth term. Choose the smoothness of the smooth term using AIC_C. Is there evidence that the relationship with cohort is nonlinear?

12. The file **AIDS** contains information gathered by the U.S. Bureau of the Census and the Centers for Disease Control and Prevention Divisions of AIDS/HIV Prevention, summarizing the number of people in the United States living with AIDS in 2000, separated by age group, racial group, and sex. I am indebted to Brendan Lim for gathering these data and sharing them with me. Model the incidence of AIDS as a function of these variables, allowing for the possibility of interaction effects. Given the very large number of people involved (roughly 220 million), consider model selection taking sample size into account (using BIC, for example). What can you say about AIDS incidence in the United States at the start of the 21st century?

13. Two link functions g_1 and g_2 are said to form a *complementary pair* if $g_1(p) = -g_2(1 - p)$. That is, if g_1 is used to model successes and g_2 is used to model failures, the regression coefficients will be identical in absolute value with opposite sign, and the fitted probabilities $\hat{p}$ (based on g_1) and $\widehat{1 - p}$ (based on g_2) sum to one.

    (a) Show that logit(p) and logit$(1 - p)$ form a complementary pair, as do probit(p) and probit$(1 - p)$.

    (b) Show that cloglog(p) and loglog$(1 - p)$ form a complementary pair, as do loglog(p) and cloglog$(1 - p)$.

14. (a) Show that the estimated median lethal dose (page 395) when fitting a model using the logistic or probit link functions equals $-\hat{\beta}_0/\hat{\beta}_1$.

    (b) Show that the estimated median lethal dose when fitting a model using the complementary log–log link function equals $-\hat{\beta}_0/\hat{\beta}_1 + \log(\log 2)/\hat{\beta}_1$.

    (c) Confirm the median lethal dose value given earlier in the discussion of the beetle dose-response data (page 397), and compare it to the values for logit and probit regression. Are they similar to each other?

    When the response in a dose response study is something other than death, such as response to a drug, the median lethal dose is referred to as the *median effective dose*, ED_{50}.

15. Limiting dilution analysis (LDA) is a type of analysis used to determine in a population of cells the unknown frequency of discrete clones of cells that respond with a defined response. Limiting dilution analysis methods can be used to allow detection, quantification, and analysis of individual (rare) cell types by investigating their clonal growth as a function, for example, of the number of cells in the sample. The file **lda** gives LDA data from StemCell Technologies (2001), representing the number of positive responses (the number of wells of

cells with a positive response) out of the total number of replicates at different levels of cells per well, for three experimental protocols.

(a) Fit a logistic regression model to these data. Is there a significant difference in the effectiveness of the three protocols?

(b) LDA models are often fit based on a complementary log–log link function. Does that provide a better fit to these data than does the logistic model?

16. Pregibon (1980) described a test for the adequacy of the link function in a generalized linear model using a linear approximation to the link function. The test is based on creating two artificial variables that represent symmetric link violations (as in the logit versus the probit link) and asymmetric link violations (as in the logit versus the complementary log–log link), and testing the significance of these variables when added to the original model. The proposed variables in the logistic regression context are

$$z_{1i} = [\log^2(\hat{p}_i) - \log^2(1 - \hat{p}_i)]/2$$

(symmetric link violation) and

$$z_{2i} = -[\log^2(\hat{p}_i) + \log^2(1 - \hat{p}_i)]/2$$

(asymmetric link violation). Construct this test for the Bliss beetle dose-response data (page 397). Does it support the use of the complementary log–log link over the logit link?

17. Consider again the *Challenger* space shuttle data, given in the file **challenger**.

(a) Fit a probit model for the probability of primary O-ring damage as a function of ambient temperature, addressing the outlying October 30, 1985 flight if necessary. How does the probit fit to the data compare to that of the logit model derived in the text?

(b) Now fit a complementary log–log model to these data. How does its fit compare to those of the logit and probit models?

(c) Fit logit, probit, and complementary log–log models to these data that are quadratic functions of temperature. Does adding the quadratic term improve the fit noticeably?

(d) Estimate the probability of O-ring damage at an ambient temperature of 31 ° for each of the models constructed in parts (a)–(c). How do these values compare to the logistic regression estimate of .96, and to each other? Use each of the probability estimates to estimate the probability that more than half of the six O-rings would be damaged (using the underlying binomial distribution) at a temperature of 31 °. Are these estimates very different from each other?

(*Note*: Draper, 1995, used an analysis of this type to illustrate the importance of the effects of model selection uncertainty on the precision of predictions.)

18. Consider again the data on menstruation patterns of Warsaw girls, graphed in Figure 9.1 and given in the file **menarche**.

   (a) Fit a logistic regression model to these data, using age as the predictor. Does this model seem to provide an adequate fit to the data?

   (b) Does the goodness-of-link test discussed in the previous question suggest that a different link might be a better choice for these data?

   (c) Now fit a probit regression to these data. Does the probit link result in a better fit than the logit link? Why do you think that the former link function might be better for these data than the latter one?

   (d) Draw a plot of the empirical logits

   $$\log\left[\frac{\hat{p}_i + 1/(2n_i)}{1 - \hat{p}_i + 1/(2n_i)}\right]$$

   versus age. Does this suggest any nonlinearity in the relationship between probability of menstruation and age in a logit scale?

   (e) Fit a model that is quadratic in age using logit and probit link functions. Do either of these provide a better fit to these data?

   (f) Fit a logistic regression model based on $\log(\text{Age} - 9)$ and $\log(18 - \text{Age})$ to these data (Lloyd, 1999, pages 196–198). Is this a better choice than any of the other models? What does this model say about the relative odds of menstruation as age decreases to 9 compared to that as age increases to 18?

(*Note*: See Stukel, 1988, for discussion of application of generalizations of the logistic regression family to these data.)

19. The file **hiroshima** gives the number of cells with exchange chromosome aberrations, among 100 cells examined per subject, for 649 survivors of the atomic bomb in Hiroshima, along with the radiation dose in rads (Prentice, 1986). The dose given is actually the average of values from ranges of radiation values.

   (a) Fit a logistic regression model to these data, treating dose as a numerical variable. What does the model say about the relationship between radiation dose and chromosome aberration rate? Does the model provide a reasonable fit to the data? If not, why not?

(b) Now fit a model treating dose as nominal. Is there evidence that a nonlinear relationship (in the logit scale) is a better choice here?

(c) Is there evidence of overdispersion in these data? How do your inferences in parts (a) and (b) change if you take this into account? Does a beta-binomial regression model provide a better fit to the data, treating dose as numerical? What if dose is treated as nominal? Would separate beta-binomial fits for each exposure group be a more effective summary of the patterns in the data?

20. The discussion of the Broadway shows data starting on page 96 focused on the number of major Tony Awards won, based on the number of nominations for each show. Fit a regression model for these data, along the lines of what was done for Tony nominations in Section 9.5.2, using show characteristics that might be relevant, such as type of show, newspaper reviews, and so on. How does the chosen model for estimating the probability of winning a Tony given being nominated differ from that for estimating the probability of getting a nomination? In particular, how do the newspaper reviews relate to the probability of winning a major Tony Award given being nominated for it? How do these relationships compare to those when estimating the probability of being nominated for a Tony Award?

21. The file ossification summarizes the results of an experiment where 81 pregnant rats were randomly allocated to either a control, 100 mg/kg of trichloropropene oxide (TCPO), 60 mg/kg of phenytoin (PHT), or a combination of the two drugs (Morel and Neerchal, 1997). After 18 days each rat was sacrificed, and the presence or absence of ossification (the hardening of soft tissue into a bonelike material) of the left middle phalanx was determined for each fetus.

(a) Is there evidence of a drug effect on ossification rate? Does the combination of the two drugs provide an effect more or less than an additive one?

(b) The clustering of observations within each mother suggests that overdispersion might be a problem here. Is that the case? Does a beta-binomial model provide a better fit to the data? Does this model change your inferences of the drug effects?

22. As was noted on page 158, while the analysis of the 2000 Florida election data treated Buchanan's vote as a count, with logged total votes an offset, a natural way to model these data is as a set of binary random variables (vote for Buchanan out of total votes for each county).

(a) Repeat the previous Poisson regression analysis, now as a logistic regression analysis. Are the results different? Would you expect them to be?

(b) Now reanalyze the data, allowing for overdispersion by fitting a beta-binomial regression. Are the results similar to those of the negative binomial regressions constructed earlier? Would you expect them to be?

23. The file **seropositive** comes from Farrington et al. (2001), and gives age-stratified seroprevalence (the rate at which the population tests positive from a diagnostic test) data from the United Kingdom for mumps and rubella (collected in 1986–1987) and parvovirus (collected in 1991).

(a) Fit a constant shift logistic regression of seroprevalence by age to these data (that is, consistent with the same odds ratio by age for the three diseases). How does this model fit?

(b) Fit separate logistic regressions for each disease. Is this a better representation of the data?

(c) Fit a smooth predictor in the logistic regression for each disease. Do these curves indicate significant nonlinearity in the logit scale? Are the patterns similar for the three diseases?

(d) The last epidemics before the survey occurred in 1984 for mumps and 1983 for rubella. This should result in lower than expected seropositive rates for children born just after these epidemics. Is that the case here?

24. On March 13, 1999, a fight between heavyweight champions Evander Holyfield and Lennox Lewis ended in controversial draw. Many observers believed that Lewis was the clear winner of the fight. Lee et al. (2002) discussed this fight, and gave the round-by-round judging results of the fight for the three official judges, as well as those for seven unofficial observers of the fight. The data are available in the file **boxing1**. In these data rounds that were judged even are treated as being awarded to Holyfield, since Lewis was perceived as the winner by the media and the public.

(a) Construct a logistic regression model for the probability of a round being awarded to Lewis, as a function of the judge and the round (treating the round as a nominal variable). Is there evidence that the judges significantly disagreed with each other (that is, a significant judge effect)? Which judges significantly favored Holyfield? Which favored Lewis? Is there evidence of a significant round effect? One judge, Eugenia Williams, was accused of fixing the fight in favor of Holyfield; is there any evidence of this in the data? Williams awarded the fifth round to Holyfield at the time of the fight, but then said several days later that she should have awarded it to Lewis. Was her choice at the time of the fight particularly surprising, given the model fit?

(b) Classify each round as predicted to be for Holyfield or Lewis based on the estimated probability from the model. Based on this, who should have been declared the winner of the fight?

(c) It is difficult to fit this model, because of redundancies in the data. For example, all judges awarded rounds 1 and 2 to Lewis, and all awarded round 10 to Holyfield; three of the unofficial judges (HBO, Sportsticker, and the *Boxing Times*) scored the fight in exactly the same way; and one judge (ESPN) awarded every round but the tenth to Lewis. Does a conditional analysis allow you to make specific statements about the judge and row effects despite this? Recode the data so as to incorporate these redundancies in the model, so that all of the parameters and their standard errors are estimable. Can you say anything more specific about the judge and round effects now?

(d) Can the judge effect be more simply represented as just an official versus unofficial judge effect?

(e) The file boxing2, also from Lee et al. (2002), gives corresponding data for the September 18, 1999, welterweight title fight between Oscar de la Hoya and Felix Trinidad. The fight was awarded to Trinidad on a majority decision, supposedly because de la Hoya thought he had the fight won after nine rounds, and stayed away from Trinidad in the last three rounds. Fit a logistic regression model to these data. Is there evidence of a judge effect in this fight? Do the data support the idea that de la Hoya lost the fight because of his performance in the last three rounds? Overall, who does this model predict should have been declared the winner of the fight?

25. Hutto et al. (1991) describe a small prospective study on the usefulness of CD4 and CD8 serum levels (each at one of two blood levels) in infants as predictors of HIV infection. The data are given in the file hivcd4cd8.

(a) Attempt to fit a logistic regression for probability of HIV infection as a function of the other variables. What is going wrong here? Does fitting a simpler model solve the problem?

(b) Perform a conditional (exact) analysis of these data. Which variables provide significant predictive power?

(c) Can you effectively approximate the exact analysis using saddlepoint methods? (See Kolassa, 1997.)

26. The file **fraud** contains data from Mehta et al. (2000) regarding 127 automobile accident insurance claims in Massachusetts in 1987. The ten predictors are indicator variables that fall into broad categories relating to either accident, claimant, or injury, while the target is an assessment by consensus among four independent claims adjusters as to whether the claim was fraudulent.

(a) Fit a logistic regression model using all of the predictors. Does the model fit the data? According to the Wald test p-values, which variables provide predictive power for the probability of fraud?

(b) Use AIC_C to choose your "best" model(s). How do your conclusions compare to those in part (a)?

(c) Perform a conditional analysis of these data. How do the exact p-values compare to those in part (a)? How does the assessment of which variables are important here compare to your results in part (b)?

10
Regression Models for Multiple Category Response Data

The generalization of categorical data from two categories (the binomial random variable) to multiple categories (the multinomial random variable) is a fundamental step in the analysis of contingency tables, allowing (for example) generalization of analysis for 2×2 tables to $I \times J$ tables. The need for such generalizations carries over to regression analysis as well. In a clinical trial context, for example, investigation of the effectiveness of a treatment might be the goal of the analysis, but the effects of a treatment might have several levels, such as ineffective, moderately effective with side effects, moderately effective without side effects, completely effective with side effects, and completely effective without side effects. This multinomial target variable is obviously different from a binary success/failure target, but the goal remains the same: to develop a model relating predictors to the probability of falling in each of the levels of the target. Depending on the structure of the response variable (nominal or ordinal), different generalizations are possible.

10.1   Nominal Response Variable

10.1.1   Multinomial Logistic Regression

Consider the situation where there is no natural ordering to the levels of the target variable. *Multinomial* (or *polytomous*) *logistic regression* models (or, more simply, multinomial logit models) are derived in the following way. Let $p_j, j = 1, \ldots, J$ be the probability of falling in the jth level of a multinomial target variable Y. The goal is to construct a model for p_j as a function of a set of predictor values $\mathbf{x}$, where $\sum_j p_j = 1$. The model is again based on logits, but the difficulty is that there is no single "success" category to use to construct the logit. The solution is to construct all of the logits relative to one of the levels, which is termed the baseline category. If one of the target categories can be viewed as different from the others, it is natural to choose it as the baseline. For example, in the clinical trial example mentioned earlier, "ineffective" would be a natural baseline, while in a multicategory case/control study, the level corresponding to the control group is the best choice. If no level is an obvious baseline, one is chosen arbitrarily.

Say the baseline category is the Jth category. The logistic regression model is then

$$\log\left(\frac{p_j}{p_J}\right) = \beta_{0j} + \beta_{1j}x_1 + \cdots + \beta_{kj}x_k, \qquad j = 1, \ldots, J-1. \qquad (10.1)$$

The model incorporates $J-1$ separate equations, each of which is based on a distinct set of parameters $\boldsymbol{\beta}$. Obviously, for the baseline category J, $\beta_{0J} = \beta_{1J} = \cdots = \beta_{kJ} = 0$. The arbitrariness of the choice of the baseline category is apparent from the model itself. Say category I is taken to be the baseline category. From (10.1),

$$
\begin{aligned}
\log\left(\frac{p_j}{p_I}\right) &= \log\left(\frac{p_j/p_J}{p_I/p_J}\right) \\
&= \log\left(\frac{p_j}{p_J}\right) - \log\left(\frac{p_I}{p_J}\right) \\
&= (\beta_{0j} + \beta_{1j}x_1 + \cdots + \beta_{kj}x_k) - (\beta_{0I} + \beta_{1I} + \cdots + \beta_{kI}x_k) \\
&= (\beta_{0j} - \beta_{0I}) + (\beta_{1j} - \beta_{1I})x_1 + \cdots + (\beta_{kj} - \beta_{kI})x_k \qquad (10.2)
\end{aligned}
$$

That is, the logit coefficients for level j relative to a baseline category I are the difference between the coefficients for level j relative to baseline J and the coefficients for level I relative to baseline J. Note that there is nothing in this derivation that requires category I to be a baseline category; equation (10.2) applies for any pair of categories.

This result can be useful if the odds for a particular pair of levels is of interest. Since the choice of the baseline is arbitrary, the estimated coefficients (and, perhaps more usefully, their standard errors) can be easily obtained using any multinomial logistic regression software by changing the baseline category to one of the two levels, with the guarantee that the estimated probabilities of falling in each level will not change. Model (10.1) implies a simple functional form for these probabilities. The form is the familiar S-shape for a logistic relationship,

$$p_j = \frac{\exp(\beta_{0j} + \beta_{1j}x_1 + \cdots + \beta_{kj}x_k)}{\sum_{\ell=1}^{J} \exp(\beta_{0\ell} + \beta_{1\ell}x_1 + \cdots + \beta_{k\ell}x_k)}. \qquad (10.3)$$

The estimates $\{\beta_1, \ldots, \beta_{J-1}\}$ are obtained using maximum likelihood, where the log-likelihood is

$$L = \sum_{j=1}^{J} \sum_{y_i=j} \log p_{j(i)},$$

where the second summation is over all observations i with response level j, and $p_{j(i)}$ is the probability (10.3), substituting in the predictor values for the ith observation. Exponentiating the fitted regression coefficient $\hat{\beta}_{mj}$ gives the estimated odds ratio of level j relative to baseline level J for predictor values $\mathbf{x}_0^*$ versus $\mathbf{x}_0$, where the two sets of predictor values differ only in that x_m is one unit higher in $\mathbf{x}_0^*$ than in $\mathbf{x}_0$. The estimated odds of level j relative to baseline level J when all predictor values equal zero is obtained from the constant term as $\exp(\hat{\beta}_{0j})$. Adequacy of the model fit can be assessed using χ^2 goodness-of-fit tests, as long as there are enough replications for each observation (that is, each observation has as a response a multinomial vector based on n_i trials, with n_i large enough). Similarly, Pearson residuals can help identify poorly fitting probabilities if there are enough replications.

When all of the predictors are categorical the multinomial logistic regression model is equivalent to a loglinear model. The model includes the highest-order interaction among all of the predictors, and the interactions of the predictor terms with the variable that corresponds to the target, just as was true for binary logistic regression models.

An obvious question when considering model (10.1) is to wonder about the need for these models at all. Why not fit the $J - 1$ logit functions in (10.1) using $J - 1$ separate logistic regressions? There are several reasons. First, the estimates of β are more efficient when fit using the multinomial logistic regression model (that is, they have smaller standard errors than they would if separate logistic regressions were fit). Second, if the multinomial regression model is fit, the estimated coefficients for the logit comparing any two categories as given in (10.2) will be the same no matter which level is taken as the baseline category. This is not the case if separate binary logistic regressions are fit, which means that in the latter case estimated probabilities and odds ratios are dependent on the (possibly arbitrary) choice of baseline category. Finally, fitting a multinomial regression allows for a test of whether a predictor, or set of predictors, provides significant predictive power for discrimination between all of the levels of the response, rather than a specific pair of them. For example, consider the test for whether the variable x_m is necessary given the others, a test of the null hypothesis

$$H_0 : \beta_{m1} = \cdots = \beta_{mJ} = 0$$

versus the alternative that at least one of these coefficients is not zero. The likelihood ratio statistic for testing this null is the difference between $-2L$

for the model that includes x_m and $-2L$ for the model that does not, and can be compared to a χ^2 critical value on $J - 1$ degrees of freedom.

There is, however, a price that must be paid for the latter capability. If different subsets of the predictors are useful in comparing level i to baseline level J than when comparing level j to level J, separate logistic regressions allow for the appropriate simplifications for each of the individual pairs. The multinomial regression model requires that if a variable x_m is useful in modeling the odds for one level versus the baseline, it must appear in the model for all of the levels, even if it does not provide any useful predictive power for other levels (although in principle it is possible to construct such models with individual coefficients constrained to be zero).

Analyzing Literary Styles

It is generally recognized that authors have inherent literary styles that are unique to them, and serve as a way to identify them from their written works. Such identification allows researchers to ascribe authorship to works with disputed or unknown authors. Author identification is also becoming increasingly important in criminal investigation, as ransom notes, threatening letters, suicide notes, and so on, provide a paper trail leading to suspects (Chaski, 1997). Quantifying authorship styles is not easy — it is challenging to determine and summarize appropriate numerical characteristics of the style of an author's work. Such quantifications have been based on the frequencies of each word in a text, the lengths of words, vocabulary richness (the use of new words given the number of distinct words in earlier works), and sentence length. The famous study of Mosteller and Wallace (1963, 1964) on authorship of *The Federalist* papers focused on frequencies of function words (words with very little contextual meaning), the idea being that authors do not think consciously about the use of these words, and they are thus likely to be used similarly in a variety of works by the same author.

Peng and Hengartner (2002) used several multivariate analysis techniques, including canonical discriminant analysis and principal component analysis, to identify authorship using function words counts in a set of works by several different authors. The file **authorship** contains word counts (from blocks of text that each contain 1700 total words) from several sets of works by Jane Austen, Jack London, John Milton, and William Shakespeare. I am indebted to Roger Peng for making these data available here. The data include counts for 69 function words, such as "a," "by," "no," "that," and "with." The task is to develop a model that distinguishes between the authors based on these word counts. That is, each block of text has as the response a multinomial vector with four categories (the four authors), based on one trial (each block of text was written by one and only one author).

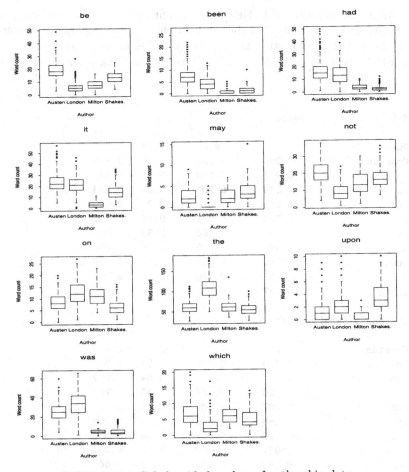

FIGURE 10.1. Side-by-side boxplots of authorship data.

With 69 potential predictors, it is clear that variable selection is key here. Peng and Hengartner (2002) found 11 words to load highly on the first three principal components in their data set (which included other authors besides those studied here), and we will treat those as the set of potential predictors: "be," "been," "had," "it," "may," "not," "on," "the," "upon," "was," and "which." Figure 10.1 gives side-by-side boxplots for each of these 11 words, separated by author. It is apparent that these word counts differentiate the authors from one another. Jane Austen is characterized by high usage of "be" and "not," and low usage of "the" and "upon"; Jack London used "on," "the," and "was" often, but "be," "may," "not," and "which" rarely; John Milton's counts were generally low (except for "on" and "not"), especially of "been," "had," "it," and "was"; and William Shakespeare used "upon" relatively often, and "been," "had," "on," "the," and "was" rarely.

An initial multinomial logistic regression fit (with Shakespeare being the baseline category) finds that the word "been" adds only marginally to any of the underlying logistic fits, and "may" adds nothing given the other words. This leaves a model based on three logistic fits, each based on ten parameters, as follows:

Milton vs. Shakespeare

Word	Coef	s.e.	Wald	p
Constant	-2.223	4.594	-0.48	0.628
be	-0.070	0.150	-0.47	0.639
had	0.174	0.206	0.84	0.398
it	-0.820	0.216	-3.80	0.000
not	0.054	0.121	0.45	0.653
on	0.536	0.190	2.83	0.005
the	0.074	0.045	1.66	0.096
upon	-1.763	0.625	-2.82	0.005
was	0.293	0.160	1.83	0.067
which	-0.087	0.193	-0.45	0.650

Austen vs. Shakespeare

Constant	-16.557	4.723	-3.51	0.000
be	0.460	0.129	3.56	0.000
had	0.670	0.220	3.04	0.002
it	0.026	0.087	0.30	0.762
not	-0.037	0.088	-0.42	0.674
on	0.468	0.195	2.40	0.016
the	-0.002	0.029	-0.06	0.950
upon	-1.951	0.562	-3.47	0.001
was	0.654	0.153	4.29	0.000
which	0.064	0.147	0.43	0.664

London vs. Shakespeare

Constant	-16.065	5.203	-3.09	0.002
be	-0.134	0.159	-0.84	0.399
had	0.605	0.231	2.62	0.009
it	0.105	0.101	1.04	0.299
not	-0.279	0.116	-2.41	0.016
on	0.496	0.207	2.39	0.017
the	0.129	0.027	4.69	0.000
upon	-1.644	0.622	-2.64	0.008
was	0.642	0.151	4.24	0.000
which	-0.547	0.209	-2.62	0.009

It is intriguing, although hardly definitive, that as an author's lifetime gets farther from that of Shakespeare, the strength of the discrimination from Shakespeare increases. So, while the model for Milton (who was born eight years before Shakespeare's death) has only three coefficients statistically significant at a .05 level, the model for Austen (born 159 years after Shakespeare's death) has five such coefficients, and the model for London (born 260 years after Shakespeare's death) has seven such coefficients. The coefficients in each logistic equation follow the patterns in the side-by-side boxplots, demonstrating that the marginal and conditional relationships are similar for these data.

How well does this model distinguish these authors? Very well, in fact, as the classification table shows:

Predicted result

		Austen	London	Milton	Shakespeare	
	Austen	308	5	0	4	317
Actual	London	1	294	1	0	296
result	Milton	0	1	52	2	55
	Shakespeare	3	1	1	168	173
		312	301	54	174	

Almost 98% of the blocks of text are correctly classified, far greater than $C_{max} = 38\%$ and $C_{pro} = 39\%$. As would be expected, the overall likelihood ratio test of the strength of the relationships is overwhelmingly significant, with $LR = 1982.4$ on 27 degrees of freedom.

In situations where there is only one predictor, a plot of the estimated probabilities for each group as a function of that predictor provides a useful counterpoint to side-by-side boxplots. Figure 10.2 gives such a plot for a model based on only counts of the word "been" (recall that this variable was omitted in the multiple regression, but the side-by-side boxplots in Figure 10.1 show that it has separating ability alone).

These probability curves make more precise the patterns in the side-by-side boxplots. Jane Austen's counts of "been" were the highest for all authors, so the probability that she is the author of a work approaches one as the count for "been" increases. Shakespeare and Milton used "been" rarely, so small counts favor them. Note, however, that the probability for Shakespeare is always greater than that for Milton. Moderate values of the count for "been" are consistent with Jack London's works, and the probability for London is thus nonmonotonic, peaking at around five counts per 1700-word block. A model based only on counts of "been" would classify works with zero or one occurrence as being by Shakespeare, two through

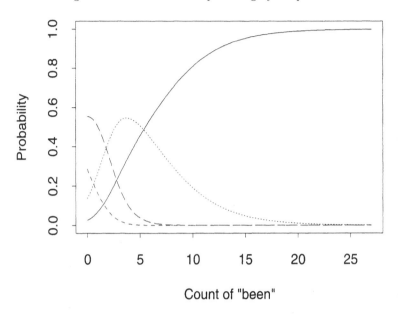

Count of "been"

FIGURE 10.2. Estimated probabilities of each author when using only count of "been" as a predictor. The solid line is the probability for Jane Austen; the dotted line is the probability for Jack London; the short dashed line is the probability for John Milton; and the long dashed line is the probability for William Shakespeare.

five occurrences as being by London, and six or more occurrences as being by Austen.

10.1.2 Independence of Irrelevant Alternatives

As was shown in (10.2), under the multinomial logit model the (log-)odds of one level of the response versus another does not depend on any of the other levels; that is, other possible outcomes are not relevant. This is known as the *independence of irrelevant alternatives* (IIA) property. This is a very reasonable property in classification problems of the sort described in the previous section, as there is no reason to think that the odds of a manuscript being written by Shakespeare versus Milton (for example) would depend on whether Austen is a possible alternative.

Discrete choice situations, where individuals decide between choices of, for example, different products or different travel options, are different. The famous red bus/blue bus discrete travel choice example illustrates the problem. Say a traveler has three choices of transportation: a red bus, car, or train. Further, say that the traveler had no preference between the three alternatives. This means, of course, that the traveler's probability of choosing each of the three alternatives is 1/3. Now, say an indistinguishable alternative to the red bus, a blue bus, becomes available. A reasonable representation of consumer behavior would be that the two bus types would split the "bus market" (each bus line then having probability 1/6 of being chosen), while the other two possibilities would be unchanged. The multinomial logit model (and any other model with the IIA property), however, implies that the odds of choosing the red bus versus car or train is unchanged by the presence or absence of the blue bus, which leads to a uniform probability distribution over the four alternatives (each alternative having probability 1/4), and hence an increase in the probability of taking a bus from 1/3 to 1/2. Thus, the multinomial logit model should only be used in situations where IIA is reasonable, such as when the different response categories are distinct and dissimilar.

It is possible to formally test the IIA assumption using observed sample data. The IIA property implies that the odds between any pair of responses do not change if an additional alternative response is present or absent, and the test is based on comparing the estimated regression coefficients using all of the responses to those after omitting one response level. The so-called *Hausman test* has the form

$$H_{IIA} = (\hat{\boldsymbol{\beta}}_R - \hat{\boldsymbol{\beta}}_F)'[\hat{V}(\hat{\boldsymbol{\beta}}_R) - \hat{V}(\hat{\boldsymbol{\beta}}_F)]^{-1}(\hat{\boldsymbol{\beta}}_R - \hat{\boldsymbol{\beta}}_F),$$

where $\hat{\boldsymbol{\beta}}_R$ is the vector of estimated regression coefficients from fitting the model with a response level omitted, $\hat{\boldsymbol{\beta}}_F$ is the vector of regression coefficients from fitting the model with all response levels included, and then omitting the coefficient for the level that was omitted in estimating $\hat{\boldsymbol{\beta}}_R$, and $\hat{V}(\cdot)$ is the estimated covariance matrix of the appropriate vector of estimated coefficients. H_{IIA} is compared to a χ^2_d critical value, where d is the length of $\hat{\boldsymbol{\beta}}_R$. If H_{IIA} is negative, the IIA hypothesis is not rejected. Note that a drawback of this test is that the choice of which response level is omitted can change the inference from the test statistic.

10.2   Ordinal Response Variable

When the levels of the target variable are ordered, models that take this ordering into account can provide a more parsimonious representation of the relationship between the target and the predictors. A common approach to handling data of this type is to just use ordinary least squares regression, with the group membership identifier as the target variable. This often doesn't work so badly, but there are several problems with its use in general:

1. An integral target variable clearly is inconsistent with continuous (and hence Gaussian) errors, violating one of the assumptions when

constructing hypothesis tests and confidence intervals. Further, an ordinal variable tends to exhibit less variability near the limits of its scale than in the middle, violating the usual constant variance assumption in least squares fitting.

2. Predictions from an ordinary regression model are of course nonintegral in general, resulting in difficulties in interpretation: what does a prediction to group 2.763 (for example) mean, exactly?

3. The least squares regression model doesn't answer the question that is of most interest in this situation — what is the estimated probability that an observation falls in a particular group given its predictor values?

4. The least squares regression model implicitly assumes that the groups are "equally distant" from each other. It can easily be argued that this is not sensible in many situations.

The solution to these problems is to generalize the logistic regression model to ordinal data. There are several ways to do this. While these approaches are based on different assumptions, they generally result in similar inferences regarding the ordinal structure in the data. The first of the models, the *proportional odds model*, is the most commonly used approach.

10.2.1   The Proportional Odds Model

The proportional odds model is based on logits, but now they are *cumulative* logits. Consider a model with predictors $\mathbf{x}$ and target group identifier Y. Let

$$L_j(\mathbf{x}) \equiv \text{logit}[F_j(\mathbf{x})] = \log\left[\frac{F_j(\mathbf{x})}{1 - F_j(\mathbf{x})}\right], \qquad j = 1, \ldots, J - 1,$$

where $F_j(\mathbf{x}) = P(Y \leq j \mid \mathbf{x})$ is the cumulative probability for response category j, and J is the total number of response categories (so Y takes on values $\{1, \ldots, J\}$). That is, $L_j(\mathbf{x})$ is the log-odds of $Y \leq j$ versus $Y > j$. The proportional odds model says that

$$L_j(\mathbf{x}) = \alpha_j + \beta_1 x_1 + \cdots + \beta_k x_k, \qquad j = 1, \ldots, J - 1. \qquad (10.4)$$

Note that this means that a positive β_ℓ is associated with increasing odds of being less than a given value j with increasing x_ℓ; that is, a positive coefficient implies increasing probability of being in lower-numbered categories with increasing x_ℓ, holding all else fixed. For this reason, the model is sometimes parameterized as

$$L_j(x) = \alpha_j - \beta_1 x_1 - \cdots - \beta_k x_k,$$

so that a positive slope means that large values of the predictor imply increasing probability of being in higher-numbered response categories.

This model can be motivated using an underlying (latent) continuous response variable Y^* representing an "actual" value that has been discretized to obtain Y. Consider, for example, a survey taken before an election, where respondents are asked their opinion of a particular candidate. The actual feelings of each respondent presumably fall somewhere along a continuum of opinions from extremely unfavorable to extremely favorable, but this is not what is ultimately reported. Rather, responses are given on an ordinal scale, such as "Strongly unfavorable" — "Unfavorable" — "Neutral" — "Favorable" — "Strongly favorable" (a Likert scale). This ordinal scale is a categorization of the underlying latent variable of actual feelings. Say Y^* (the latent variable) has a distribution with location $\beta_0 + \beta_1 x_1 + \cdots + \beta_k x_k$ for given value of $\mathbf{x}$. Then, for $-\infty = \alpha_0 < \alpha_1 < \cdots < \alpha_J = \infty$, the observed response Y satisfies

$$Y = j \text{ if } \alpha_{j-1} < Y^* \leq \alpha_j.$$

That is, we observe a response in category j when the underlying Y^* falls in the jth interval of $\boldsymbol{\alpha}$ values. This is a generalization of the threshold model for dose response modeling discussed in Section 9.4.1 to more than two groups, and is the reason that models of this type are sometimes called *grouped continuous* models.

Figure 10.3 gives a graphical representation of the model for one predictor. The points in the plot represent the relationship between the latent variable Y^* and the predictor X. What is actually observed, however, are values on the ordinal scale, depending on which of the horizontal regions contains the observed value of Y^*. So, for example, if the actual latent value is between α_1 and α_2, $Y = 2$ ("Unfavorable") is the response given.

This construction shows how the probabilities $p_j(x) \equiv P(Y = j \mid x)$ are defined. Since it represents the probability of the latent variable falling between α_{j-1} and α_j,

$$p_j(x) = F(\alpha_j) - F(\alpha_{j-1}).$$

Graphically, the probability is the area under the appropriate density curve between α_{j-1} and α_j. Note that the distances between the entries in $\boldsymbol{\alpha}$ have direct implications for the probabilities, and there is no restriction that they be (close to) equal. As x changes, the location of the distribution changes, and thus the probabilities of falling at each level change. If the latent variable given the predictor variable values follows a logistic distribution, as in (9.7) and Figure 10.3, the probabilities $p_j(x)$ satisfy the proportional odds form (10.4). Under this model $L_j(x+1) - L_j(x) = \beta$ for all j and all x; that is, the odds of a response below a category multiply by $\exp(\beta)$ when x increases by one unit, for any category j, which accounts for the "proportional odds" in the name of the model. The assumption

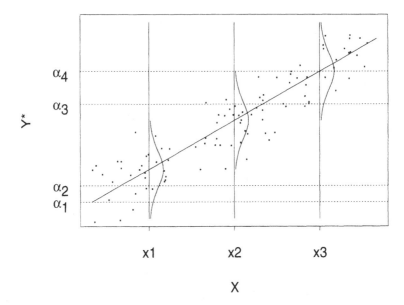

FIGURE 10.3. Graphical representation of the latent variable model for an ordinal regression relationship.

that the slope β is the same for all categories j is often called the *constant slopes* assumption. Figure 10.4 illustrates the implications of this in terms of the cumulative probabilities of each response occurring for a five-response model. The shape of the logistic cumulative distribution function for each category is the same, and is determined by β, with the curves for different categories shifted right or left, with the shifts determined by the relative values of $\boldsymbol{\alpha}$.

The latent variable representation of the proportional odds model, as represented in Figure 10.3, illustrates an important property of the model. Removal of one of the α_j values, which corresponds to merging two neighboring categories, does not change the underlying relationship. Similarly, adding a new α_j value, which corresponds to splitting a category into two parts, does not change the relationship. So, for example, if the model fits well using a "Strongly unfavorable" — "Unfavorable" — "Neutral" — "Favorable" — "Strongly favorable" scale, the estimated effects will be roughly similar based on a "Unfavorable" — "Neutral" — "Favorable" scale. Collapsing categories leads to loss of efficiency of the parameter estimates, but this is usually not serious unless the scale is collapsed down to two levels.

10.2.2   Other Link Functions

The proportional odds formulation (10.4) can be generalized to allow for linear relationships with functions of the cumulative probability other than

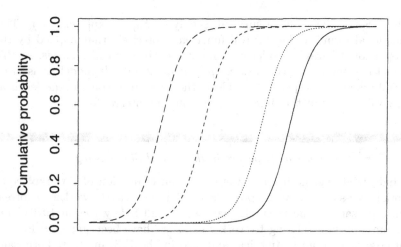

FIGURE 10.4. Cumulative probabilities for a five-group proportional odds model. The solid line refers to $P(Y \leq 1)$; the dotted line refers to $P(Y \leq 2)$; the short dashed line refers to $P(Y \leq 3)$; and the long dashed line refers to $P(Y \leq 4)$.

the logit. The general form is

$$G^{-1}[P(Y \leq j \mid \mathbf{x})] = \alpha_j + \beta_1 x_1 + \cdots + \beta_k x_k,$$

where G^{-1} is a link function, so these are termed *cumulative link* models. If the logit link is replaced by a probit, so that $G = \Phi$, the normal cumulative distribution function, the resultant model

$$\Phi^{-1}[P(Y \leq j \mid \mathbf{x})] = \alpha_j + \beta_1 x_1 + \cdots + \beta_k x_k$$

is the *cumulative probit* model. This model is consistent with an appealing conditional Gaussian distribution for the latent variable, but it sacrifices the easy interpretation of the coefficients in terms of odds ratios. In any event, just as in the binary case, the proportional odds (cumulative logit) and cumulative probit model generally give very similar fits.

An asymmetric distribution for the latent variable can be addressed using an appropriate link function. Recall from Section 9.4.3 that the complementary log-log link corresponds to an underlying extreme value distribution. In this case the model is

$$\log\{-\log[1 - P(Y \leq j \mid \mathbf{x})]\} = \alpha_j + \beta_1 x_1 + \cdots + \beta_k x_k. \tag{10.5}$$

Define $S_j(\mathbf{x}) = P(Y > j \mid \mathbf{x})$ to be the survivor function. Exponentiating both sides of the model equation shows that it implies that

$$S_j(\mathbf{x}) = S_{0j}^{\exp(\beta_1 x_1 + \cdots + \beta_k x_k)},$$

where S_{0j} is the baseline survival, satisfying $S_{0j} = \exp[-\exp(\alpha_j)]$. This functional form for the survival function is precisely that implied by the proportional hazards model commonly used in survival analysis, so this model is often called the *discrete proportional hazards* model. It is often useful when the response variable is the time until some event (such as survival time), but the times are given only in interval form.

Asbestos Exposure in the United States Navy

Industrial hygiene involves the study and evaluation of a workplace or work process for any and all possible health hazards to the workers involved. Since passage of the Occupational Health and Safety Act in 1970, large amounts of data relating to worker exposure have been gathered for this purpose in corporate America, and also in the U.S. military. Formisano et al. (2001) describe and discuss one such database, the Occupational Exposure Database of the United States Navy (NOED). The data described here come from that database, and I am indebted to Jerry Formisano for sharing them with me.

Exposure to asbestos dust and fibers is known to increase the risk of several serious diseases, including asbestosis (a chronic lung ailment that can produce shortness of breath and permanent lung damage and increase the risk of dangerous lung infections), lung cancer, mesothelioma (a cancer of the thin membranes that line the chest and abdomen), and other cancers, including those of the larynx and of the gastrointestinal tract. For this reason, the Occupational Safety and Health Administration (OSHA) has issued regulations governing worker exposure to asbestos, with the legal limit being .1 fibers per cubic centimeter for an 8-hour workday. In addition, general industrial hygiene practice is to consider one-half the legal limit to be the "action level," where intervention to reduce exposure would be explored. Exposure is determined by taking air samples in the worker's breathing zone (a circle of diameter 2.5 feet around the worker's mouth and nose), capturing asbestos fibers using multicellulose ester filters.

The file `asbestos` contains data from the NOED on asbestos exposure for 83 Navy workers. These workers were engaged in jobs involving potential asbestos exposure, removing either asbestos tile or asbestos insulation. It would be expected that exposure would be different in these two tasks: while insulation is friable (it easily crumbles and releases fibers into the air), tiles are not. The workers either worked with ordinary (general) ventilation (only a fan or natural occurring wind current being the methods of dispersing fibers) or negative pressure (where a pump with a High Efficiency Particulate Air filter is used to draw air (and fibers) from the work area). For each worker, the duration of the sampling period (in minutes) was recorded, and their asbestos exposure was measured and classified into one of the three exposure categories, low exposure (less than .05 fibers per

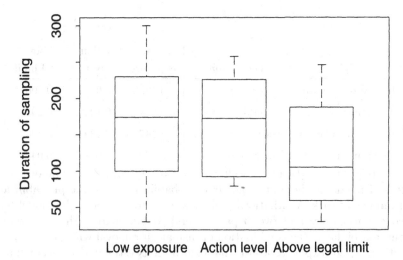

Low exposure   Action level Above legal limit

Exposure level

FIGURE 10.5. Side-by-side boxplots of job duration by exposure level for asbestos data.

cubic centimeter), action level (between .05 and .1 fibers per cubic centimeter), and above the legal limit (more than .1 fibers per cubic centimeter).

There is strong evidence in the data that, as expected, the type of task and type of ventilation are related to asbestos exposure. More than 86% of the tile removal jobs resulted in low exposure, with only 8% above the legal limit. In contrast, 63% of the insulation removal jobs resulted in exposure above the legal limit, with almost 9% more at the action level. Negative pressure ventilation was associated with far lower exposure (80% at a low level, 4% at the action level) than general ventilation (12% at the action level, 71% above the legal limit). The time spent sampling is much more weakly related to exposure (Figure 10.5), with high exposure associated with shorter sampling periods.

A proportional odds fit to these data finds the duration variable to be highly insignificant (Wald statistic $z = 0.20$), so the model can be refit using only task and ventilation. Note that from an investigative quality control point of view this is a desirable result, since it means that the sampling scheme is not related to the (hopefully objective) exposure measurement. The two remaining variables have highly significant predictive power for exposure level ($LR = 47.7$, $df = 2$), with the Wald statistics for each variable large ($z = 3.7$ for task and $z = 3.8$ for ventilation, respectively). The resultant structure for the data is then a $2 \times 2 \times 3$ table of estimated probabilities, as follows.

| | | Exposure level | | |
| | | Low | Action | Above |
Task	*Ventilation*	exposure	level	legal limit
Insulation removal	General	0.122	0.072	0.806
Tile removal	General	0.578	0.125	0.297
Insulation removal	Negative pressure	0.547	0.129	0.324
Tile removal	Negative pressure	0.922	0.031	0.047

The model fits the data very well, with $G^2 = 2.00$ and $X^2 = 1.86$ on 4 degrees of freedom (there are eight free probabilities, since the probabilities within each (Task, Ventilation) pair must sum to zero, and the model has two α parameters and two β parameters). It is apparent that the worst situation by far is the combination of insulation removal with general ventilation, where exposure above the legal limit is estimated to occur more than 80% of the time. Tile removal with general ventilation and insulation removal with negative pressure ventilation result in very similar probability profiles, with about a 30% chance of exposure above the legal limit, while tile removal with negative pressure is by far the safest situation, with a very high probability of low exposure.

10.2.3   Adjacent-Categories Logit Model

An alternative to the grouped continuous approach to modeling an ordinal variable is to focus directly on specific probability relationships through the logits. The *adjacent-categories logit* model hypothesizes a linear relationship with predictors for the logit of probabilities in adjacent categories,

$$\log\left(\frac{p_{j+1}}{p_j}\right) = \alpha_j + \beta_1 x_1 + \cdots + \beta_k x_k, \qquad j = 1, \ldots, J - 1.$$

This model is a special case of the multinomial logit model (10.1). Each specifies $J - 1$ logit equations, and it is possible to reexpress the logit equations above in terms of baseline-category logits using the identity

$$\log\left(\frac{p_{j+1}}{p_j}\right) = \log\left(\frac{p_{j+1}}{p_J}\right) - \log\left(\frac{p_j}{p_J}\right).$$

Whereas the multinomial logit model incorporates $J - 1$ different values of each β_ℓ, in the adjacent-categories model, a single β_ℓ summarizes the effect of x_ℓ: holding all else fixed, a one-unit increase in x_ℓ multiplies the odds of being in category $j+1$ relative to category j by $\exp(\beta_\ell)$, for any categories j and $j+1$. Note that since

$$
\begin{aligned}
\log\left(\frac{p_{j+m}}{p_j}\right) &= \log\left(\frac{p_{j+m}}{p_{j+m-1}}\right) + \log\left(\frac{p_{j+m-1}}{p_{j+m-2}}\right) + \cdots + \log\left(\frac{p_{j+1}}{p_j}\right) \\
&= (\alpha_{j+m-1} + \beta_1 x_1 + \cdots + \beta_k x_k) \\
&\quad + (\alpha_{j+m-2} + \beta_1 x_1 + \cdots + \beta_k x_k) + \cdots \\
&\quad + (\alpha_j + \beta_1 x_1 + \cdots + \beta_k x_k) \\
&= \sum_{r=j}^{j+m-1} \alpha_r + m(\beta_1 x_1 + \cdots + \beta_k x_k),
\end{aligned}
$$

the effect of an increase in x_ℓ on the odds of being in one category m levels higher than another is based on parameter $m\beta_\ell$. The appropriateness of the assumption of a single slope for each variable can be assessed by comparing the fit of the adjacent-categories model to that of the general multinomial logit model.

The adjacent-categories model is equivalent to certain models for ordinal contingency tables considered earlier. For example, consider a two-way contingency table with columns corresponding to the response variable and rows corresponding to the distinct values of the predictor x. The linear-by-linear (uniform) association model for the expected cell counts was given in (7.1),

$$
\log e_{ij} = \lambda + \lambda_i^X + \lambda_j^Y + \theta u_i v_j,
$$

where $\mathbf{u}$ and $\mathbf{v}$ are the row and columns scores, respectively (the scores have not been centered in this form, but that contribution can be absorbed into λ). Define the column scores to be $v_j = j$, and the row score for each row to be the value of x for that row. Then the adjacent-categories logit for row i satisfies

$$
\log\left(\frac{p_{j+1}}{p_j}\right) = \lambda_{j+1}^Y - \lambda_j^Y + \theta u_i. \tag{10.6}
$$

This is equivalent to the adjacent-categories logit model with $\alpha_j = \lambda_{j+1}^Y - \lambda_j^Y$ and $\beta = \theta$. Models with multiple predictors correspond to association models for multidimensional contingency tables, as in Section 8.3.

10.2.4 Continuation Ratio Models

Another approach to modeling ordinal regression data is based on so-called continuation ratios, the conditional probability of being in category j given being in category j or higher,

$$\delta_j(\mathbf{x}) = P(Y = j \mid Y \geq j, \mathbf{X} = \mathbf{x})$$
$$= \frac{p_j(\mathbf{x})}{p_j(\mathbf{x}) + \cdots + p_J(\mathbf{x})}.$$

If Y represents a categorized survival time, the continuation ratio represents the probability of survival to time level j given survival at least that long, which is the hazard rate.

Continuation ratio models hypothesize a linear function for a function of the continuation ratios,

$$G^{-1}[\delta_j(\mathbf{x})] = \alpha_j + \beta_1 x_1 + \cdots + \beta_k x_k.$$

The continuation ratio logit model assumes this linear relationship for the logit of $\delta_j(\mathbf{x})$. Another possibility is to use the complementary log-log link,

$$\log\{-\log[1 - \delta_j(\mathbf{x})]\} = \alpha_j + \beta_1 x_1 + \cdots + \beta_k x_k.$$

In fact, this model is equivalent to the grouped continuous model with complementary log-log link (10.5), providing another justification of its use for grouped survival data. As was true for grouped continuous data, a positive coefficient implies a shorter survival with larger value of the predictor, since in that case the hazard is increasing.

The Success of Broadway Shows (continued)

Recall again the data on Broadway shows examined in earlier chapters. The data file includes an ordered categorical variable representing how long the show remained open: less than six months, six months to one year, one to two years, two to three years, or more than three years. These data constitute a categorized survival time, and as such, the proportional hazards model is appropriate; that is, the continuation ratio model with complementary log-log link, or equivalently, the grouped continuous model with complementary log-log link. Potential predictors for survival are the type of show, whether or not it is a revival, *New York Times* and *Daily News* ratings, the number of major Tony Award nominations and wins, and the opening-week attendance as a percentage of the available seats.

Figure 10.6 shows that several of the available characteristics of the shows are related to survival time. Not surprisingly, earning more major Tony nominations and winning more major Tony Awards (presumably reflecting both inherent quality and a positive signal to the ticket-buying public) is associated with a longer show run. Higher first-week attendance (an indicator of both initial public reaction to the show and the effectiveness of pre-opening publicity) is also associated with a longer run.

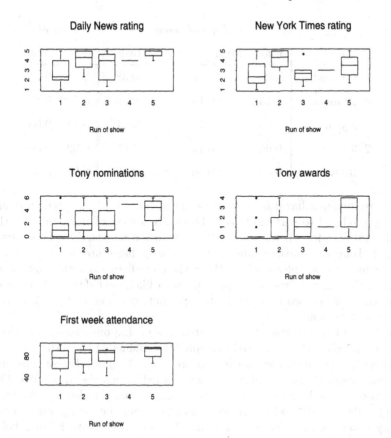

FIGURE 10.6. Side-by-side boxplots of predictors by run of Broadway shows. The categories of show run correspond to less than six months (1), six to twelve months (2), one to two years (3), two to three years (4), and more than three years (5).

The relationship between newspaper review longevity is (surprisingly) more complex. While better reviews in the *Daily News* are associated with a longer run, there is no apparent relationship between the review in *The New York Times* and show longevity. This is surprising; given the reputation and perceived importance of the newspaper in the Broadway theater industry, it might be expected that the views of this newspaper would be an important factor on survival time of a show. While there does not appear to be a relationship between survival and whether the show is a revival or not, there is one with the type of show, as musicals have generally longer survival times than musical revues and especially plays. The following tables summarize these marginal relationships as observed proportions in each of the five survival categories:

Run	Type of show			Revival	
	Play	Musical revue	Musical	No	Yes
< 6 months	0.82	0.57	0.58	0.70	0.73
6-12 months	0.13	0.14	0.09	0.07	0.17
1-2 years	0.05	0.14	0.21	0.17	0.05
2-3 years	0.00	0.14	0.00	0.02	0.00
≥ 3 years	0.00	0.00	0.12	0.04	0.05

A proportional hazards fit to these data shows that the predictors are strongly related to survival ($LR = 47.9$ on 8 degrees of freedom), but the Wald statistics for whether or not the show was a revival, *New York Times* rating, Tony nominations, and first-week attendance are not statistically significant. The simpler model without these predictors is highly significant ($LR = 45.4$ on 4 degrees of freedom), and a likelihood ratio test confirms that the omitted variables add little predictive power ($LR = 2.5$ on 4 degrees of freedom).

Table 10.1 summarizes the estimated model. Exponentiating the slope coefficients gives the estimated multiplicative change of the hazard associated with a one-unit increase in the predictor, holding all else constant. So, for example, one additional Tony Award is associated with a 39% decrease in the hazard of the show closing, given all else is held fixed ($\exp(-.496) = .61$), while an additional point on the rating scale of the *Daily News* review is associated with a 31% decrease in the hazard, holding all else fixed ($\exp(-.369) = .69$). Musicals and musical revues have much better survival prospects than plays, with musicals having 72% lower hazard of closing, and musical revues having 67% lower hazard, holding all else fixed.

Unfortunately, this model does a poor job of classifying shows into survival categories from predictor values. The classification table on the next page shows that almost all shows are classified as either closing in less than six months, or lasting more than three years. Two shows are particularly poorly classified: *Fosse!*, which ran for more than $2\frac{1}{2}$ years, and *Jekyll and Hyde*, which ran for more than three years, each of which is classified to the less than six months category. A total of 75% of the shows are correctly classified, which is not very impressive compared to $C_{\max} = 71\%$ and $C_{\mathrm{pro}} = 82\%$, reinforcing the ineffectiveness of the model for classification.

	Musical	Musical revue	Tony awards	Daily News review
(a) Proportional hazards model				
Coefficient	−1.270	−1.096	−0.496	−0.369
Wald	−4.01	−2.31	−3.45	−3.08
p	< .001	.021	.001	.002
(b) Proportional odds model				
Coefficient	−2.219	−2.399	−0.878	−0.782
Wald	−3.58	−2.63	−3.66	−3.16
p	< .001	.009	< .001	.002

TABLE 10.1. Ordinal regression estimation for Broadway performances data.

		Predicted result					
		< 6 months	6–12 months	1–2 years	2–3 years	≥ 3 years	
	< 6 months	65	0	0	0	0	65
Actual	6–12 months	9	0	0	0	2	11
result	1–2 years	9	0	1	0	1	11
	2–3 years	1	0	0	0	0	1
	≥ 3 years	1	0	0	0	3	4
		85	0	1	0	6	

Table 10.1 also summarizes the results of a proportional odds fit to these data. The implications of this model are very similar to those of the proportional hazards model, although the estimated slopes naturally have a different interpretation. So, for example, the model implies that (holding all else fixed) an additional Tony award is associated with an estimated 58% decrease in the odds of surviving up to any given survival category $(\exp(-.878) = .42)$. The significance of the regression is slightly less than that of the proportional hazards model ($LR = 44.0$ on 4 degrees of freedom), but the estimated probabilities yield more appealing classifications, wherein only *Jekyll and Hyde* is seriously misclassified, as shown in the classification table on the next page.

Predicted result

		< 6 months	6–12 months	1–2 years	2–3 years	≥ 3 years	
	< 6 months	64	0	1	0	0	65
Actual	6–12 months	8	0	3	0	0	11
result	1–2 years	9	0	2	0	0	11
	2–3 years	0	0	1	0	0	1
	≥ 3 years	1	0	1	0	2	4
		82	0	8	0	2	

Simonoff and Ma (2003) analyzed the uncategorized survival data (in which some of the survival values are censored, since the shows had not closed at the time of writing) using the semiparametric Cox proportional hazards model. The results obtained there were similar to those of the discrete proportional hazards model, as would be expected.

A noteworthy aspect of these analyses is that while the *Daily News* review was previously seen as not being predictive for winning Tony awards, and that in *The New York Times* was, the opposite pattern occurs here. The difference in results suggests (but, of course, doesn't prove) a potential "disconnect" between the awards and the audience — while the Tony voters pay more attention to the *Times* (the newspaper of the elite?), the ticket-buyers pay more attention to the *Daily News* (the newspaper of the people?). Another possibility is that the newspaper reviews reflect the opinions of very different audiences — critics and theater professionals (the *Times*) versus the ticket-buying public (the *Daily News*).

10.3   Background Material

Lloyd (1999), pages 315–325, discusses multinomial logit models, and their connection to Poisson regression models. McFadden (1974) described the red bus/blue bus example of violation of the IIA property in discrete choice problems. Hausman and McFadden (1984) proposed the test H_{IIA} for the independence of irrelevant alternatives hypothesis. Greene (2002, Section 21.7) describes the application of multinomial logit models to discrete choice problems. In that context it is often the case that some of the predictor values vary across the choices (for example, the price of a product or cost of a travel choice), while others vary across individuals (for example, income), which is not consistent with (10.1). McFadden (1973) proposed an appropriate model that is commonly called the *conditional logit* model (note that the word "conditional" is not being used here in the sense of exact or small-sample inference), which can accommodate both choice-varying and individual-varying predictors. There has been a good deal of research on discrete choice modeling that is not discussed here; Manski and McFadden (1981) describes some of this work. Chen and Kuo (2001) described how the multinomial logit model can be generalized to allow for overdispersion from correlation in the responses.

Bender and Benner (2000) discussed how ordinal regression models can be fit using SAS and S-PLUS, including describing how continuation ratio models can be fit using binary regression techniques (see also Berridge and Whitehead, 1991). Hosmer and Lemeshow (1999) discusses survival analysis in general, and the proportional hazards model in particular. Laara and Matthews (1985) demonstrated the equivalence of the grouped continuous and continuation ratio models when using the complementary log-log link. Harrell (2001, Chapters 13 and 14) discusses ordinal regression models in detail, including plots and tests designed to assess the fit of the different models. The Dirichlet-multinomial distribution is a generalization of the beta-binomial distribution that can be used to model overdispersion in multinomial regression models; see Poortema (1999) for discussion. Agresti (1999) discusses extensions of ordinal regression models to repeated measurement data, among other applications. Simonoff and Tutz (2000) discuss smoothing for ordinal categorical targets with either single or multiple predictors.

10.4 Exercises

1. As was noted on page 418, members of the Baseball Hall of Fame can be elected in one of two ways: by a vote of the members of the BBWAA during the first 20 years of eligibility, or selected by a vote of the Veterans Committee if not elected by the members of the BB-WAA. Construct a multinomial logistic regression model for Hall of Fame membership that includes both types of induction (as well as not being a member). Are the results noticeably different from those when pooling the two types of membership together? Are the two types of Hall of Famers very different from each other?

2. The earlier analysis of train accidents in the United Kingdom in Section 5.2 (pages 145–147) analyzed the rates of different types of accidents (SPAD-preventable, other preventable, and nonpreventable) per billion train-kilometers. A different question is how the relative probabilities of the three types of accidents has changed over time, and as a function of the size of the Mark I fleet and amount of train

travel. Construct a multinomial logistic regression model for the annual distribution of type of accident as a function of time, train-kilometers, and size of Mark I fleet. How can you summarize how the relative likelihood of the three types of accidents has changed over time?

3. Previous analyses of the Broadway show data on pages 400 and 423 examined the probability of a show being nominated for a major Tony Award, and the probability of winning given a nomination. That is, for each show there are three possibilities for the six major nominations: no nomination, nomination but no award, and award. Analyze the data in this form using multinomial logistic regression. Do you learn anything new from this formulation? Examine the residuals from the model fit. Is overdispersion still apparently a problem? Is an analysis treating these categories as ordinal rather than nominal appropriate?

4. The file **votesurvey** summarizes the results of a survey taken in October 2000 of first-year MBA students at New York University's Leonard N. Stern School of Business. I am indebted to Amy Lefkowitz and Matthieu Vacarie for gathering these data and sharing them with me. The survey gathered information on characteristics of the students, including gender, age, salary before coming to business school, and expected salary after graduating. The response variable is the respondent's preferred presidential candidate in the 2000 election, with four responses: George W. Bush, Al Gore, undecided, or refuse to answer.

   (a) Fit a multinomial logistic regression to these data. Which variable(s) appear useful as predictors of voting choice?

   (b) The appropriateness of the IIA assumption is open to question here. For example, if the respondents were not given the option of answering "undecided," and people who were undecided were leaning more to one candidate or the other, the IIA assumption would be violated. Fit a logistic regression model to these data, omitting the people who were undecided. Does the IIA assumption seem reasonable here? That is, do the estimated coefficients change very much if the undecideds are omitted? If you have access to a statistical language with matrix programming capabilities, construct the Hausman test H_{IIA} for these data based on omitting the undecideds. Does the test suggest any violation of IIA?

   (c) Another way that IIA could be violated would be if people who refused to answer were forced to pick one candidate or the other, and chose "undecided" instead. Repeat the analysis in part (b), now dropping out people who refused to answer. Does IIA seem reasonable?

(d) Fit a multinomial logit regression using only gender as a predictor. Is there evidence of a "gender gap" in these data (that is, a significant difference in preferences between men and women)? Note that if only gender is used as a predictor, the data form a 2×4 contingency table. Would the Hausman test be useful in this context? Why or why not?

5. Show that (10.6) is true; that is, that the adjacent-categories logit model is equivalent to the linear-by-linear association model for a model with a single predictor variable.

6. Abzug et al. (2000) gathered data relating the bond rating of a city's general obligation bonds to various factors (see page 240). Available variables include population, per capita income, average total household income, average household discretionary income, the number of public entities, nonprofits, for profits, and utilities among the top ten employers in the city, the geographic region, and whether the city is a state capital, for 57 large cities that issued general obligation bonds in 1994. The data are given in the file **bondrate**.

(a) Fit a proportional odds model for bond rating to these data. Consider reasonable transformations of variables. Simplify the model as appropriate. What are the implications for the fiscal health of large cities?

(b) Simonoff (1998b) smoothed the contingency table that cross-classifies bond rating by the number of nonprofit organizations among the top ten employers, ultimately arguing that one nonprofit in the top ten is best (in terms of highest bond rating), with fewer or more nonprofits being associated with a lower bond rating. Is this supported by the proportional odds model? (Hint: try fitting a model that is quadratic in the number of nonprofits.)

7. The file **currency** contains information on the 2000 Standard and Poor's currency rating for 31 countries, along with the national income level and debt service ratio (debt principal and interest divided by total exports). I am indebted to Elizabeth Brock for gathering these data and sharing them with me.

(a) Fit a proportional odds model for currency rating on debt service ratio. Does this variable provide significant predictive power for currency rating?

(b) A way to assess the validity of the constant slopes assumption of the proportional odds model is to fit separate logistic regressions for each of the underlying cumulative logits, and compare the resultant slope coefficients to the estimated slope in the proportional odds model. Is there evidence of violation of the constant slopes assumption here?

(c) Fit a multinomial logistic regression to these data. Based on AIC_C, which model provides a better representation of the data — the (unstructured) multinomial logit model or the proportional odds model?

(d) Now include the income level as a predictor, and repeat parts (a)–(c). Does your answer change? Are there any difficulties in fitting the models? Why would that be the case?

(e) Argentina is a middle-income country with a debt service ratio of 85.5. At what currency level would these models classify Argentina? In fact, Standard and Poor's rated Argentina's currency as in selective default. Is this surprising to you?

8. Not surprisingly, organizational behavior theory states that the willingness of one partner in a negotiation to do business with the other partner in the future depends on the perceptions of how favorable the outcome of the negotiation was and of how fair the negotiation procedure was. It also states that the willingness to engage in future business also depends on the person's view of their own status, and on their general level of trust in people. A study of MBA students at New York University's Leonard N. Stern School of Business in 2001 was designed to illustrate these points. The students filled out a questionnaire designed to measure their perception of status and their level of trust, participated in an employment negotiation game in which they were either a recruiter or a job candidate, and then reported on their perceptions of outcome favorability, procedural fairness, and whether they would engage in business with their partner in the negotiation in the future. The willingness to do future business variable was coded on a Likert scale from 1 (not at all willing) to 7 (very much willing), and the other variables are scaled to be between 0 and 100, with the high limit referring to high status, high trust, high outcome favorability, and high procedural fairness, respectively. The results are given in the file **negotiation**. I am indebted to Yaru Chen for sharing these data with me.

(a) Fit a proportional odds model for willingness to do future business, simplifying the model as necessary. Does the role played (recruiter or candidate) matter? Are the estimated coefficients consistent with what you would expect?

(b) Fit separate cumulative logit regressions to these data (see the previous exercise). Is the assumption of constant slopes supported here?

(c) Fit a proportional hazards model to these data. Do you prefer the proportional odds model or the proportional hazards model?

(d) Theory suggests that there could be an interaction effect of status, outcome favorability, and procedural fairness on the willingness for future business. Specifically, for people with higher perceived status, outcome favorability will not matter if there is a perception of procedural unfairness, but it will matter if there is a perception of procedural fairness; on the other hand, for people with lower perceived status, outcome favorability will not matter if there is a perception of procedural fairness, but it will matter if there is a perception of procedural unfairness. Explore this possibility for these data by categorizing the status, outcome favorability, and procedural fairness variables into low, medium, and high ranges, and fitting the appropriate interaction effects. Do the data support the existence of this interaction effect?

(e) Fit a least squares regression to these data. Are the results very different from the ordinal regression results?

9. Clogg (1982) reported results from the General Social Survey relating the happiness level of 1517 respondents ("Not too happy" — "Pretty happy" — "Very happy") to the person's level of schooling and number of siblings. The data are given in the file **happiness**.

(a) Fit an appropriate ordinal regression model to these data, treating the schooling and siblings variables as numerical. Can you find a model that fits the data well?

(b) Fit an appropriate ordinal regression model to these data, treating the schooling and siblings variables as categorical. Is there evidence of an interaction effect of the two predictors on happiness level? Does this model formulation provide meaningful improvement over the model in part (a)?

(c) Repeat parts (a) and (b), ignoring the ordering in the happiness variable.

(d) Beh and Davy (1999) analyzed these data as a three-dimensional contingency table, treating the happiness variable as nominal. They found evidence (using partitions of the Pearson X^2 statistic) that the relationship between schooling and happiness was reflected in differences in both the location and variation in years of schooling for different happiness levels, while this was less true for the relationship between the number of siblings and happiness level. Are these patterns consistent with your results in parts (a)–(c)? What if you analyze the table using the association models of Section 8.3?

10. Recall the proportional hazards and proportional odds analyses of the Broadway show run data in Section 10.2.4. In those analyses, no shows were predicted to last between 6 and 12 months, or between 2 and 3 years. What happens to the analyses if the run of the show is collapsed to the three levels of less than 1 year, 1–2 years, and more than 2 years? Would you expect the coefficients of either the proportional hazards or proportional odds models to change very much? What if a continuation ratio logit model was fit using the two categorization schemes? Would you expect the results of the two analyses to be similar?

11. Brillinger (1996) examined the performance of the Toronto Maple Leafs hockey team during the 1993–1994 season. The data are given in the file **mapleleafs**, which gives the result of each game (win — tie — loss) and site (home or away).

(a) Is there evidence of a site (home/away) effect on game outcome? Answer this question treating the tie outcome as "between" a win and a loss.

(b) Brillinger proposed investigating the presence of autocorrelation in the data by including the previous game's outcome (treated as a categorical variable) as a predictor, resulting in an autoregressive form. Such autocorrelation could be viewed as evidence of "streakiness" in the performance of the team. Is there evidence of streakiness in the Maple Leafs' performance in 1993–1994? Are your results surprising, given that the Maple Leafs started the season with 10 straight wins? Do your results differ if you change link functions or model forms?

(c) Consider again the data on home runs hit by Mark McGwire, Sammy Sosa, and Barry Bonds, respectively, in their record-setting seasons, given in the file **homerun**. Examine the streakiness in their home run hitting in the way described in part (b). Is there any evidence of streakiness based on this model for any of the players?

Appendix A
Some Basics of Matrix Algebra

Matrix algebra is a multivariate generalization of usual algebra. While a thorough understanding of matrix algebra is not necessary to analyze categorical data, it is useful to have a basic understanding of it to understand some of the subtleties in analysis.

1. A set of numbers can be represented as a *vector*, as follows:

$$\mathbf{a} = \begin{pmatrix} 4 \\ 2 \\ 0 \\ 6 \\ 1 \end{pmatrix}$$

By convention, vectors are represented as column vectors. The vector above has dimension 5×1. Each element of a vector can be identified by a subscript; for example, $a_1 = 4$; $a_2 = 2$; etc.

2. A set of vectors taken together forms a *matrix*, as follows:

$$A = \begin{pmatrix} 4 & 6 & 3 \\ 2 & 7 & 2 \\ 0 & 5 & 6 \\ 6 & 1 & 2 \\ 1 & 2 & 1 \end{pmatrix}$$

The matrix above has dimension 5×3 (rows $\times$ columns). Each element of a matrix can be identified by a paired subscript; for example, $a_{11} = 4$; $a_{13} = 3$; $a_{31} = 0$; etc. Note that a vector is a special case of a matrix, with the number of columns being one.

3. A constant k can be added to a matrix; just add the constant to each element:

$$A + k = \begin{pmatrix} 4+k & 6+k & 3+k \\ 2+k & 7+k & 2+k \\ 0+k & 5+k & 6+k \\ 6+k & 1+k & 2+k \\ 1+k & 2+k & 1+k \end{pmatrix}$$

Similarly, multiplying (or dividing) a matrix by a constant means multiplying or dividing each element by the constant.

4. Say we have two matrices of the same dimensions. Then, the two matrices can be added by adding them element by element:

$$A + B = \begin{pmatrix} 4 & 6 & 3 \\ 2 & 7 & 2 \\ 0 & 5 & 6 \\ 6 & 1 & 2 \\ 1 & 2 & 1 \end{pmatrix} + \begin{pmatrix} 1 & 4 & 2 \\ 0 & 2 & 8 \\ 2 & 9 & 4 \\ 1 & 6 & 3 \\ 0 & 1 & 5 \end{pmatrix} = \begin{pmatrix} 5 & 10 & 5 \\ 2 & 9 & 10 \\ 2 & 14 & 10 \\ 7 & 7 & 5 \\ 1 & 3 & 6 \end{pmatrix}$$

5. The *transpose* of a matrix changes the (i, j)th element to the (j, i)th element:

$$A' = \begin{pmatrix} 4 & 2 & 0 & 6 & 1 \\ 6 & 7 & 5 & 1 & 2 \\ 3 & 2 & 6 & 2 & 1 \end{pmatrix}$$

6. Multiplication of matrices is more complicated, but important. Let A be $n \times p$ (n rows, p columns) and B be $p \times k$ (p rows, k columns). Then A and B can be multiplied, as follows. If $C = AB$, then

$$c_{ij} = \sum_{\ell=1}^{p} a_{i\ell} b_{\ell j}.$$

For example:

$$A = \begin{pmatrix} 4 & 6 & 3 \\ 2 & 7 & 2 \\ 0 & 5 & 6 \\ 6 & 1 & 2 \\ 1 & 2 & 1 \end{pmatrix}, B = \begin{pmatrix} 2 & 5 & 1 \\ 4 & 1 & 9 \\ 6 & 1 & 0 \end{pmatrix}$$

Then, the $(2, 1)$ entry of AB is $(2)(2) + (7)(4) + (2)(6) = 44$. The full product matrix is calculated the same way, yielding

$$C = AB = \begin{pmatrix} 50 & 29 & 58 \\ 44 & 19 & 65 \\ 56 & 11 & 45 \\ 28 & 33 & 15 \\ 16 & 8 & 19 \end{pmatrix}$$

Note that if A is $n \times p$ and B is $p \times k$, then AB is $n \times k$. Note also that two vectors can be multiplied; for example,

$$\mathbf{a}'\mathbf{a} = \sum_{i=1}^{k} a_i^2$$

if $\mathbf{a}$ is $k \times 1$.

7. Division is multiplication by the reciprocal of a number ($a/b = a \times (1/b) = a \times b^{-1}$); this is also true for matrices. We want to find the matrix such that when you multiply it with A, you get the *identity matrix* (I), which is the equivalent of the number one:

$$I = \begin{pmatrix} 1 & 0 & 0 & 0 \\ 0 & 1 & 0 & 0 \\ 0 & 0 & 1 & 0 \\ 0 & 0 & 0 & 1 \end{pmatrix}.$$

8. So, A^{-1} is defined as the matrix such that

$$AA^{-1} = I = A^{-1}A.$$

This is defined for square matrices only (the same number of rows and columns).

References

Abzug, R., Simonoff, J.S., and Ahlstrom, D. (2000), "Nonprofits as Large Employers: A City-Level Geographical Inquiry," *Nonprofit and Voluntary Sector Quarterly*, **29**, 455–470.

Aerts, M., Augustyns, I., and Janssen, P. (1997), "Local Polynomial Estimation of Contingency Table Cell Probabilities," *Statistics*, **30**, 127–148.

Agresti, A. (1984), *Analysis of Ordinal Categorical Data*, John Wiley and Sons, New York.

Agresti, A. (1996), *An Introduction to Categorical Data Analysis*, John Wiley and Sons, New York.

Agresti, A. (1999), "Modelling Ordered Categorical Data: Recent Advances and Future Challenges," *Statistics in Medicine*, **18**, 2191–2207.

Agresti, A. (2002), *Categorical Data Analysis*, 2nd. ed., John Wiley and Sons, New York.

Agresti, A. and Caffo, B. (2000), "Simple and Effective Confidence Intervals for Proportions and Differences of Proportions Result From Adding Two Successes and Two Failures," *American Statistician*, **54**, 280–288.

Agresti, A. and Coull, B.A. (1998a), "Approximate Is Better Than 'Exact' for Interval Estimation of Binomial Proportions," *American Statistician*, **52**, 119–126.

Agresti, A. and Coull, B.A. (1998b), "Order-Restricted Inference for Monotone Trend Alternatives in Contingency Tables," *Computational Statistics and Data Analysis*, **28**, 139–155.

Agresti, A. and Winner, L. (1997), "Evaluating Agreement and Disagreement Among Movie Reviewers," *Chance*, **10(2)**, 10–14.

Agresti, A. and Yang, M.–C. (1987), "An Empirical Investigation of Some Effects of Sparseness in Contingency Tables," *Computational Statistics and Data Analysis*, **5**, 9–21.

Ahmed, F., McRae, J.A., and Ahmed, N. (1990), "Factors Associated With Not Receiving Adequate Prenatal Care in an Urban Black Population: Program Planning Implications," *Social Work in Health Care*, **14**, 107–123.

Aickin, M. (1990), "Maximum Likelihood Estimation of Agreement in the Constant Predictive Probability Model, and Its Relation to Cohen's Kappa," *Biometrics*, **46**, 293–302.

Aitkin, M. (1992), "Evidence and the Posterior Bayes Factor," *Mathematical Scientist*, **17**, 15–25.

Altman, D.G. (1991), *Practical Statistics for Medical Research*, Chapman and Hall, London.

Andersen, E.B. (1997), *Introduction to the Statistical Analysis of Categorical Data*, Springer-Verlag, Berlin.

Anderson, T.W., Reid, D.B.W., and Beaton, G.H. (1972), "Vitamin C and the Common Cold," *Canadian Medical Association Journal*, **107**, 503–508.

Andrews, D.F. and Herzberg, A.M. (1985), *Data: A Collection of Problems from Many Fields for the Student and Research Worker*, Springer-Verlag, New York.

Anscombe, F.J. (1973), "Graphs in Statistical Analysis," *American Statistician*, **27**, 17–21.

Appleton, D.R., French, J.M., and Vanderpump, M.P.J. (1996), "Ignoring a Covariate: An Example of Simpson's Paradox," *American Statistician*, **50**, 340–341.

Arnold, B.C. and Strauss, D.J. (1991), "Bivariate Distributions with Conditionals in Prescribed Exponential Families," *Journal of the Royal Statistical Society, Ser. B*, **53**, 365–375.

Associated Press and Media General Financial Services (1998), "Third Quarter Stock Market Review," *Newsday*, October 4, 1998, F8.

Bailey, B.J.R. (1990), "A Model for Function Word Counts," *Applied Statistics*, **39**, 107–114.

Baker, S.G. and Kramer, B.S. (2001), "Good for Women, Good for Men, Bad for People: Simpson's Paradox and the Importance of Sex-Specific Analysis in Observational Studies," *Journal of Women's Health and Gender-Based Medicine*, **10**, 867–872.

Barker, L. (2002), "A Comparison of Nine Confidence Intervals for a Poisson Parameter When the Expected Number of Events is $\leq$ 5," *American Statistician*, **56**, 85–89.

Barnard, G.A. (1947), "Significance Tests for 2×2 Tables," *Biometrika*, **34**, 123–138.

Barnett, V. and Lewis, T. (1994), *Outliers in Statistical Data*, 3rd. ed., John Wiley and Sons, Chichester.

Basu, A. and Basu, S. (1998), "Penalized Minimum Disparity Methods for Multinomial Models," *Statistica Sinica*, **8**, 841–860.

Basu, A., Harris, I.R., and Basu, S. (1996), "Tests of Hypotheses in Discrete Models Based on the Penalized Hellinger Distance," *Statistics and Probability Letters*, **27**, 367–373.

Basu, A. and Sarkar, S. (1994), "On Disparity Based Goodness-of-Fit Tests for Multinomial Models," *Statistics and Probability Letters*, **19**, 307–312.

Baum, J.K., Myers, R.A., Kehler, D.G., Worm, B., Harley, S.J., and Doherty, P.A. (2003), "Collapse and Conservation of Shark Populations in the Northwest Atlantic," *Science*, **299**, 389–392.

Beh, E.J. (1997), "Simple Correspondence Analysis of Ordinal Cross-Classifications Using Orthogonal Polynomials," *Biometrical Journal*, **39**, 589–613.

Beh, E.J. and Davy, P.J. (1999), "Partitioning Pearson's Chi-Squared Statistic for a Partially Ordered Three-Way Contingency Table," *Australian and New Zealand Journal of Statistics*, **41**, 233–246.

Bender, R. and Benner, A. (2000), "Calculating Ordinal Regression Models in SAS and S-Plus," *Biometrical Journal*, **42**, 677–699.

Berndt, E., Hall, B., Hall, R., and Hausman, J. (1974), "Estimation and Inference in Nonlinear Structural Models," *Annals of Economic and Social Measurement*, **3/4**, 653–665.

Berresford, G.C. (1980), "The Uniformity Assumption in the Birthday Problem," *Mathematics Magazine*, **53**, 286–288.

Berridge, D.M. and Whitehead, J. (1991), "Analysis of Failure Time Data with Ordinal Categories of Response," *Statistics in Medicine*, **10**, 1703–1710.

Berry, S.M. (2000), "Modeling Acceptance to the Major League Baseball Hall of Fame," *Chance*, **13(1)**, 52–57.

Biller, C. (2000), "Adaptive Bayesian Regression Splines in Semiparametric Generalized Linear Models," *Journal of Computational and Graphical Statistics*, **9**, 122–140.

Bishop, Y.M.M., Fienberg, S.E., and Holland, P.W. (1975), *Discrete Multivariate Analysis*, MIT Press, Cambridge, MA.

Blane, D., Harding, S., and Rosato, M. (1999), "Does Social Mobility Affect the Size of the Socioeconomic Mortality Differential?: Evidence from the Office for National Statistics Longitudinal Study," *Journal of the Royal Statistical Society, Ser. A*, **162**, 59–70.

Bliss, C.I. (1934), "The Method of Probits," *Science*, **79**, 38–39.

Bliss, C.I. (1935), "The Calculation of the Dosage–Mortality Curve," *Annals of Applied Biology*, **22**, 134–167.

Böhning, D. (1998), "Zero-Inflated Poisson Models and C.A.MAN: A Tutorial Collection of Evidence," *Biometrical Journal*, **40**, 833–843.

Böhning, D., Dietz, E., Schlattmann, P., Mendonça, L., and Kirchner, U. (1999), "The Zero-Inflated Poisson Model and the Decayed, Missing and Filled Teeth Index in Dental Epidemiology," *Journal of the Royal Statistical Society, Ser. A*, **162**, 195–209.

Boland, P.J. and Hutchinson, K. (2000), "Student Selection of Random Digits," *The Statistician*, **49**, 519–529.

Borokhovich, K.A., Parrino, R., and Trapani, T. (1996), "Outside Directors and CEO Selection," *Journal of Financial and Quantitative Analysis*, **31**, 337–355.

Boswell, M.T. and Patil, G.P. (1970), "Chance Mechanisms Generating the Negative Binomial Distributions," in G.P. Patil, ed., *Random Counts in Models and Structures*, volume 1, Pennsylvania State University Press, University Park, PA, 3–22.

Brazzale, A.R. (1999), "Approximate Conditional Inference in Logistic and Loglinear Models," *Journal of Computational and Graphical Statistics*, **8**, 653–661.

Brillinger, D.R. (1996), "An Analysis of an Ordinal-Valued Times Series," *Proceedings of the Athens Conference on Applied Probability and Time Series Analysis. Vol. II: Time Series Analysis*, Lecture Notes in Statistics, **115**, Springer-Verlag, New York, 73–87.

Brown, L.D., Cai, T.T., and DasGupta, A. (2001), "Interval Estimation for a Binomial Proportion (with discussion)," *Statistical Science*, **16**, 101–133.

Browne, R.H. (2002), "Comment on Paper by Winkler et al.," *American Statistician*, **56**, 252.

Buck, C.E. and Sahu, S.K. (2000), "Bayesian Models for Relative Archaeological Chronology Building," *Applied Statistics*, **49**, 423–440.

Burnham, K.P. and Anderson, D.R. (2002), *Model Selection and Multimodel Inference: A Practical Information-Theoretic Approach*, 2nd. ed., Springer-Verlag, New York.

Butterfield, F. (2001), "Victim's Race Affects Decisions on Killer's Sentence," *New York Times*, April 20, 2001, A10.

Cai, Z. and Tsai, C.-L. (1999), "Diagnostics for Nonlinearity in Generalized Linear Models," *Computational Statistics and Data Analysis*, **29**, 445–469.

Cameron, A.C. and Trevidi, P.K. (1986), "Econometric Models Based on Count Data: Comparisons and Applications of Some Estimators and Tests," *Journal of Applied Econometrics*, **1**, 29–53.

Cameron, A.C. and Trevidi, P.K. (1998), *Regression Analysis of Count Data*, Cambridge University Press, Cambridge.

Cameron, A.C. and Windmeijer, F.A.G. (1997), "R-Squared Measures for Count Data Regression Models with Application to Health Care Utilization," *Journal of Business and Economic Statistics*, **14**, 209–220.

Cave, H.C., van der Werff ten Bosch, J., Suciu, S., Guidal, C., Waterkeyn, C., Otten, J., Bakkus, M., Thielmans, K., Grandchamp, B., Vilmer, E., Nelken, B., Fournier, M., Boutard, P., Lebrun, E., Méchinaud, E., Garand, R., Robert, A., Dastugne, N., Plouvier, E., Racadot, E., Ferster, A., Gyselinck, J., Fenneteau, O., Duval, M., Solbu, G., and Manel, A.-M. (1998), "Clinical Significance of Minimal Residual Disease in Childhood Acute Lymphoblastic Leukemia," *New England Journal of Medicine*, **339**, 591–598.

Cazeneuve, B. (2002), "Advantage USA," *Sports Illustrated*, February 4, 2002, 102–107.

Chamberlain, J.M. (1998), "Pacifier Thermometer Comment," *Archives of Pediatrics and Adolescent Medicine*, **152**, 206–207.

Chaski, C.E. (1997), "Who Wrote It? Steps Toward a Science of Authorship Identification," *National Institute of Justice Journal*, **233**, 15–22.

Chatterjee, S. and Hadi, A.S. (1988), *Sensitivity Analysis in Linear Regression*, John Wiley and Sons, New York.

Chatterjee, S., Hadi, A.S., and Price, B. (1999), *Regression Analysis by Example*, 3rd. ed., John Wiley and Sons, New York.

Chatterjee, S., Handcock, M.S., and Simonoff, J.S. (1995), *A Casebook for a First Course in Statistics and Data Analysis*, John Wiley and Sons, New York.

Chen, Z. and Kuo, L. (2001), "A Note on the Estimation of the Multinomial Logit Model with Random Effects," *American Statistician*, **55**, 89–95.

Chernick, M.R. and Liu, C.Y. (2002), "The Saw-Toothed Behavior of Power Versus Sample Size and Software Solutions: Single Binomial Proportion Using Exact Methods," *American Statistician*, **56**, 149–155.

Christen, W.G., Glynn, R.J., Ajani, U.A., Schaumberg, D.A., Buring, J.E., Hennekens, C.H., and Manson, J.E. (2000), "Smoking Cessation and Risk of Age-Related Cataract in Men," *Journal of the American Medical Association*, **284**, 713–716.

Clogg, C.C. (1982), "Some Models for the Analysis of Association in Multiway Cross-Classifications Having Ordered Categories," *Journal of the American Statistical Association*, **77**, 803–815.

Cochran, J.J. (2000), "Career Records for All Modern Position Players Eligible for the Major League Baseball Hall of Fame," *Journal of Statistics Education*, **8** (http://www.amstat.org/publications/jse/secure/v8 n2/datasets.cochran.new.cfm).

Cohen, J. (1960), "A Coefficient of Agreement for Nominal Tables," *Educational and Psychological Measurement*, **20**, 37–46.

Cohen, J. (1968), "Weighted Kappa: Nominal Scale Agreement with Provision for Scaled Disagreement or Partial Credit," *Psychological Bulletin*, **70**, 213–220.

Collett, D. (2003), *Modelling Binary Data*, 2nd. ed., Chapman and Hall/CRC, Boca Raton, FL.

Collett, D. and Jemain, A.A. (1985), "Residuals, Outliers and Influential Observations in Regression Analysis," *Sains Malaysiana*, **14**, 493–511.

Consumer Reports (1997), "The 1997 Cars," April, 1997.

Coory, M. (1998), "Gestational-Age-Specific Stillbirth Risk Among Australian Aborigines," *International Journal of Epidemiology*, **27**, 83–86.

Cressie, N. and Pardo, L. (2002), "Model Checking in Loglinear Models using φ–Divergences and MLEs," *Journal of Statistical Planning and Inference*, **103**, 437–453.

Cytel Software Corporation (1999), *LogXact 4 for Windows User Manual*, Cytel Software Corporation, Cambridge, MA.

Cytel Software Corporation (2001a), *Egret for Windows User Manual*, Cytel Software Corporation, Cambridge, MA.

Cytel Software Corporation (2001b), *StatXact 4 for Windows User Manual*, Cytel Software Corporation, Cambridge, MA.

Dalal, S.R., Fowlkes, E.B., and Hoadley, B. (1989), "Risk Analysis of the Space Shuttle: Pre-*Challenger* Prediction of Failure," *Journal of the American Statistical Association*, **84**, 945–957.

Darroch, J.N., Lauritzen, S.L., and Speed, T.P. (1980), "Markov Fields and Log-Linear Interaction Models for Contingency Tables," *Annals of Statistics*, **8**, 522–539.

Darroch, J.N. and McCloud, P.I. (1986), "Category Distinguishability and Observer Agreement," *Australian Journal of Statistics*, **28**, 371–388.

Dawson, R.J.M. (1995), "The 'Unusual Episode' Data Revisited," *Journal of Statistics Education*, **3** (http://www.amstat.org/publications/jse/v3n3/datasets.dawson.html).

Dean, C.B. (1992), "Testing for Overdispersion in Poisson and Binomial Regression Models," *Journal of the American Statistical Association*, **87**, 451–457.

Dean, C. and Lawless, J.F. (1989), "Tests for Detecting Overdispersion in Poisson Regression Models," *Journal of the American Statistical Association*, **84**, 467–472.

De Hertog, S.A.E., Wensveen, C.A.H., Castiaens, M.T., Kielich, C.J., Berkhout, M.J.P., Westendorp, R.G.J., Vermeer, B.J., and Bouwes Bavinck, J.N. (2001), "Relation Between Smoking and Skin Cancer," *Journal of Clinical Oncology*, **19**, 231–238.

Department of Justice (2001), "The Federal Death Penalty System: Supplementary Data, Analysis and Revised Protocols for Capital Case Review," Washington, DC (http://www.usdoj.gov/dag/pubdoc/deathpenalty study.htm).

Dieckmann, A. (1981), "Ein Einfaches Stochastisches Modell zur Analyse von Häufigkeitsverteilungen Abweichenden Verhaltens," *Zeitschrift für Soziologie*, **10**, 319–325.

Dietz, E. and Böhning, D. (2000), "On Estimation of the Poisson Parameter in Zero-Modified Poisson Models," *Computational Statistics and Data Analysis*, **34**, 441–459.

Dolezal, J.M., Perkins, E.S., and Wallace, R.B. (1989), "Sunlight, Skin Sensitivity, and Senile Cataract," *American Journal of Epidemiology*, **129**, 559–568.

Draper, D. (1995), "Assessment and Propagation of Model Uncertainty (with discussion)," *Journal of the Royal Statistical Society, Ser. B*, **57**, 45–97.

Draper, N.R. and Smith, H. (1998), *Applied Regression Analysis*, 3rd. ed., John Wiley and Sons, New York.

Edwards, D. (2000), *Introduction to Graphical Modelling*, 2nd. ed., Springer-Verlag, New York.

Eliason, S.R. (1986), "A Manual for Clogg's ANOAS: PC Version 1.2," Technical Report, Department of Sociology, Pennsylvania State University, University Park, PA.

Elms, L. and Brady, H.E. (2001), "Mapping the Buchanan Vote Escarpment in Palm Beach County, Florida" (http://ucdata.berkeley.edu/new_web/VOTE2000/MapBuchanan.PDF).

Emerson, J.D. and Hoaglin, D.C. (1983), "Analysis of Two-Way Tables by Medians," in *Understanding Robust and Exploratory Data Analysis*, eds. D.C. Hoaglin, F. Mosteller, and J.W. Tukey, John Wiley and Sons, New York, 166–210.

Englin, J.E., Holmes, T.P., and Sills, E.O. (2003), "Estimating Forest Recreation Demand Using Count Data Models," in E.O. Sills and K. Abt, eds., *Forests in a Market Economy*, Kluwer Academic Publishers, Dordrecht, Netherlands, to appear.

Eubank, R.L. (1997), "Testing Goodness-of-Fit With Multinomial Data," *Journal of the American Statistical Association*, **92**, 1084–1093.

Evans, A.W. (2000), "Fatal Train Accidents on Britain's Mainline Railways," *Journal of the Royal Statistical Society, Ser. A*, **163**, 99–119.

Fan, J. and Gijbels, I. (1996), *Local Polynomial Modelling and Its Applications*, Chapman and Hall, London.

Fan, J., Heckman, N.E., and Wand, M.P. (1995), "Local Polynomial Kernel Regression for Generalized Linear Models and Quasi-Likelihood Functions," *Journal of the American Statistical Association*, **90**, 141–150.

Farrington, C.P., Kanaan, M.N., and Gay, N.J. (2001), "Estimation of the Basic Reproduction Number for Infectious Diseases from Age-Stratified Serological Survey Data (with discussion)," *Applied Statistics*, **50**, 251–292.

Finkelstein, M.O. and Levin, B. (2001), *Statistics for Lawyers*, 2nd. ed., Springer-Verlag, New York.

Fisher, R.A. (1925), *Statistical Methods for Research Workers*, Oliver and Boyd, London.

Fisher, R.A. (1935), *The Design of Experiments*, Oliver and Boyd, London.

Fleiss, J.L. (1981), *Statistical Methods for Rates and Proportions*, 2nd. ed., John Wiley and Sons, New York.

Fleiss, J.L., Cohen, J., and Everitt, B.S. (1981), "Large-Sample Standard Errors of Kappa and Weighted Kappa," *Psychological Bulletin*, **72**, 323–327.

Forcheh, N. (2002), "Ehrenberg Law–Like Relationship and Anthropometry," *Journal of the Royal Statistical Society, Ser. A*, **165**, 155–172.

Formisano, J.A., Jr., Still, K., Alexander, W., and Lippmann, M. (2001), "Application of Statistical Models for Secondary Data Usage of the U.S. Navy's Occupational Exposure Database (NOED)," *Applied Occupational and Environmental Hygiene*, **16**, 201–209.

Franklin, M.S. (2002), "Comparative Efficiency of Insect Repellents Against Mosquito Bites," *New England Journal of Medicine*, **347**, 13–18.

Friedberg, M., Saffran, B., Stinson, T.J., Nelson, W., and Bennett, C.L. (1999), "Evaluation of Conflict of Interest in Economic Analyses of New Drugs Used in Oncology," *Journal of the American Medical Association*, **282**, 1453–1457.

Friendly, M. (1999), "Extending Mosaic Displays: Marginal, Conditional, and Partial Views of Categorical Data," *Journal of Computational and Graphical Statistics*, **8**, 373–395.

Friendly, M. (2002), "A Brief History of the Mosaic Display," *Journal of Computational and Graphical Statistics*, **11**, 89–107.

García-Pérez, M.A. (1999a), "MPROB: Computation of Multinomial Probabilities," *Behavior Research Methods, Instruments, and Computers*, **31**, 701–705.

García-Pérez, M.A. (1999b), "Exact Finite-Sample Significance and Confidence Regions for Goodness-of-Fit Statistics in One-Way Multinomials," *British Journal of Mathematical and Statistical Psychology*, **53**, 193–207.

Garwood, F. (1936), "Fiducial Limits for the Poisson Distribution," *Biometrika*, **28**, 437–442.

Gelberg, L., Stein, J., and Neumann, C.G. (1995), "Determinants of Undernutrition Among Homeless Adults," *Public Health Reports*, **110**, 448–454.

Glass, D.V. (ed.), (1954), *Social Mobility in Britain*, Free Press, Glencoe, IL.

Goldenberg, S.B., Landsea, C.W., Mestas–Nuñuez, A.M., and Gray, W.M. (2001), "The Recent Increase in Atlantic Hurricane Activity: Causes and Implications," *Science*, **293**, 474–479.

Golder, P.N. (2000), "Historical Method in Marketing Research with New Evidence on Long-Term Market Share Stability," *Journal of Marketing Research*, **37**, 156–172.

Good, P.I. (2001), *Applying Statistics in the Courtroom: A New Approach for Attorneys and Expert Witnesses*, Chapman and Hall, Boca Raton, FL.

Goodman, L.A. (1979), "Simple Models for the Analysis of Association in Cross-Classifications Having Ordered Categories," *Journal of the American Statistical Association*, **74**, 537–552.

Greenacre, M.J. (1984), *Theory and Applications of Correspondence Analysis*, Chapman and Hall, London.

Greenacre, M.J. (2000), "Correspondence Analysis of Square Asymmetric Matrices," *Applied Statistics*, **49**, 297–310.

Greene, W.H. (1994), "Accounting for Excess of Zeroes and Sample Selection in Poisson and Negative Binomial Regression Models," *Working Paper 94-10*, Department of Economics, Leonard N. Stern School of Business, New York University.

Greene, W.H. (2000a), *LIMDEP, Version 7.0*, Econometric Software, Bellport, NY.

Greene, W.H. (2000b), *Econometric Analysis*, 4th. ed., Prentice Hall, Upper Saddle River, NJ.

Greene, W.H. (2002), *Econometric Analysis*, 5th. ed., Prentice Hall, Upper Saddle River, NJ.

Grogger, J.T. and Carson, R.T. (1991), "Models for Truncated Counts," *Journal of Applied Econometrics*, **6**, 225–238.

Grossman, D.C., Reay, D.T., and Baker, S.A. (1999), "Self-Inflicted and Unintentional Firearm Injuries Among Children and Adolescents," *Archives of Pediatric and Adolescent Medicine*, **153**, 875–878.

Guggenmoos-Holzmann, I. and Vonk, R. (1998), "Kappa-like Indices of Observer Agreement Viewed from a Latent Class Perspective," *Statistics in Medicine*, **17**, 797–812.

Gwiazda, J., Ong, E., Held, R., and Thorn, F. (2000), "Vision — Myopia and Ambient Night-time Lighting," *Nature*, **904**, 144.

Haberman, S.J. (1977), "Loglinear Models and Frequency Tables with Small Expected Counts," *Annals of Statistics*, **5**, 1148–1169.

Hadi, A.S. and Simonoff, J.S. (1993), "Procedures for the Identification of Multiple Outliers in Linear Models," *Journal of the American Statistical Association*, **88**, 1264–1272.

Hall, D.B. (2000), "Zero-Inflated Poisson and Binomial Regression with Random Effects: A Case Study," *Biometrics*, **56**, 1030–1039.

Hall, D.B. and Berenhaut, K.S. (2002), "Score Tests for Heterogeneity and Overdispersion in Zero-Inflated Poisson and Binomial Regression Models," *Canadian Journal of Statistics*, **30**, 415–430.

Hardin, J. and Hilbe, J. (2001), *Generalized Linear Models and Extensions*, Stata Press, College Station, TX.

Harrell, F.E., Jr. (2001), *Regression Modeling Strategies*, Springer–Verlag, New York.

Harris, I.R. and Basu, A. (1994), "Hellinger Distance as a Penalized Log Likelihood," *Communications in Statistics — Simulation and Computation*, **23**, 1097–1113.

Hartigan, J.A. and Kleiner, B. (1981), "Mosaics for Contingency Tables," in W.F. Eddy, ed., *Computer Science and Statistics: Proceedings of the 13th Symposium on the Interface*, Springer-Verlag, New York, 268–273.

Hastie, T.J. and Tibshirani, R.J. (1990), *Generalized Additive Models*, Chapman and Hall, London.

Hausman, J.A. and McFadden, D. (1984), "Specification Tests for the Multinomial Logit Model," *Econometrica*, **52**, 909–938.

Hayes, B. (2002), "Statistics of Deadly Quarrels," *American Scientist*, **90**, 10–15.

Hinde, J. and Demétrio, C.G.B. (1998), "Overdispersion: Models and Estimation," *Computational Statistics and Data Analysis*, **27**, 151–170.

Holcomb, Z.C. (2002), *Interpreting Basic Statistics: A Guide and Workbook Based on Excerpts from Journal Articles*, 3rd. ed., Pyrczak Publishing, Los Angeles.

Hosmer, D.W., Hosmer, T., le Cessie, S., and Lemeshow, S. (1997), "A Comparison of Goodness-of-Fit Tests for the Logistic Regression Model," *Statistics in Medicine*, **16**, 965–980.

Hosmer, D.W. and Lemeshow, S. (1999), *Applied Survival Analysis*, John Wiley and Sons, New York.

Hosmer, D.W. and Lemeshow, S. (2000), *Applied Logistic Regression*, 2nd. ed., John Wiley and Sons, New York.

Hsu, K. (2000), "Effect of Nightlights on Children Disputed," *Boston Globe*, March 9, 2000, A21.

Hueston, W.J. and Slott, K. (2000), "Improving Quality or Shifting Diagnoses?," *Archives of Family Medicine*, **9**, 933–935.

Hurvich, C.M., Simonoff, J.S., and Tsai, C.-L. (1998), "Smoothing Parameter Selection in Nonparametric Regression Using an Improved Akaike Information Criterion," *Journal of the Royal Statistical Society, Ser. B*, **60**, 271–293.

Hurvich, C.M. and Tsai, C.-L. (1989), "Regression and Time Series Model Selection in Small Samples," *Biometrika*, **76**, 297–307.

Hutto, C., Parks, W.P., and Lai, S. (1991), "A Hospital Based Prospective Study of Perinatal Infection with HIV-1," *Journal of Pediatrics*, **118**, 347–353.

Ihaka, R. and Gentleman, R. (1996), "R: A Language for Data Analysis and Graphics," *Journal of Computational and Graphical Statistics*, **5**, 299–314.

Infuso, A., Hubert, B., and Etienne, J. (1998), "Underreporting of Legionnaires' Disease in France: The Case for More Active Surveillance," *Eurosurveillance*, **3**, 48–50.

Insightful Corporation (2001), *S-PLUS for Windows Release 6*, Insightful Corporation, Seattle, WA.

Jackson, C.L., Hirst, L., de Jong, I.C., and Smith, N. (2002), "Can Australian General Practitioners Effectively Screen for Diabetic Retinopathy? A Pilot Study," *BMC Family Practice*, **3**, 4 (http://www.biomed central.com/1471-2296/3/4).

Jeon, J.W., Ching, H.Y., and Bae, J.S. (1987), "Chances of Simpson's Paradox," *Journal of the Korean Statistical Society*, **16**, 117–125.

Johnson, V.E. and Albert, J.H. (1999), *Ordinal Data Modeling*, Springer-Verlag, New York.

Kagan, A. (2001), "A Note on the Logistic Link Function," *Biometrika*, **88**, 599–601.

Karlis, D. and Xekalaki, E. (2000), "A Simulation Comparison of Several Procedures for Testing the Poisson Assumption," *The Statistician*, **49**, 355–382.

Kass, N.E., Faden, R.R., Fox, R., and Dudley, J. (1991), "Loss of Private Health Insurance Among Homosexual Men with AIDS," *Inquiry*, **28**, 249–254.

Kauermann, G. and Carroll, R.J. (2001), "A Note on the Efficiency of Sandwich Covariance Matrix Estimation," *Journal of the American Statistical Association*, **96**, 1387–1396.

Kim, D. and Agresti, A. (1997), "Nearly Exact Tests of Conditional Independence and Marginal Homogeneity for Sparse Contingency Tables," *Computational Statistics and Data Analysis*, **24**, 89–104.

King, E.N. and Ryan, T.P. (2002), "A Preliminary Investigation of Maximum Likelihood Logistic Regression Versus Exact Logistic Regression," *American Statistician*, **56**, 163–170.

Kocherlakota, S. and Kocherlakota, K. (1993), *Bivariate Discrete Distributions*, Marcel Dekker, New York.

Kolassa, J.E. (1997), "Infinite Parameter Estimates in Logistic Regression, with Application to Approximate Conditional Inference," *Scandinavian Journal of Statistics*, **24**, 523–530.

Kotze, T.J.vW. and Hawkins, D.M. (1984), "The Identification of Outliers in Two-Way Contingency Tables Using 2×2 Subtables," *Applied Statistics*, **33**, 215–223.

Kupper, L.L. and Haseman, J.K. (1978), "The Use of a Correlated Binomial Model for the Analysis of Certain Toxicological Experiments," *Biometrics*, **34**, 69–76.

Kvam, P.H. (2000), "The Effect of Active Learning Methods on Student Retention in Engineering Statistics," *American Statistician*, **54**, 136–140.

Laara, E. and Matthews, J.N.S. (1985), "The Equivalence of Two Models for Ordinal Data," *Biometrika*, **72**, 206–207.

LaBarbera, P.A. and Simonoff, J.S. (1999), "Toward Enhancing the Quality and Quantity of Marketing Majors," *Journal of Marketing Education*, **21**, 4–13.

Lambert, D. (1992), "Zero-Inflated Poisson Regression with an Application to Defects in Manufacturing," *Technometrics*, **34**, 1–14.

Lancaster, H.O. (1949), "The Combination of Probabilities Arising from Data in Discrete Distributions" *Biometrika*, **36**, 370–382.

Landis, J.R. and Koch, G.G. (1977), "The Measurement of Observer Agreement for Categorical Data," *Biometrics*, **33**, 159–174.

Lapp, K., Molenberghs, G., and Lesaffre, E. (1998), "Models for the Association Between Ordinal Variables," *Computational Statistics and Data Analysis*, **28**, 387–411.

Lee, H.K.H., Cork, D.L., and Algranati, D.J. (2002), "Did Lennox Lewis Beat Evander Holyfield? Methods for Analysing Small Sample Interrater Agreement Problems," *The Statistician*, **51**, 129–146.

Lehr, R.G. (2000), "Comment on Paper of Irony et al.," *American Statistician*, **54**, 325.

Lenk, P.J. (1999), "Bayesian Inference for Semiparametric Regression Using a Fourier Representation," *Journal of the Royal Statistical Society, Ser. B*, **61**, 863–879.

Ley, E. (1996), "On the Peculiar Distribution of the U.S. Stock Indexes' Digits," *American Statistician*, **50**, 311–313.

Lindsey, J.K. (1995), *Modelling Frequency and Count Data*, Oxford University Press, Oxford.

Lindsey, J.K. (1997), *Applying Generalized Linear Models*, Springer-Verlag, New York.

Lindsey, J.K. and Altham, P.M.E. (1998), "Analysis of the Human Sex Ratio by Using Overdispersion Models," *Applied Statistics*, **47**, 149–157.

Lindsey, J.K. and Mersch, G. (1992), "Fitting and Comparing Probability Distributions with Log Linear Models," *Computational Statistics and Data Analysis*, **13**, 373–384.

Linhart, H. and Zucchini, W. (1986), *Model Selection*, John Wiley and Sons, New York.

Little, R.J.A. and Rubin, D.R. (2002), *Statistical Analysis with Missing Data*, 2nd. ed., John Wiley and Sons, New York.

Lloyd, C.J. (1999), *Statistical Analysis of Categorical Data*, John Wiley and Sons, New York.

Long, J.S. (1997), *Regression Models for Categorical and Limited Dependent Variables*, Sage, Thousand Oaks, CA.

Louis, T.A. (1981), "Confidence Intervals for a Binomial Parameter After Observing No Successes," *American Statistician*, **35**, 154.

Manski, C.F. and McFadden, D.L., eds. (1981), *Structural Analysis of Discrete Data and Econometric Applications*, MIT Press, Cambridge, MA.

Mantel, N. (1987), "Understanding Wald's Test for Exponential Families," *American Statistician*, **41**, 147–148.

Margolis, L.H., Foss, R.D., and Tolbert, W.G. (2000), "Alcohol and Motor Vehicle–Related Deaths of Children as Passengers, Pedestrians, and Bicyclists," *Journal of the American Medical Association*, **283**, 2245–2248.

Marsh, L.C. and Mukhopadhyay, K. (1999), "Discrete Nonparametric Regression," *Proceedings of the Business and Economic Statistics Section of the American Statistical Association*, 233–238.

Martin Andrés, A. and Tapia García, J.M. (1999), "Optimal Unconditional Test in 2 × 2 Multinomial Trials," *Computational Statistics and Data Analysis*, **31**, 311–321.

McCullagh, P. (1978), "A Class of Parametric Models for the Analysis of Square Contingency Tables with Ordered Categories," *Biometrika*, **65**, 413–418.

McCullagh, P. (1980), "Regression Models for Ordinal Data (with discussion)," *Journal of the Royal Statistical Society, Ser. B*, **42**, 109–142.

McCullagh, P. and Nelder, J.A. (1989), *Generalized Linear Models*, 2nd. ed., Chapman and Hall, London.

McCurdy, S.A. (1994), "Epidemiology of Disaster. The Donner Party (1846–1847)," *Western Journal of Medicine*, **160**, 338–342.

McDonald, J.M. and Diamond, I. (1983), "Fitting Generalized Linear Models with Linear Inequality Constraints," *GLIM Newsletter*, **6**, 29–36.

McDonald, J.W., DeRoure, D.C., and Michaelides, D.T. (1998), "Exact Tests for Two-Way Symmetric Contingency Tables," *Statistics and Computing*, **8**, 391–399.

McFadden, D. (1973), "Conditional Logit Analysis of Qualitative Choice Behavior," in P. Zarembka, ed., *Frontiers in Econometrics*, Academic Press, New York, 105–142.

McFadden, D. (1974), "The Measurement of Urban Travel Demand," *Journal of Public Economics*, **3**, 303–328.

McNutt, L.-A., Hafner, J.-P., and Xue, X. (1999), "Correcting the Odds Ratio in Cohort Studies of Common Outcomes," *Journal of the American Medical Association*, **282**, 529.

McQuarrie, A.D.R. and Tsai, C.-L. (1998), *Regression and Time Series Model Selection*, World Scientific, Singapore.

Mehta, C.R., Patel, N.R., and Senchaudhuri, P. (2000), "Efficient Monte Carlo Methods for Conditional Logistic Regression," *Journal of the American Statistical Association*, **95**, 99–108.

Menard, S. (2000), "Coefficients of Determination for Multiple Logistic Regression Analysis," *American Statistician*, **54**, 17–24.

Milicer, H. and Szczotka, F. (1966), "Age at Menarche in Warsaw Girls in 1965," *Human Biology*, **38**, 199–203.

Mittlböck, M. and Waldhör, T. (2000), "Adjustments for R^2-Measures for Poisson Regression Models," *Computational Statistics and Data Analysis*, **34**, 461–472.

Moore, S. (1999), "Cases vs. Lectures: A Comparison of Learning Outcomes in Undergraduate Principles of Finance," *Journal of Financial Education*, **25** (Fall 1999), 37–49.

Morel, J.G. and Neerchal, N.K. (1997), "Clustered Binary Logistic Regression in Teratology Data Using a Finite Mixture Distribution," *Statistics in Medicine*, **16**, 2843–2853.

Morrell, C.H. (1999), "Simpson's Paradox: An Example From a Longitudinal Study in South Africa," *Journal of Statistics Education*, **7** (http://www.amstat.org/publications/jse/secure/v7n3/datasets .morrell.cfm)

Mosteller, F. and Parunak, A. (1985), "Identifying Extreme Cells in a Sizable Contingency Table: Probabilistic and Exploratory Approaches," in D.C. Hoaglin, F. Mosteller, and J.W. Tukey, eds., *Exploring Data Tables, Trends and Shapes*, John Wiley and Sons, New York, 189–224.

Mosteller, F. and Wallace, D.L. (1963), "Inference in an Authorship Problem," *Journal of the American Statistical Association*, **58**, 275–309.

Mosteller, F. and Wallace, D.L. (1964), *Applied Bayesian and Classical Inference: The Case of the Federalist Papers*, Springer-Verlag, New York.

Nair, V. (1986), "Testing in Industrial Experiments with Ordered Categorical Data," *Technometrics*, **28**, 283–311.

Nakashima, E. (1997), "Some Methods for Estimation in a Negative-Binomial Model," *Annals of the Institute of Statistical Mathematics*, **49**, 101–115.

Neeman, T. (1998), "Statistical Review, MEDI–493 (Palivizumab) Humanized Monoclonal Antibody to RSV F Protein," memorandum, April 3, 1998, Department of Health and Human Services, Food and Drug Administration, Division of Biostatistics and Epidemiology (http://www.fda.gov/cber/review/palimed061998r6.pdf).

Nelder, J.A. (2000), "Quasi-Likelihood and Pseudo-Likelihood Are Not the Same Thing," *Journal of Applied Statistics*, **27**, 1007–1011.

Nelder, J.A. and Wedderburn, R.W.M. (1972), "Generalized Linear Models," *Journal of the Royal Statistical Society, Ser. A*, **135**, 370–384.

Nelson, B. (2000), "A Stir Over Screening," *Newsday*, September 17, 2000, A7, A38–A40.

Nigrini, M.J. (2000), *Digital Analysis Using Benford's Law: Tests and Statistics for Auditors*, Global Audit Publications, Vancouver.

Norman, J.M. (1998), "Soccer," in J. Bennett, ed., *Statistics in Sport*, Arnold, London, 105–120.

Ochs, R. (1998), "Prostate Study Fuels Debate," *Newsday*, May 19, 1998, A39.

Ochs, R. (2000), "To Change Our Water Ways, Drink Two Quarts Every Day," *Newsday*, August 8, 2000, C4.

O'Connell, A.A. (2002), "Student Perceptions of Assessment Strategies in a Multivariate Statistics Course," *Journal of Statistics Education*, **10** (http://www.amstat.org/publications/jse/v10n1/oconnell.html).

Ohland, M., Zhang, G., Foreman, F., and Haynes, F. (2000), "The Engineering Concepts Institute: The Foundation of a Comprehensive Minority Student Development Program at the FAMU–FSU College of Engineering," *Proceedings of the 30th ASEE/IEEE Frontiers in Education Conference*, Kansas City, MO, F1F-17–F1F-20.

Pan, W. (2002), "Approximate Confidence Intervals for One Proportion and Difference of Two Proportions," *Computational Statistics and Data Analysis*, **40**, 143–157.

Park, T. (2002), "Is the Exact Test Better Than the Asymptotic Test for Testing Marginal Homogeneity in 2 × 2 Tables?," *Biometrical Journal*, **44**, 571–583.

Parzen, M., Lipsitz, S., Ibrahim, J., and Klar, N. (2002), "An Estimate of the Odds Ratio That Always Exists," *Journal of Computational and Graphical Statistics*, **11**, 420–436.

Pavlidis, I., Eberhardt, N.L., and Levine, J.A. (2002), "Seeing Through the Face of Deception," *Nature*, **415**, 35.

Paternoster, R., Brame, R., Bacon, S., Ditchfield, A., Biere, D., Beckman, K., Perez, D., Strauch, M., Frederique, N., Gawkoski, K., Zeigler, D., and Murphy, K. (2003), "An Empirical Analysis of Maryland's Death Sentencing System With Respect to the Influence of Race and Legal Jurisdiction" (http://www.urhome.umd.edu/newsdesk/pdf/finalrep.pdf).

Peng, R.D. and Hengartner, N.W. (2002), "Quantitative Analysis of Literary Styles," *American Statistician*, **56**, 175–185.

Pierce, D.A. and Schafer, D.W. (1986), "Residuals in Generalized Linear Models," *Journal of the American Statistical Association*, **81**, 977–986.

Pievatolo, A. and Rotondi, R. (2000), "Analysing the Interevent Time Distribution to Identify Seismicity Phases: a Bayesian Nonparametric Approach to the Multiple-Changepoint Problem," *Applied Statistics*, **49**, 543–562.

Pigeon, J.G. and Heyse, J.F. (1999), "A Cautionary Note About Assessing the Fit of Logistic Regression Models," *Journal of Applied Statistics*, **26**, 847–853.

Poortema, K. (1999), "On Modelling Overdispersion of Counts," *Statistica Neerlandica*, **53**, 5–20.

Pregibon, D. (1980), "Goodness of Link Tests for Generalized Linear Models," *Applied Statistics*, **29**, 15–24.

Prentice, R.L. (1986), "Binary Regression Using an Extended Beta-Binomial Distribution, with Discussion of Correlation Induced by Covariate Measurement Errors," *Journal of the American Statistical Association*, **81**, 321–327.

Press, S. and Quinn, B.J. (1997), "The Pacifier Thermometer: Comparison of Supralingual with Rectal Temperatures in Infants and Young Children," *Archives of Pediatrics and Adolescent Medicine*, **151**, 551–554.

Price, S.L. (2002), "The Indian Wars," *Sports Illustrated*, **96**(10), 66–72.

Quinn, G.E., Shin, C.H., Maguire, M.G., and Stone, R.A. (1999), "Myopia and Ambient Lighting at Night," *Nature*, **399**, 113–114.

Raftery, A.E. (1986a), "A Note on Bayes Factors for Log-linear Contingency Table Models With Vague Prior Information," *Journal of the Royal Statistical Society, Ser. B*, **48**, 249–250.

Raftery, A.E. (1986b), "Choosing Models for Cross–Classifications," *American Sociological Review*, **51**, 145–146.

Ramsey, F.L. and Schafer, D.W. (1997), *The Statistical Sleuth: A Course in Methods of Data Analysis*, Duxbury, Belmont, CA.

Rayner, J.C.W. and Best, D.J. (2001), *A Contingency Table Approach to Nonparametric Testing*, Chapman and Hall/CRC, Boca Raton, FL.

Read, T.R.C. and Cressie, N. (1988), *Goodness-of-Fit Statistics for Discrete Multivariate Data*, Springer-Verlag, New York.

Reed, K.P., van den Berg, S.S., Rudolph, A., Albright, J.A., Casey, H.W., Marino, A.A. (1994), "Treatment of Tendon Injuries in Thoroughbred Racehorses Using Carbon–Fiber Implants," *Journal of Equine Veterinary Science*, **14**, 371–377.

Reifman, A. (1999), "U.S. Senate Votes on Clinton Removal," *Journal of Statistics Education*, **7** (http://www.amstat.org/publications/jse/datasets/impeach.txt).

Ridout, M., Hinde, J., and Demétrio, C.G.B. (2001), "A Score Test for Testing a Zero-Inflated Poisson Regression Model Against Zero-Inflated Negative Binomial Alternatives," *Biometrics*, **57**, 219–223.

Robbins, A.S., Chao, S.Y., and Fonseca, V.P. (2002), "What's the Relative Risk? A Method to Directly Estimate Risk Ratios in Cohort Studies of Common Outcomes," *Annals of Epidemiology*, **12**, 452–454.

Romaniuk, H., Skinner, C.J., and Cooper, P.J. (1999), "Modelling Consumers' Use of Products," *Journal of the Royal Statistical Society, Ser. B*, **162**, 407–421.

Ross, K. (2001), "A Freshman Seminar: Statistics and Mathematics of Baseball," *Newsletter of the Statistics in Sports Section of the American Statistical Association*, **3(1)**, 6.

Russell, M. and Haney, W. (1997), "Testing Writing on Computers: An Experiment Comparing Student Performance on Tests Conducted via Computer and via Paper-and-Pencil," *Education Policy Analysis Archives*, **5(3)** (http://olam.ed.asu.edu/epaa/v5n3.html).

Russell, J.M., Simonoff, J.S., and Nightingale, J. (1997), "Nursing Behaviors of Beluga Calves (*Delphinapterus Leucas*) Born in Captivity," *Zoo Biology*, **16**, 247–262.

Ryan, T.P. (1997), *Modern Regression Methods*, John Wiley and Sons, New York.

Salama, P., Spiegel, P., Van Dyke, M., Phelps, L., Wilkinson, C. (2000), "Mental Health and Nutritional Status Among the Adult Serbian Minority in Kosovo," *Journal of the American Medical Association*, **284**, 578–584.

Salsburg, D. (2001), *The Lady Tasting Tea: How Statistics Revolutionized Science in the Twentieth Century*, W.H. Freeman, New York.

Santner, T.J. (1998), "A Note on Teaching Binomial Confidence Intervals," *Teaching Statistics*, **20**, 20–23.

Santner, T.J. and Duffy, D.E. (1989), *The Statistical Analysis of Discrete Data*, Springer-Verlag, New York.

SAS Institute (2000), *SAS/STAT User's Guide, Version 8*, SAS Publishing, Cary, NC.

Scheidegger, K. (2003), "Maryland Study, When Properly Analyzed, Supports Death Penalty" (`http://www.cjlf.org/deathpenalty/MdMora torium.htm`).

Schriefer, M.E., Dennis, D.T., Gubler, D.J., Hayes, E.B., Johnson, B.J.B., and Chu, M.C. (2000), "Serologic Testing for Lyme Disease," *Journal of the American Medical Association*, **284**, 695–696.

Schuster, C. and von Eye, A. (2001), "Models for Ordinal Agreement Data," *Biometrical Journal*, **43**, 795–808.

Schutzer, S.E., Coyle, P.K., Reid, P., and Holland, B. (1999), "*Borrelia burgdorferi*-Specific Immune Complexes in Acute Lyme Disease," *Journal of the American Medical Association*, **282**, 1942–1946.

Seikel, K., Primm, P.A., Elizondo, B.J., and Remley, K.L. (1999), "Hand-held Metal Detector Localization in Ingested Metallic Foreign Bodies," *Archives of Pediatric and Adolescent Medicine*, **153**, 853–857.

Sekar, C.C. and Deming, W.E. (1949), "On a Method of Estimating Birth and Death Rates and the Extent of Registration," *Journal of the American Statistical Association*, **44**, 101–115.

Seneta, E. and Phipps, M.C. (2001), "On the Comparison of Two Observed Proportions," *Biometrical Journal*, **43**, 23–43.

Shane, K.V. and Simonoff, J.S. (2001), "A Robust Approach to Categorical Data Analysis," *Journal of Computational and Graphical Statistics*, **10**, 135–157.

Shekarriz, B., Lu, H.-F., and Stoller, M.L. (2001), "Correlation of Unilateral Urolithiasis," *Journal of Urology*, **165**, 1085–1087.

Simonoff, J.S. (1983), "A Penalty Function Approach to Smoothing Large Sparse Contingency Tables," *Annals of Statistics*, **11**, 208–218.

Simonoff, J.S. (1985), "An Improved Goodness-of-Fit Statistic for Sparse Multinomials," *Journal of the American Statistical Association*, **80**, 671–677.

Simonoff, J.S. (1987), "Probability Estimation via Smoothing in Sparse Contingency Tables with Ordered Categories," *Statistics and Probability Letters*, **5**, 55–63.

Simonoff, J.S. (1988), "Detecting Outlying Cells in Two-Way Contingency Tables via Backwards-Stepping," *Technometrics*, **30**, 339–345.

Simonoff, J.S. (1995), "Smoothing Categorical Data," *Journal of Statistical Planning and Inference*, **47**, 41–69.

Simonoff, J.S. (1996), *Smoothing Methods in Statistics*, Springer-Verlag, New York.

Simonoff, J.S. (1997), "The 'Unusual Episode' and a Second Statistics Course," *Journal of Statistics Education*, **5** (http://www.amstat.org/publications/jse/v5n1/simonoff.html).

Simonoff, J.S. (1998a), "Move Over, Roger Maris: Breaking Baseball's Most Famous Record," *Journal of Statistics Education*, **6** (http://www.amstat.org/publications/jse/v6n3/datasets.simonoff.html).

Simonoff, J.S. (1998b), "Three Sides of Smoothing: Categorical Data Smoothing, Nonparametric Regression, and Density Estimation," *International Statistical Review*, **66**, 137–156.

Simonoff, J.S. (1998c), "Logistic Regression, Categorical Predictors and Goodness-of-Fit: It Depends on Who You Ask," *American Statistician*, **52**, 10–14.

Simonoff, J.S. and Hurvich, C.M. (1993), "A Study of the Effectiveness of Simple Density Estimation Methods," *Computational Statistics*, **8**, 259–278.

Simonoff, J.S. and Ma, L. (2003), "An Empirical Study of Factors Relating to the Success of Broadway Shows," *Journal of Business*, **76**, 135–150.

Simonoff, J.S. and Tsai, C.-L. (1991), "Higher Order Effects in Log-Linear and Log-Non-Linear Models for Contingency Tables with Ordered Categories," *Applied Statistics*, **40**, 449–458.

Simonoff, J.S. and Tsai, C.-L. (1999), "Semiparametric and Additive Model Selection Using an Improved Akaike Information Criterion," *Journal of Computational and Graphical Statistics*, **8**, 22–40.

Simonoff, J.S. and Tutz, G. (2000), "Smoothing Methods for Discrete Data," in M.G. Schimek, ed., *Smoothing and Regression: Approaches, Computation, and Application*, John Wiley and Sons, New York, 193–228.

Simpson, D.G. (1987), "Minimum Hellinger Distance Estimation for the Analysis of Count Data," *Journal of the American Statistical Association*, **82**, 802–807.

Simpson, D.G. (1989), "Hellinger Deviance Tests: Efficiency, Breakdown Points, and Examples," *Journal of the American Statistical Association*, **84**, 107–113.

Simpson, E.H. (1951), "The Interpretation of Interaction in Contingency Tables," *Journal of the Royal Statistical Society, Ser. B*, **13**, 238–241.

Smith, C.A.B. (1951), "A Test for Heterogeneity of Proportions," *Annals of Eugenics*, **16**, 16–25.

Smith, T. and Schwertman, N.C. (1999), "Can the NCAA Basketball Tournament Seeding Be Used to Predict Margin of Victory?," *American Statistician*, **53**, 94–98.

Sommers, P.M. (2003), "The Writing on the Wall," *Chance*, **16(1)**, 35–38.

SPSS, Inc. (2001), *SPSS for Windows, Release 11.0*, SPSS, Inc., Chicago, IL.

Standard and Poor's (1998), "Special Report: Ratings Performance 1997," August 1998.

Starr, N. (1997), "Nonrandom Risk: The 1970 Draft Lottery," *Journal of Statistics Education*, **5** (http://www.amstat.org/publications/jse/v5n2/datasets.starr.html).

Stein, C.E., Bennett, S., Crook, S., and Maddison, F. (2001), "The Cluster That Never Was: Germ Warfare Experiments and Health Authority Reality in Dorset," *Journal of the Royal Statistical Society, Ser. A*, **164**, 23–27.

Stelfox, H.T., Chua, G., O'Rourke, K., and Detsky, A.S. (1998), "Conflict of Interest in the Debate over Calcium-Channel Antagonists," *New England Journal of Medicine*, **338**, 101–106.

Steltner, H., Staats, R., Timmer, J., Vogel, M., Guttmann, J., Matthys, H., and Virchow, J.C. (2002), "Diagnosis of Sleep Apnea by Automatic Analysis of Nasal Pressure and Forced Oscillation Impedance," *American Journal of Respiratory and Critical Care Medicine*, **165**, 940–944.

StemCell Technologies (2001), *L-Calc^{TM} Version 1.1: A Statistical Software Program for Limiting Dilution Analysis*, StemCell Technologies, Vancouver, Canada.

Stern, H.S. (1998), "How Accurate Are the Posted Odds?," *Chance*, **11(4)**, 17–21.

Stigler, S.M. (1995), "Galton and Identification by Fingerprints," *Genetics*, **140**, 857–860.

Stigler, S.M. (1999), *Statistics on the Table: The History of Statistical Concepts and Methods*, Harvard University Press, Cambridge, MA.

Stone, E. (2001), "Year-End Rash of Killings," *Burlington Free Press*, January 11, 2001, 1A.

Strand, L.I. and Wie, S.L. (1999), "The Sock Test for Evaluating Activity Limitation in Patients with Musculoskeletal Pain," *Physical Therapy*, **79**, 136–145.

Stukel, T.A. (1988), "Generalized Logistic Models," *Journal of the American Statistical Association*, **83**, 426–431.

Tapia García, J.M. and Martin Andrés, A. (2000), "Optimal Unconditional Critical Regions for 2 × 2 Multinomial Trials," *Journal of Applied Statistics*, **27**, 689–695.

Tarone, R.E. (1979), "Testing the Goodness of Fit of the Binomial Distribution," *Biometrika*, **66**, 585–590.

Tarone, R.E. (1985), "On Heterogeneity Tests Based on Efficient Scores," *Biometrika*, **72**, 91–95.

Theus, M. and Lauer, S.R.W. (1999), "Visualizing Loglinear Models," *Journal of Computational and Graphical Statistics*, **8**, 396–412.

Trice, A.D., Holland, S.A., and Gagné, P.E. (2000), "Voluntary Class Absences and Other Behaviors in College Students: An Exploratory Analysis," *Psychological Reports*, **87**, 179–182.

Tufte, E.R. (1997), *Visual Explanations: Images and Quantities, Evidence and Narrative*, Graphics Press, Cheshire, CT.

Tukey, J.W. (1947), "Nonparametric Estimation II. Statistically Equivalent Blocks and Tolerance Regions. The Continuous Case," *Annals of Mathematical Statistics*, **18**, 529–539.

Tukey, J.W. (1961), "Curves as Parameters and Touch Estimation," in J. Neyman, ed., *Proceedings of the 4th Berkeley Symposium on Mathematical Statistics and Probability*, 681–694.

Tukey, J.W. (1977), *Exploratory Data Analysis*, Addison-Wesley, Reading, MA.

Tunaru, R. (2001), "Models of Association Versus Causal Models for Contingency Tables," *The Statistician*, **50**, 257–269.

Vieira, A.M.C., Hinde, J.P., and Demétrio, C.G.B. (2000), "Zero-Inflated Proportion Data Models Applied to a Biological Control Assay," *Journal of Applied Statistics*, **27**, 373–389.

Wainer, H. (2002), "The BK-Plot: Making Simpson's Paradox Clear to the Masses," *Chance*, **15(3)**, 60–62.

Walker, G., Greenfield, J., Fox, J., and Simonoff, J.S. (1985), "The Yale Survey: A Large-Scale Study of Book Deterioration in the Yale University Library," *College and Research Libraries*, **46**, 111–132.

Wand, J.N., Shotts, K.W., Sekhon, J.S., Mebane, W.R., Jr., Herron, M.C., and Brady, H.E. (2001), "The Butterfly Did It: The Aberrant Vote for Buchanan in Palm Beach County, Florida," *American Political Science Review*, **95**, 793–810.

Webber, W.F. (2003), "Comment on Paper by King and Ryan," *American Statistician*, **57**, 147–148.

Weisberg, S. (1985), *Applied Linear Regression*, 2nd. ed., John Wiley and Sons, New York.

Wermuth, N. and Cox, D.R. (1998), "On the Application of Conditional Independence to Ordinal Data," *International Statistical Review*, **66**, 181–199.

Wermuth, N. and Lauritzen, S.L. (1990), "On Substantive Research Hypotheses, Conditional Independence Graphs and Graphical Chain Models (with discussion)," *Journal of the Royal Statistical Society, Ser. B*, **52**, 21–72.

Williams, D.A. (1982), "Extra-Binomial Variation in Logistic Linear Models," *Biometrics*, **31**, 144–148.

Winkelmann, R. (2000), *Econometric Analysis of Count Data*, 3rd. ed., Springer-Verlag, Berlin.

Woodworth, G.C. (1988), "Statistics and the Death Penalty," *Stats*, **2**, 9–12.

World Health Organization (1998), "Gender and Health: Technical Paper," World Health Organization, Geneva, WHO/FRH/WHD/98.16.

Xie, M., He, B., and Goh, T.N. (2001), "Zero-Inflated Poisson Model in Statistical Process Control," *Computational Statistics and Data Analysis*, **38**, 191–201.

Yates, F. (1934), "Contingency Tables Involving Small Numbers and the χ^2 Test," *Journal of the Royal Statistical Society Supplement*, **1**, 217–235.

Ye, J. (1998), "On Measuring and Correcting the Effects of Data Mining and Model Selection," *Journal of the American Statistical Association*, **93**, 120–131.

Yick, J.S. and Lee, A.H. (1998), "Unmasking Outliers in Two-Way Contingency Tables," *Computational Statistics and Data Analysis*, **29**, 69–79.

Yokota, F. and Thompson, K.M. (2000), "Violence in G-Rated Animated Films," *Journal of the American Medical Association*, **283**, 2716–2720.

Yule, G.U. (1903), "Notes on the Theory of Association of Attributes in Statistics," *Biometrika*, **2**, 121–134.

Zadnik, K., Jones, L.A., Irvin, B.C., Kleinstein, R.N., Manny, R.E., Shin, J.A., and Mutti, D.O. (2000), "Vision — Myopia and Ambient Night-time Lighting," *Nature*, **904**, 143–144.

Zelterman, D. (1999), *Models for Discrete Data*, Oxford University Press, Oxford.

Zhang, J. and Yu, K.F. (1998), "What's the Relative Risk? A Method of Correcting the Odds Ratio in Cohort Studies of Common Outcomes," *Journal of the American Medical Association*, **280**, 1690–1691.

Zhang, X. (1999), "Investigating the Association of Inbreeding and Schizophrenia," unpublished paper, Department of Statistics, Carnegie Mellon University.

Zheng, B. and Agresti, A. (2000), "Summarizing the Predictive Power of a Generalized Linear Model," *Statistics in Medicine*, **19**, 1771–1781.

Zorn, C.J.W. (1998), "An Analytic and Empirical Examination of Zero-Inflated and Hurdle Poisson Specifications," *Sociological Methods and Research*, **26**, 368–400.

Index

Springer Texts in Statistics *(continued from page ii)*

ALSO AVAILABLE FROM SPRINGER!

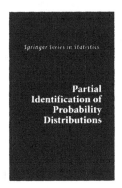

SMOOTHING METHODS IN STATISTICS

JEFFREY S. SIMONOFF

This book surveys the uses of smoothing methods in statistics. The coverage has an applied focus, and is very broad, including simple and complex univariate and multivariate density estimation, nonparametric regression estimation, categorical data smoothing, and applications of smoothing to other areas of statistics.

1996/338 PP./HARDCOVER/ISBN 0-387-94716-7
SPRINGER SERIES IN STATISTICS

PRINCIPAL COMPONENT ANALYSIS

SECOND EDITION

I.T. JOLLIFFE

Principal component analysis is central to the study of multivariate data. Although one of the earliest multivariate techniques it continues to be the subject of much research, ranging from new model-based approaches to algorithmic ideas from neural networks. It is extremely versatile with applications in many disciplines. The first edition of this book was the first comprehensive text written solely on principal component analysis. The second edition updates and substantially expands the original version, and is once again the definitive text on the subject. It includes core material, current research and a wide range of applications. Its length is nearly double that of the first edition. Researchers in statistics, or in other fields that use principal component analysis, will find that the book gives an authoritative yet accessible account of the subject. It is also a valuable resource for graduate courses in multivariate analysis. The book requires some knowledge of matrix algebra.

2002/502 PP./HARDCOVER/ISBN 0-387-95442-2
SPRINGER SERIES IN STATISTICS

PARTIAL IDENTIFICATION OF PROBABILITY DISTRIBUTIONS

CHARLES F. MANSKI

Sample data alone never suffice to draw conclusions about populations. Inference always requires assumptions about the population and sampling process. Statistical theory has revealed much about how strength of assumptions affects the precision of point estimates, but has had much less to say about how it affects the identification of population parameters. Indeed, it has been commonplace to think of identification as a binary event – a parameter is either identified or not – and to view point identification as a pre-condition for inference. Yet there is enormous scope for fruitful inference using data and assumptions that partially identify population parameters. This book explains why and shows how.

2003/196 PP./HARDCOVER/ISBN 0-387-00454-8
SPRINGER SERIES IN STATISTICS

To Order or for Information:

In the Americas: CALL: 1-800-SPRINGER or
FAX: (201) 348-4505 • WRITE: Springer-Verlag New York, Inc., Dept. S5433, PO Box 2485, Secaucus, NJ 07096-2485 • VISIT: Your local technical bookstore • E-MAIL: orders@springer-ny.com

Outside the Americas: CALL: +49 (0) 6221 345-217/8 • FAX: + 49 (0) 6221 345-229 • WRITE: Springer Customer Service, Haberstrasse 7, 69126 Heidelberg, Germany • E-MAIL: orders@springer.de
PROMOTION: S5433

Springer